STRATEGIES
FOR
BUSINESS
AND
TECHNICAL
WRITING

STRATEGIES FOR BUSINESS AND TECHNICAL WRITING

Fourth Edition

KEVIN J. HARTY

La Salle University

Allyn and Bacon

Boston • London • Toronto • Sydney • Tokyo • Singapore

Vice President, Eben W. Ludlow
Series Editorial Assistant: Linda M. D'Angelo
Marketing Manager: Lisa Kimball
Composition and Prepress Buyer: Linda Cox
Manufacturing Buyer: Suzanne Lareau
Cover Administrator: Jenny Hart
Editorial–Production Service: Shepherd, Inc.

Library of Congress Cataloging-in-Publication Data

Harty, Kevin J.
    Strategies for business and technical writing / Kevin J. Harty. —
4th ed.
        p. cm.
    Includes bibliographical references and index.
    ISBN 0-205-26120-5
    1. Commercial correspondence. 2. Business report writing.
3. Technical writing. I. Title.
HF5721.H37 1998
808'.066651—dc21   98–25025
                                                                      CIP

Printed in the United States of America

10 9 8 7 6 5 4 3 2 1   02 01 00 99 98

For
Jim Butler, Gabe Fagan, John Keenan, John Kleis,
Fran Lottier, and Linda Merians

—friends and colleagues all—
merito et tempore.

Contents

PART FOUR: REPORTS AND OTHER LONGER DOCUMENTS

Preface

This fourth edition of *Strategies for Business and Technical Writing* represents a thorough reworking of the previous edition. I have increased the number of selections from thirty to thirty-two, with thirteen appearing for the first time or in revised form, guided by the same simple principle that informed the selection of materials for inclusion in the three previous editions: *Strategies* should present the best advice from the best sources about the most important issues in business and technical writing.

The changes I have made in preparing this fourth edition broaden the coverage in *Strategies* of key issues in both business and technical writing and increase the usefulness of *Strategies* as a textbook and as a reference. In particular, the new and revised essays:

- provide additional perspectives on the *process* of writing,
- offer an analysis of the failures in communication that led to the Challenger disaster,
- increase the coverage of graphics by adding materials on a greater variety of visual aids including computer-generated graphics,
- expand the discussion of such topics as audience analysis, word choice—especially in terms of bias-free and nonsexist usage—and business and technical correspondence including e-mail,
- discuss a greater variety of approaches to the job search in terms of different kinds of resumes and cover letters,
- extend coverage to such important new topics as proposals, methods of revising, the dynamics that should inform written communications within business and industry, and the ways in which established office practices can prove counterproductive to effective business and technical writing.

This fourth edition of *Strategies* can, like its predecessors, be used in three ways: (1) as a supplement to a business or technical writing textbook, (2) as a supplement to a composition textbook or handbook, and (3) on its own as a textbook or as a reference source.

I have designed *Strategies* to appeal to practical-minded instructors, students, and people already working in business and technical occupations. The contributors to this volume write with the kind of purpose and understanding that come only from a career dedicated to improving the effectiveness of business and technical communications. As a result, the selections in *Strategies* not only teach professional writing but also are fine examples of the genre.

In selecting essays for this fourth edition of *Strategies,* I have been guided by the belief that all courses in business and technical writing should firmly implant one ideal in the minds of students (and of those in the world of work). *All* successful writing consists of clear and effective prose. However, students should recognize that business and technical writing differs in important ways from the expository prose usually taught in freshman or advanced composition courses. Because business and technical writers communicate with multiple audiences and with a variety of intentions, learning to write well for the world of work requires some adjustments in technique. Therefore, in addition to touching on such broad problems as style, jargon, and diction, *Strategies* includes comprehensive discussions of specific forms of writing—for example, chronological as well as functional resumes, letters of application, letters that sell, memos, e-mail, reports, and proposals. Students learn the specific techniques used by successful executives in business and industry who communicate information as part of their jobs.

The essays in this fourth edition of *Strategies* have, like those in the three previous editions, survived the most rigorous scrutiny—that of my students in seminars and courses in business and technical writing at La Salle University, Temple University, Rhode Island College, the Federal Reserve Bank of Philadelphia, RCA, Philadelphia Life Insurance Company, the Board of Pensions of the United Presbyterian Church (USA), First Pennsylvania Bank (now CoreStates Bank), Fidelity Bank (now First Union Bank), Blue Cross of Greater Philadelphia, and Philadelphia Newspapers, Inc. I am grateful to these former students for being such willing and helpful critics.

I am also grateful to those who have helped in the preparation of this anthology in other ways. Professors Stuart C. Brown of New Mexico State University, Barry Batorsky of the DeVry Institute, Phillip Vassallo of Middlesex County College, and Carol M. Barnum of Southern Polytechnic State University wrote detailed reviews of the third edition and the proposed changes for the fourth edition from which I benefited greatly. For permission to reprint the selections included in this fourth edition of *Strategies,* I am happy to thank the authors, agents, and editors who handled my permissions requests with such efficiency and good cheer. At La Salle University, I owe a continuing debt to my many friends and colleagues, especially those to whom this volume is dedicated. My thanks too to Professor Steven J. Patterson of St. Joseph's University for help with proofreading the manuscript of this book.

Finally, I am indebted to the editorial and production staffs at Allyn & Bacon, especially to Eben Ludlow and his assistant Linda M. D'Angelo.

Introduction

Writing consumes a substantial portion of the working day for almost all college-educated workers.[1]

Ability to express ideas cogently and goals persuasively—in plain English—is the most important skill to leadership. I know of no greater obstacle to the progress of good ideas and good people than the inability to compose a plain English sentence.[2]

Whether you are planning a career in business, in industry, or in some technical field, or whether you already have such a career, you may be ill-prepared for what is perhaps the most difficult part of your job. Although you may be an excellent accountant, salesperson, manager, engineer, or scientist, all of your training may not help you when you sit down to write. If this is the case, you are not alone in being frustrated by the writing demands of your job. Nevertheless, writing is essential to most technical and professional occupations; the ability to communicate effectively both in person and on paper can help you advance in your profession. Good writing can mean the difference between winning or losing a major sale, a pleased customer, or a challenging new position with a higher salary.

The selections in this book will help you make your letters, reports, memos, and other professional documents more effective. The selections represent some of the best advice veteran teachers and practitioners of business and technical writing have to offer. None of the selections, however, offers a quick cure for writing ills. There is no such cure. Like any worthwhile skill, good professional writing requires time, practice, and, most of all, discipline.

To begin with, we need to realize that there are only two kinds of professional writing:

1. clear, effective writing that meets the combined needs of the reader and the writer, and
2. bad writing.

Bad writing is unclear and ineffective; it wastes time and money. Even worse, it ignores its readers, usually raising more questions than answers.

---

[1] Paul V. Anderson, "What Survey Research Tells Us about Writing at Work," in *Writing in Nonacademic Settings*, ed. Lee Odell and Dixie Goswarmi (New York: Guilford, 1985), p. 30.

[2] The late John D. deButts, former Chairman, AT&T. Quoted as the epigraph to the third edition of the Bell Laboratories' *Editorial Style Guide* (Whippany, NJ: 1979).

To be effective writers, each of us needs to develop a variety of strategies for the different writing situations we face. We must also respect writing as something more than words typed or written on the page or fed into a dictaphone or word processor. Writing takes thought and planning.

THREE KEY QUESTIONS

Business and technical writing situations present us with a variety of challenges. One of the ways to meet these challenges is to begin by asking ourselves these three key questions:

1. Exactly who is my audience?
2. What is the most important thing I want to tell my audience?
3. What is the best way of making sure my audience understands what I have to say?

Exactly Who Is My Audience?

Whatever we write on the job, somebody—across the hall, across the city, across the country, or even across the world—will eventually read what we have written. Most likely, our readers will receive our message in their own offices or homes. Since we won't be on hand to explain anything that is unclear or that doesn't make sense, our message will have to speak for itself.

Whenever we sit down to write, we become experts. That is in large part why we are the writers and not the readers. Our readers depend on us for clear, effective writing, so that they may share in our expertise. Although this seems obvious, business, industrial, and technical practice shows that the single most common cause of bad writing is ignoring the readers. There are any number of ways to ignore our audience: we can talk over their heads; we can talk down to them; we can beat around the bush; we can leave out important details—and, as a result, we can undermine our credibility.

When you next sit down to write, therefore, it is important to ask yourself what you know about your audience—you must determine your audience's situation and anticipate your audience's potential reaction to your message. To paraphrase the golden rule, make sure you are prepared to write unto others the way you would have them write unto you.

What Is the Most Important Thing I Want to Tell My Audience?

Whether you write one page or one hundred pages, you should be able to condense your most important idea into about 25 words. If you can't, you aren't ready to write, since you haven't figured out what you are trying to say. You may not be able to explain every reason for a decision or a proposal in 25 words, but 25 words should be enough for a statement of the decision or proposal itself. For instance, you may have several reasons for recommending that your company replace its Brand X word processors with Brand Y word processors. You can make the recommendation in 22 words flat:

> I recommend that we immediately replace all our Brand X word processors with new Brand Y units for the reasons presented below.

In short, you can cut through all the bull and come right out with your recommendation. Then, in the body of your memo or report, you can present the relevant facts. Busy people dislike having their time wasted, and they hate surprises. So, in professional writing, the bottom line should also be the top line.

What Is the Best Way of Making Sure My Audience Understands What I Have To Say?

Different writing situations require different strategies. After answering the first two questions, you may decide that you shouldn't write at all—that you should phone instead or make a personal visit. But once you do commit yourself to writing, you need to determine how to put the particular document together. You will need to use one strategy for good news and a different one for bad news, one for giving information and another for changing someone's mind, one for accepting a job offer and another for turning the offer down. But in each case you will need to get your main idea out as soon as possible in as few words as possible, and then later defend or elaborate on that idea as clearly and as effectively as possible.

To illustrate the importance of answering these three key questions properly, let's look at a brief excerpt from a booklet formerly used by the New York State Income Tax Bureau:

> The return for the period before the change of residence must include all items of income, gain, loss or deduction accrued to the taxpayer up to the time of his change of residence. This includes any amounts not otherwise includable in the return because of an election to report income on an installment basis.
>
> Stated another way, the return for the period prior to the change of residence must be made on the accrual basis whether or not that is the taxpayer's established method of reporting. However, in the case of a taxpayer changing from nonresident to resident status, these accruals need not be made with respect to items derived from or connected with New York sources.

How did you respond to this information? If you were a taxpayer in need of quick and concise information to solve a particular problem, you would surely be disappointed and probably even annoyed. Tax forms need not be complicated; since ordinary people normally fill out these forms, they should be written for the average taxpayer, not for a trained accountant. In fact, all professional writing should be reader-directed.

Furthermore, the result of such convoluted writing is that these two paragraphs do not fulfill their purpose; they do not tell taxpayers how to adjust their returns when changing residence. Although the grammar is correct, the message is certainly not effective. Among other faults, the writer forgot to ask, "Who is my audience?" Unfamiliar words or usages such as *includable, accrual basis,* and *election* appear without explanation. If taxpayers cannot understand such language, they will complete their returns incorrectly and cheat either themselves or the state out of money. In the end, these two paragraphs—and others like them in the tax booklet—will have missed their mark, and unnecessary complications for both taxpayers and the state will be the result.

GUIDELINES FOR EFFECTIVE WRITING

To make sure you answer the three key questions correctly, keep these guidelines in mind:

- *Put your main idea up front.* Then give whatever background or evidence your reader needs to understand your main idea.
- *Avoid using technical terms with nontechnical readers.* If you are an accountant writing to another accountant, use all the technical terminology you need. But if you are an accountant writing to a layperson, use ordinary language that the layperson will understand. If you must use technical terms, define them carefully.
- *Limit the length and complexity of your sentences.* Whenever possible, try to include only one main idea in each sentence. More mechanically, try also to limit your sentences to 25 to 35 words (or about 2½ typed lines). Generally speaking, the shorter the sentences, the fewer the difficulties you and your reader will get into. To ensure that your sentences are easily understood, move the main subject and verb to the front of the sentence and cut out any unnecessary words.
- *Limit the length and content of your paragraph.* Nothing intimidates a reader more than a half-page-long block of dense prose. Rather than using one paragraph to discuss every aspect of a topic, use a separate paragraph for each aspect of your topic. In that way, you can limit your paragraphs to five or six sentences. If you really need a longer paragraph, use indentation or enumeration so that your subpoints stand out from the pack.
- *Give your reader all necessary information in a clear and unmistakable way.* If your reader needs to let you know something by June 12, tell the reader so plainly. If you want your reader to check back with you for additional information, make sure you provide your address or telephone number, and your travel or vacation schedule if you are going to be out of the office. If you are attaching or enclosing anything, specify exactly what. If your letter is a follow-up to some previous correspondence or to a recurring problem, make sure you give the reader this information too.

The selections that follow suggest many ways in which you can make your professional writing more effective. They also suggest ways you can make the process more enjoyable. If you view your professional writing as a chore, its effectiveness will be limited. But the more you get into the habit of asking the three key questions before you write your memos, letters, reports, and other documents, the easier, the more effective—and the more enjoyable—writing will become for you.

Part 1

Process as Well as Product

The old adage tells us to think before we speak. We should also think before we write. Actually, we should think before we write, while we write, and after we write. Thinking is what the writing process is all about. A process approach to writing is simply a writing strategy that asks writers to make sure they carefully plan each written message, rather than simply letting words fall where they may in the hope they will somehow organize themselves effectively and make sense to their intended audience.

People in business and industry are, however, trained to think in terms of finished products, bottom lines, and deadlines, so at first they may react with skepticism—if not with antagonism—to a process that requires them to slow down and think, especially when it is 4:30 and they need to get a memo to the boss by 5:00. Sometimes, sheer momentum and adrenaline will actually help, and a writer will produce an acceptable, even an effective, document. More times than not, though, the document will fail to achieve its purpose. The writer who balked at spending an extra ten minutes thinking a document through ends up spending an additional hour writing a follow-up document to clarify what should have been clear in the first document or, even worse, to mend fences because the original document created new problems.

In the introduction, I suggested an approach to the writing process that involved having writers ask themselves three key questions:

1. Exactly who is the audience?
2. What is the most important thing they want to tell this audience?
3. What is the best way of making sure this audience understands what they have to say?

These three key questions are simply one of any number of approaches to the writing process. The essays that follow in this section of *Strategies* offer other approaches and additional insights into a process approach to writing.

In the first essay, John Keenan offers a five-part approach to the writing process that asks writers to consider purpose, audience, format, evidence, and organization. Michael Adelstein follows Keenan with another five-part approach to writing that lays down a schedule for any writing task: 15% worrying, 10% planning, 25% writing, 45% revising, and 5% proofreading. The number of steps involved is, of course, unimportant. What is important is that there be a conscious effort to understand that, in business and technical writing, the process is as important as the goal. Keenan's approach is more open-ended, allowing writers considerable freedom in the amount of time they spend on the different steps in the writing process. While allowing for some variation among writing situations, Adelstein suggests a more regimented approach that will appeal to other writers.

Peter Elbow takes a different tack than Keenan or Adelstein by simply dividing writing time in half. Using what Elbow calls "the direct writing process," writers spend the first half of their available time "fast writing" in an attempt to get their ideas on the page before worrying about organization, correctness, or precision. Elbow suggests that these latter issues are best left to the second step in his approach to the writing process, revision. Revision is the key to successful writing, and Linda Flower and John Ackerman suggest ways in which the process of revision can help writers produce more effective documents.

In an often reprinted and adapted essay, John S. Harris next examines the writing process from a different perspective. The fault in poor quality documents may lie as much with the managers who assign them as with those who produce them. To counter problems in the management of writing, Harris offers a project worksheet designed to facilitate efficient writing management.

Finally, in the last essay in this section of *Strategies*, Dorothy A. Winsor examines the public documents available on the Challenger explosion to show how a "history of miscommunication" contributed to the accident. Her aim is to show how the dynamics at work in the Challenger case can help business and technical writers and managers reduce miscommunication in their own companies.

John Keenan

Using PAFEO Planning

A former technical writer for SmithKline, John Keenan is now Professor Emeritus of English at La Salle University in Philadelphia and coauthor with Kevin J. Harty of Writing for Business and Industry: Process and Product *(Macmillan, 1987).*

Surveys tell us that "lack of clarity" is the most frequently cited weakness when executives evaluate their employees' writing. But since lack of clarity results from various bad habits, it is often used as a convenient catchall criticism. Pressed to explain what they meant by lack of clarity, readers would probably come up with comments such as these:

"I had to read the damn thing three times before I got what he was driving at. At least I *think* I got his point; I'm not sure."

"I couldn't follow her line of thought."

"I'm just too busy to wade through a lot of self-serving explanation and justification. If he has something to say, why didn't he just say it without all that buildup?"

Most unclear writing results from unclear thinking. But it is also true that habits of clear writing help one to think clearly. When a writer gets thoughts in order and defines the purpose and goal, what he wants to say may no longer be exactly the same as it was originally. Sometimes the writer may even decide it is better *not* to write. Whatever the case, the work done *before* the first draft in the prewriting process and *after* the draft in the revising process is often the difference between successful communication and time-wasting confusion. . . .

THINK PAFEO

Some years ago I made up a nonsense word to help my students remember the important steps in constructing a piece of writing. It seemed to stick in their minds. In fact, I keep running into former students who can't remember my name, but remember PAFEO because it proved to be so useful. Here are the magic ingredients:

P stands for purpose
A stands for audience
F stands for format
E stands for evidence
O stands for organization.

Put together, they spell PAFEO, and it means the world to me!

PURPOSE

Put the question to yourself, "Why am I writing this?" ("Because I have to," is not a sufficient answer.) This writing you're going to do must aim at accomplishing something; you must be seeking a particular response from your reader. Suppose, for example, you're being a good citizen and writing to your congressman on a matter of interest to you. Your letter may encompass several purposes. You are writing to inform him of your thinking and perhaps to persuade him to vote for or against a certain bill. In explaining your position, you may find it useful to narrate a personal experience or to describe a situation vividly enough to engage his emotions. Your letter may therefore include the four main kinds of writing—exposition, argument, narration, and description—but one of them is likely to be primary. Your other purposes would be subordinated to your effort to persuade the congressman to vote your way. So your precise purpose might be stated: "I am writing to Congressman Green to persuade him to vote for increased social security benefits provided by House Bill 5883."

Write it out in a sentence just that way. Pin it down. And make it as precise as you can. You will save yourself a good deal of grief in the writing process by getting as clear a focus as you can on your purpose. You'll know better what to include and what to omit. The many choices that combine to form the writing process will be made easier because you have taken the time and the thought to determine the exact reason you have for writing this communication.

When you're writing something longer than a letter, like a report, you can make things easier for yourself and your reader by making a clear statement of the central idea you're trying to develop. This is your thesis. To be most helpful, it must be a complete sentence, not just the subject of a sentence. "Disappointing sales" may be the subject you are writing about, but it is not a thesis. Stating it that way won't help you organize the information in your report. But if you really say something specific about "disappointing sales," you will have a thesis that will provide a framework for development. For example:

Disappointing sales may be attributed to insufficient advertising, poor selection of merchandise, and inadequate staffing during peak shopping hours.

The frame for the rest of the report now exists. By the time you have finished explaining how each cause contributes to the result (disappointing sales), the report is finished. Not only finished, but clear; it follows an orderly pattern. The arguments, details, illustrations, statistics, or quotations are relevant because the thesis has given you a way of keeping your purpose clearly in focus.

Of course the ideal method is to get your purpose and your thesis down before you write a first draft, but I can tell you from experience that it doesn't always work that way. Sometimes the only way to capture a thesis statement is to sneak up on it by means of a meandering first draft. When you've finished that chore, read the whole thing over and try to ambush the thesis by saying, "OK, now just what am I trying to say?" Then *say it* as quickly and concisely as you can—out loud, so you can hear how it sounds. Then try to write it down in the most precise and concise way you can. Even if you don't capture the exact thesis statement you're looking for, there's a good chance that you will have clarified your purpose sufficiently to make your next revision better.

AUDIENCE

Most of the writing the average person does is aimed at a specific reader: the congressman, the boss, a business subordinate, a customer, a supplier. Having a limited audience can be an advantage. It enables the writer to analyze the reader and shape the writing so that it effectively achieves the purpose with that particular reader or group of readers.

Here are some questions that will help you communicate with the reader:

1. How much background do I need to give this reader, considering his or her position, attitude toward the subject, and experience with this subject?
2. What does the reader need to know, and how can I best give him or her this information?
3. How is my credibility with this reader? Must I build it gradually as I proceed, or can I assume that he or she will accept certain judgments based on my interpretations?
4. Is the reader likely to agree or disagree with my position? What tone would be most appropriate in view of this agreement or disagreement?

The last question, I think, is very important. It implies that the writer will try to see the reader's point of view, will bend every effort to look at the subject the way the reader will probably look at it. That isn't easy to do. Doing it takes both imagination and some understanding of psychology. But it is worth the considerable effort it involves; it is a gateway to true communication.

Let me briefly suggest a point here that is easily overlooked in the search for better techniques of communication. It is simply this: *better writing depends on better reading.* Technique isn't everything. One cannot learn to be imaginative and understanding in ten easy lessons, but reading imaginative literature—fiction, poetry, drama—is one way of developing the ability to put yourself in someone else's place so you can see how the problem looks from that person's angle. The writer who limits reading to his or her own technical field is building walls around the imagination. The unique value of imaginative literature is the escape that it allows from the prison of the self into the experience and emotions of persons of different backgrounds. That ability to understand why a reader is likely to think and act in a certain way is constantly useful to the writer in the struggle to communicate.

FORMAT

Having thought about your purpose and your audience, consider carefully the appropriate format for this particular communication you are writing. Of course the range of choice may be limited by company procedures, but even the usual business formats allow some room for the writer to use ingenuity and intelligence.

The key that unlocks this ingenuity is making the format as well as the words work toward achieving your purpose. You can call attention to important points by the way you arrange your material on the page.

Writers sometimes neglect this opportunity, turning the draft over to the typist without taking the trouble to plan the finished page. Remember the reader, his or her desk piled high with things to be read, and try to make your ideas as clear as possible on the page as the reader glances at it.

External signals such as headings, underlining, and numerals are ways of giving the reader a quick preview. Use them when they are appropriate to your material.

Don't be afraid of using white space as a way of drawing attention to key ideas.

To see how experts use format for clarity and emphasis, take a look at some of the sales letters all of us receive daily. Note how these persuasive messages draw the reader's attention to each advantage of the product. You may not want to go that far, but the principle of using the format to save the reader's time is worth your careful consideration.

Overburdened executives are always looking for shortcuts through the sea of paper. Many admit that they cannot possibly read every report or memo. They scan, they skip, they look for summaries. A good format identifies the main points quickly and gives an idea of the organizational structure and content. If the format looks logical and interesting, readers may be lured into spending more time on your ideas. You've taken the trouble to separate the important from the unimportant, thus saving them time. The clear format holds promise of a clear analysis, and what boss can be too busy for that?

EVIDENCE

Unless you're an authority on your subject, your opinions carry only as much weight as the evidence you can marshal to support them. The more evidence you can collect before writing, the easier the writing task will be. Evidence consists of the facts and information you gather in three ways: (1) through careful observation, (2) through intelligent fieldwork (talking to the appropriate persons), and (3) through library research.

When you follow the evidence where it leads and form a hypothesis, you are using inductive reasoning, the scientific method. If your evidence is adequate, representative, and related to the issue, your conclusion must still be considered a probability, not a certainty, since you can never possess or weigh *all* the evidence. At some point you make the "inductive leap" and conclude that the weight of evidence points to a theory as a probability. The probability is likely to be strong if you are aware of the rules of evidence.

The Rules of Evidence

Rule 1. Look at the Evidence and Follow Where It Leads.

The trick here is not to let your own bias seduce you into selecting only the evidence you agree with. If you aren't careful, you unconsciously start forcing the evidence to fit the design that seems to be emerging. When fact A and fact B both point toward the same conclusion, there is always the temptation to *make* fact C fit. Biographer Marchette Chute's warning is worth heeding: ". . . you will never succeed in getting at the truth if you think you know, ahead of time, what the truth ought to be."

A reliable generalization ought to be based on a number of verifiable, relevant facts—the more the better. Logicians tell us that evidence supporting a generalization must be (1) known or available, (2) sufficient, (3) relevant, and (4) representative. Let's see how these apply to the generalization, "Cigarette smoking is hazardous to your health."

The evidence accumulated in animal studies and in comparative studies of smokers and nonsmokers has been published in medical journals: it is therefore known and available. Is it sufficient? The government thinks so; the tobacco industry does not. Which is more likely to have a bias that might make it difficult to follow where the evidence leads?

Tobacco industry spokesmen argue that animal studies are not necessarily relevant to human beings and point to individuals of advanced age who have been smoking since childhood as living refutations of the supposed evidence. But the number of cases studied is now in the thousands, and studies have been done with representative samplings. The individuals who have smoked without apparent damage to their health are real enough, but can they be said to be numerous enough to be representative of the usual effects of tobacco on humans? "Proof" beyond doubt lies only in mathematics, but the evidence in this example is the kind on which a sound generalization can be based.

Rule 2. Look for the Simplest Explanation that Accounts for All the Evidence.

When the lights go out, the sudden darkness might be taken as evidence of a power failure. But a quick investigation turns up other evidence that must be accounted for: the streetlights are still on; the refrigerator is still functioning. So a simpler explanation may exist, and a check of the circuit breakers or fuse box would be appropriate.

Rule 3. Look at All Likely Alternatives.

Likely alternatives in the example just discussed would include such things as burned-out bulbs, loose plugs, and defective outlets.

Rule 4. Beware of Absolute Statements.

In the complexity of the real world, it is seldom possible to marshal sufficient evidence to permit an absolute generalization. Be wary of writing general statements using words like *all, never* or *always*. Sometimes these words are implied rather than stated, as in this example:

> Jogging is good exercise for both men and women.
>
> [Because it is unqualified, the statement means *all* men and women. Since jogging is not good for people with certain health problems, this statement would be better if it said *most* or *many* men and women.]

Still, caution is always necessary. Induction has its limitations, and a hypothesis is best considered a probability subject to change on the basis of new evidence.

The other kind of reasoning we do is called deduction. Instead of starting with particulars and arriving at a generalization, deduction starts with a general premise or set of premises and works toward the conclusion necessarily implied by them. If the premise is true, it follows that the conclusion must be true. This logical relationship is called an inference. A fallacy is an erroneous or unjustified inference. When you are reasoning deductively, you can get into trouble if your premise is faulty or if the route from premise to conclusion contains a fallacy. Keep the following two basic principles in mind, and watch out for the common fallacies that can act as booby traps along the path from premise to conclusion.

Two Principles for Sound Deduction

1. The ideas must be true; that is, they must be based on facts that are known, sufficient, relevant, and representative.
2. The two ideas must have a strong logical connection.

If your inductive reasoning has been sound enough to take care of the first principle, let me warn you that the second is not so simple as it sounds. We have all been exposed to so much propaganda and advertising that distort this principle that we may have trouble recognizing the weak or illogical connection. Nothing is easier than to tumble into one of the commoner logic fallacies. As a matter of fact, no barroom argument would be complete without one of these bits of twisted logic.

Twisted Logic

Begging the question. Trying to prove a point by repeating it in different words.

> Women are the weaker sex because they are not as strong as men.
> Our company is more successful because we outsell the competition.

Non sequitur. The conclusion does not follow logically.

> I had my best sales year; the company's stock should be a big gainer this year.
> I was shortchanged at the supermarket yesterday. You can't trust these young cashiers at all.

Post hoc. Because an event happened first, it is presumed to be the cause of the second event.

> Elect the Democrats and you get war; elect the Republicans and you get a depression.

Oversimplification. Treating truth as an either-or proposition, without any degrees in between.

> Better dead than Red.
> America—love it or leave it.
> If we do not establish our sales leadership in New York, we might as well close up our East Coast outlets.

False analogy. Because of one or two similarities, two different things are assumed to be *entirely* similar.

> You can lead a horse to water but you can't make him drink; there is no sense in forcing kids to go to school.
> Giving the Canal back to Panama would be like giving the Great Lakes back to the Indians.

Seen in raw form, as in the examples, these fallacies seem easy to avoid. Yet all of us fall into them with dismaying regularity. A writer must be on guard against them whenever he or she attempts to draw conclusions from data.

ORGANIZATION

Experienced writers often use index cards when they collect and organize information. Cards can be easily arranged and rearranged. By arranging them in piles, you can create an organizational plan. Here's how you play the game:

1. You can write only one point on each card—one fact, one observation, one opinion, one statistic, one whatever.
2. Arrange the cards into piles, putting all closely related points together. All evidence related to marketing goes in one pile, all evidence related to product development goes in another pile, and so on.
3. Now you can move the piles around, putting them in sequence. What kind of sequence? Consider one of the following commonly used principles:

   Chronological. From past to present to future.
   Background, present status, prospects
   Spatial. By location.
   New England territory, Middle Atlantic, Southern
   Logical. Depends on the topic.
   Classification and division according to a consistent principle (divisions of a company classified according to function)
   Cause and effect (useful in troubleshooting manuals)
   Problem-analysis-solution. Description of problem, why it exists, what to do about it.
   Order of importance. From least important to most important or from most important to least important.

   The choice of sequence will depend largely on the logic of the subject matter and the needs of your audience.
4. Go through each pile and arrange the cards in an understandable sequence. Which points need to precede others in order to present a clear picture?
5. That's it! You now have an outline that is both orderly and flexible. You can add, subtract, or rearrange whenever necessary.

SUMMARY

Think PAFEO. Use this word to remind you to clarify your purpose and analyze your audience before you write. When you have a clear sense of purpose, create a thesis that will act as a frame for your ideas.

Choose a format for your communication that will help the reader identify the main points quickly.

Collect your evidence before you write by observing, interviewing, and doing library research. Use the index card system to help you organize your evidence. Keep in mind the rules of evidence before you draw conclusions. Be alert to the common logic fallacies that can easily undermine the deductive process.

Try the file card system as a way of organizing your material. Write one point on each card and then place closely related points together in a pile. Place the piles in sequence according to a principle of organization you select. Some useful organizing principles include: chronological, spatial, logical, problem-analysis-solution, and order of importance. The sequence you choose will depend on the logic of the subject matter and the needs of your audience.

Michael E. Adelstein

The Writing Process

Michael E. Adelstein is Professor of English at the University of Kentucky and coauthor with W. Keats Sparrow of Business Communications *(Harcourt Brace Jovanovich, 1983).*

THE WRITING PROCESS

Their prevalent belief in the myths of genius, inspiration, and correctness prevents many people from writing effectively. If these individuals could realize that they can learn to write, that they cannot wait to be inspired, and that they must focus on other aspects besides correctness, then they can overcome many obstacles confronting them. But they should be aware that writing, like many other forms of work, is a process. To accomplish it well, people should plan on completing each one of its five stages. The time allocations may vary with the deadline, subject, and purpose, but the work schedule should generally follow this pattern:

1. Worrying—15%
2. Planning—10%
3. Writing—25%
4. Revising—45%
5. Proofreading—5%

Note that only 25% of the time should be spent in writing; the rest—75%—should be spent in getting ready for the task and perfecting the initial effort. Observe also that more time is spent in revising than in any other stage, including writing.

THE WRITING PROCESS. By Michael E. Adelstein. From *Contemporary Business Writing* (New York: Random House, 1971). Copyright © 1971 by Michael E. Adelstein. Reprinted by permission of Michael E. Adelstein.

Stage One—Worrying (15%)

Worrying is a more appropriate term for the first operation than *thinking* because it suggests more precisely what you must do. When you receive a writing assignment, you must avoid blocking it out of your mind until you're ready to write; instead, allow it to simmer while you try to cook up some ideas. As you stew about the subject while brushing your teeth, taking a shower, getting dressed, walking to class, or preparing for bed, jot down any pertinent thoughts. You need only note them on scrap paper or the back of an envelope. But if you fail to write them down, you'll forget them.

Another way to relieve your worrying is to read about the subject or discuss it with friends. The chances are that others have wrestled with similar problems and written their ideas. Why not benefit from their experience? But if you think that there is nothing in print about the subject or a related one—either because your assignment is restricted, localized, or topical—then you should at least discuss it with friends.

Let's say, for example, that as a member of the student government on your campus, you've been asked to investigate the possibility of opening a student-operated bookstore to combat high textbook prices. During free hours, you might drop into the library to read about the retail book business and, if possible, about college bookstores. If your student government is affiliated with the National Student Association, you might write to it for information about policies and practices on other campuses. In addition, you might start talking about the problem to the manager of your college bookstore, owners of other bookstores in your community, student transfers from other schools, managers of college bookstores in nearby communities, and some of your professors. Because you might be concerned about the public relations repercussions of a student bookstore, you should propose the venture to college officials. At this point, you will have heard many different ideas. To discuss some of them, bring up the subject of a student bookstore with friends. As you talk, you will be clarifying and organizing your thoughts.

The same technique can be applied to less involved subjects, such as a description of an ideal summer job, or an explanation of a technical matter like input-output analysis. By worrying about the subject as you proceed through your daily routine, by striving to get ideas from books and people or both, and by clarifying your thoughts in conversation with friends, you will be ready for the next step. Remember: you can proceed only if you have some ideas. You must have something to write about.

Stage Two—Planning (10%)

Planning is another term for organizing or for the task that students dread so—*outlining*. You should already have had some experience with this process. Like many students, unfortunately, you may have learned only how to outline a paper after writing it. This practice is a waste of time. If you want to operate efficiently, then it follows that you will have to specialize in each of the writing stages in turn.

When you force yourself simultaneously to conceive of ideas, to arrange them in logical order, and to express them in words, you cannot perform all of these complex activities as effectively as if you had concentrated on each one separately. By dividing the writing process into five stages, and by focusing on each one individually, the odds are that the result will be much better than you had thought possible. Of course you will hear about people who never plan their writing but who do well anyway, just as you may know someone who skis or plays golf well without having had a lesson. There are always some naturals who can flaunt the rules. But perhaps these people could have significantly improved their writing, golf, or skiing if they had received some formal instruction and proceeded in the prescribed manner.

An efficient person plans his work. Many executives take a few minutes before leaving the office or retiring at night to jot down problems to attend to the following day. Many vacationers list tasks to do before leaving home: stop the newspaper, turn down the refrigerator, inform the mailman, cut off the hot water, check the car, mail the mortgage payment, notify the police, and the like. Of course, you can go on vacation without planning—just as you can write without planning—but the chances are that the more carefully you organize, the better the results will be.

In writing, planning consists mainly in examining all your ideas, eliminating the irrelevant ones, and arranging the others in a clear, logical order. Whether you write a formal outline or merely jot down your ideas on scrap paper is up to you. The outline is for your benefit; follow the procedure that helps you the most. But realize that the harder you work on perfecting your outline, the easier the writing will be. As you write, therefore, you can concentrate on formulating sentences instead of also being concerned with thinking of ideas and trying to organize them. For the present, realize the importance of planning. . . .

Stage Three—Writing (25%)

When you know what you want to say and have planned how to say it, then you are ready to write. This third step in the writing process is self-explanatory. You need only dash away your thoughts, using pen, pencil, or typewriter, whichever you prefer. Let your mind flow along the outline. Don't pause to check spelling, worry about punctuation, or search for exact words. If new thoughts pop into your mind, as they often do, check them with your outline, and work them into your paper if they are relevant. Otherwise discard them. Keep going until either the end or fatigue halts you. And don't worry if you write slowly: many people do.

Stage Four—Revising (45%)

Few authors are so talented that they can express themselves clearly and effectively in a first draft. Most know that they must revise their papers extensively. So they roll up their sleeves and begin slashing away, cutting out excess verbiage, tearing into fuzzy sentences, stabbing at structural weaknesses, and knifing into obfuscation. Revision is painful: removing pet phrases and savory sentences is like getting rid of cherished possessions. Just as we dislike to discard old magazines, books, shoes, and clothes, so we hate to get rid of phrases and sentences. But no matter how distasteful the process may be, professional writers know that revising is crucial; it is usually the difference between a mediocre paper and an excellent one.

Few students revise their work well; they fail to realize the importance of this process, lack the necessary zeal, and are unaware of how to proceed. Once you understand why you seem to be naturally adverse to revision, you may overcome your resistance.

Because writing requires concentrated thought, it is about the most enervating work that humans perform. Consequently, upon completing a first draft, we are so relieved at having produced something that we tidy it up slightly, copy it quickly, and get rid of it immediately. After all, why bother with it anymore? But there's the rub. The experienced writer knows that he has to sweat through it again and again, looking for trouble, willing to rework cumbersome passages, and striving always to find a better word, a more felicitous phrase, and a smoother sentence. John Galbraith, for example, regularly writes five drafts, the last one reserved for inserting "that note of spontaneity" that ironically makes his work appear natural and effortless. The record for revision is probably held by Ernest Hemingway, who wrote the last page of *A Farewell to Arms* thirty-nine times before he was satisfied!

To revise effectively, you must not only change your attitude, but also your perspective. Usually we reread a paper from our own viewpoint, feeling proud of our accomplishment. We admire our cleverness, enjoy our graceful flights, and glow at our fanciful turns. How easy it is to delude ourselves! The experienced writer reads his draft objectively, looking at it from the standpoint of his reader. To accomplish this, he gets away from the paper for a while, usually leaving it until the following morning. You may not be able to budget your time this ideally, but you can put the paper aside while you visit a friend, grab a bite to eat, or phone someone. Unless you divorce yourself from the paper, you will probably be under its spell; that is, you will reread it with your mind rather than with your eyes. You will see only what you think is on the page instead of what is actually there. And you will not be able to transport yourself from your role of writer to that of critic.

Only by attacking your paper from the viewpoint of another person can you revise it effectively. You must be anxious to find fault and you must be honest with yourself. If you cannot or will not realize the weaknesses in your writing, then you cannot correct them. This textbook should open your eyes to many things that can go wrong, but above all you must *want* to find them. If you blind yourself to the bad, then your work will never be good. You must convince yourself that good writing depends on good rewriting. This point cannot be repeated too frequently or emphasized too much. Tolstoy revised *War and Peace*—one of the world's longest and greatest novels—five times. The brilliant French stylist Flaubert struggled for hours, even days, trying to perfect a single sentence. But question your professors about their own writing. Many—like me—revise papers three, four, or five times before submitting them for publication. You may not have this opportunity, your deadlines in school and business may not permit this luxury; but you can develop a respect for revision and you can devote more time and effort to this crucial stage in the writing process. . . .

Stage Five—Proofreading (5%)

Worrying, planning, writing, and revising are not sufficient. The final task, although not as taxing as the others, is just as vital. Proofreading is like a quick check in the mirror before leaving for a date. A sloppy appearance can spoil a favorable acceptance of you or your paper. If, through your fault or your typist's, words are missing or repeated, letters are transposed, or sentences juxtaposed, a reader may be perturbed enough to ignore or resist your ideas or information. Poor proofreading of an application letter can cost you a job. Such penalties may seem unfair, but we all react in this fashion. What would you think of a doctor with dirty hands? His carelessness would not only disgust you but would also raise questions about his professional competence. Similarly, carelessness in your writing antagonizes readers and raises questions in their minds about your competence. The merit of a person or paper should not depend on appearances, but frequently it does. Being aware of this possibility, you should keep yourself and your writing well groomed.

Poor proofreading results from failure to realize its importance, from inadequate time, and from improper effort. Unless you are convinced that scrutinizing your final copy is important, you cannot proofread effectively. If you realize the significance of this fifth step in the writing process, then you will not only have the incentive to work hard at proofreading but also will allow time for it in planning.

Like most things, proofreading takes time. We usually run out of time for things that we don't care about. But there's always time for what we enjoy or what is important to us. So with proofreading; lack of time is seldom a legitimate excuse for doing it badly.

But lack of technique is. To proofread well, you must forget the ideas in a paper and focus on words. Slow down your reading speed, stare hard at the black print, and search for trouble rather than trying to finish quickly. . . . Some professional proofreaders start with the last sentence and read backward to the first. Whatever technique you adopt, work painstakingly, so that a few careless errors will not spoil your efforts. If you realize the importance of proofreading, view it as one of the stages in the writing process, and labor at it conscientiously, your paper will reflect your care and concern. . . .

Each Stage Is Important

These then are the five stages in the writing process: worrying, planning, writing, revising, and proofreading. Each is vital. You need to become proficient at each of them, although, at some later point in your life, an editor or efficient secretary may relieve you of some proofreading chores. But in your career as a student and in your early business years, you will be on your own to guide your paper through the five stages. Since it is your paper, no one else will be as interested in it as you. It's your offspring—to nourish, to cherish, to coddle, to bring up, and finally, to turn over to someone else. You will be proud of it only if you have done your best throughout its growth and development.

Peter Elbow

The Direct Writing Process
for Getting Words on Paper

Peter Elbow teaches at the University of Massachusetts at Amherst. He is also the author of Writing without Teachers *(Oxford, 1973), an innovative introduction to the writing process.*

The direct writing process is most useful if you don't have much time or if you have plenty to say about your topic. It's a kind of let's-get-this-thing-over-with writing process. I think of it for tasks like memos, reports, somewhat difficult letters, or essays where I don't want to engage in much new thinking. It's also a good approach if you are inexperienced or nervous about writing because it is simple. . . .

Unfortunately, its most common use will be for those situations that aren't supposed to happen but do: when you have to write something you *don't* yet understand, but you also don't have much time. The direct writing process may not always lead to a satisfactory piece of writing when you are in this fix, but it's the best approach I know.

The process is very simple. Just divide your available time in half. The first half is for fast writing without worrying about organization, language, correctness, or precision. The second half is for revising.

Start off by thinking carefully about the audience (if there is one) and the purpose for this piece of writing. Doing so may help you figure out exactly what you need to say. But if it doesn't, then let yourself put them out of mind. You may find that you get the most benefit from ignoring your audience and purpose at this early stage of the writing process. . . .

In any event spend the first half of your time making yourself write down everything you can think of that might belong or pertain to your writing task: incidents that come to mind for your story, images for your poem, ideas and facts for your essay or report. Write fast. Don't waste any time or energy on how to organize it, what to start with, paragraphing, wording, spelling, grammar, or any other matters of presentation. Just get things down helter-skelter. If you can't find the right word just leave a blank. If you can't say it the way you want to say it, say it the wrong way. (If it makes you feel better, put a wavy line under those wrong bits to remind you to fix them.)[1]

I'm not saying you must never pause in this writing. No need to make this a frantic process. Sometimes it is very fruitful to pause and return in your mind to some productive feeling or idea that you've lost. But don't stop to worry or criticize or correct what you've already written.

While doing this helter-skelter writing, don't allow too much digression. Follow your pencil where it leads, but when you suddenly realize, "Hey, this has nothing to do with what I want to write about," just stop, drop the whole thing, skip a line or two, and get yourself back onto some aspect of the topic or theme.

Similarly, don't allow too much repetition. As you write quickly, you may sometimes find yourself coming back to something you've already treated. Perhaps you are saying it better or in a better context the second or third time. But once you realize you've done it before, stop and go on to something else.

When you are trying to put down everything quickly, it often happens that a new or tangentially related thought comes to mind while you are just in the middle of some train of thought. Sometimes two or three new thoughts crowd in on you. This can be confusing: you don't want to interrupt what you are on, but you fear you'll forget the intruding thoughts if you don't write them down. I've found it helpful to note them without spending much time on them. I stop right at the moment they arrive—wherever I am in my writing—and jot down a couple of words or phrases to remind me of them, and then I continue on with what I am writing. Sometimes I jot the reminder on a separate piece of paper. When I write at the typewriter I often just put the reminder in caps inside double parentheses ((LIKE THIS)) in the middle of my sentence. Or I simply start a new line LIKE THIS
and then start another new line to continue my old train of thought. But sometimes the intruding idea seems so important or fragile that I really want to go to work on it right away so I don't lose it. If so, I drop what I'm engaged in and start working on the new item. I know I can later recapture the original thought because I've already written part of it. The important point here is that what you produce during this first half of the writing cycle can be very fragmented and incoherent without any damage at all.

There is a small detail about the physical process of writing down words that I have found important. Gradually I have learned not to stop and cross out something

---

[1] An excerpt from a letter giving me feedback on an earlier draft: "I tried the direct writing process. Though it sounds simple enough, I . . . see now that in the past I've often interrupted the flow of writing by spending disproportionate time on spelling, punctuation, etc. I can spend hours on an opening paragraph stroking the words to death; then, if there's a deadline, have to rush through the remainder." (Joanne Turpin, 7/24/78.)

I've just written when I change my mind. I just leave it there and write my new word or phrase on a new line. So my page is likely to have lots of passages that look like

Many of my pages
Still I don't mean that you should stop and rewrite every passage till you are happy with it.
This kind of appearance.

What is involved here is developing an increased tolerance for letting mistakes show. If you find yourself crumpling up your sheet of paper and throwing it away and starting with a new one every time you change your mind, you are really saying, "I must destroy all evidence of mistakes." Not quite so extreme is the person who scribbles over every mistake so avidly that not even the tail of the *y* is visible. Stopping to cross out mistakes doesn't just waste psychic energy, it distracts you from full concentration on what you are trying to say.

What's more, I've found that leaving mistakes uncrossed out somehow makes it easier for me to revise. When I cross out all my mistakes I end up with a *draft*. And a draft is hard to revise because it is a complete whole. But when I leave my first choices there littering my page along with some second and third choices, I don't have a draft, I just have a succession of ingredients. Often it is easier to whip that succession of ingredients into something usable than, as it were, to *undo* that completed draft and turn it into a better draft. It turns out I can just trundle through that pile of ingredients, slash out some words and sections, rearrange some bits, and end up with something quite usable. And quite often I discover in retrospect that my original "mistaken phrase" is really better than what I replaced it with: more lively or closer to what I end up saying.

* * *

If you only have half an hour to write a memo, you have now forced yourself in fifteen minutes to cram down every hunch, insight, and train of thought that you think might belong in it. If you have only this evening to write a substantial report or paper, it is now 10:30 P.M., you have used up two or two and a half hours putting down as much as you can, and you only have two or more hours to give to this thing. You must stop your raw writing now, even if you feel frustrated at not having written enough or figured out yet exactly what you mean to say. If you started out with no real understanding of your topic, you certainly won't feel satisfied with what is probably a complete mess at this point. You'll just have to accept the fact that of course you will do a poor job compared to what you could have done if you'd started yesterday. But what's more to the point now is to recognize that you'll do an even crummier job if you steal any of your revising time for more raw writing. Besides, you will have an opportunity during the revising process to figure out what you want to say—what all these ingredients add up to—and to add a few missing pieces. It's important to note that when I talk about revising . . . I mean something much more substantial than just tidying up your sentences.

So if your total time is half gone, stop now no matter how frustrated you are and change to the revising process. That means changing gears into an entirely different consciousness. You must transform yourself from a fast-and-loose-thinking person who is open to every whim and feeling into a ruthless, tough-minded, rigorously logical editor. Since you are working under time pressure, you will probably use quick revising or cut-and-paste revising. . . .

* * *

Direct writing and quick revising are probably good processes to start with if you have an especially hard time writing. They help you to prove yourself that you *can* get things written quickly and acceptably. The results may not be the very best you can do, but they work, they get you by. Once you've proved you can get the job done you will be more willing to use other processes for getting words down on paper and for revising—processes that make greater demands on your time and energy and emotions. And if writing is usually a great struggle, you have probably been thrown off balance many times by getting into too much chaos. The direct writing process is a way to allow a limited amount of chaos to occur in a very controlled fashion.

It's easiest to explain the direct writing process in terms of pragmatic writing: you are in a hurry, you know most of what you want to say, you aren't trying for much creativity or brilliance. But I also want to stress that the direct writing process can work well for very important pieces of writing and ones where you haven't yet worked out your thinking at all. But one condition is crucial: you must be confident that you'll have no trouble finding lots to say once you start writing. . . .

As I wrote many parts of this book [i.e., *Writing with Power*], for example, I didn't have my thinking clear or worked out by any means, I couldn't have made an outline at gunpoint, and I cared deeply about the results. But I knew that there was lots of *stuff* there swirling around in my head ready to go down on paper. I used the direct process. I just wrote down everything that came to mind and went on to revise.

But if you want to use the direct writing process for important pieces of writing, you need plenty of time. You probably won't be able to get them the way you want them with just quick revising. You'll need thorough revising or revising with feedback. . . . For important writing I invariably spend more time revising than I do getting my thoughts down on paper the first time.

MAIN STEPS IN THE DIRECT WRITING PROCESS

- If you have a deadline, divide your total available time: half for raw writing, half for revising.
- Bring to mind your audience and purpose in writing but then go on to ignore them if that helps your raw writing.

- Write down as quickly as you can everything you can think of that pertains to your topic or theme.
- Don't let yourself repeat or digress or get lost, but don't worry about the order of what you write, the wording, or about crossing out what you decide is wrong.
- Make sure you stop when your time is half gone and change to revising, even if you are not done.
- The direct writing process is most helpful when you don't have difficulty coming up with material or when you are working under a tight deadline.

Linda Flower and John Ackerman

Evaluating and Testing As You Revise

Both Linda Flower, a member of the faculty at Carnegie Mellon University, and John Ackerman, a member of the faculty at the University of Utah, are nationally recognized authorities on the teaching of writing.

IMAGINING A READER'S RESPONSE

Even if you have a model or text to work from, as a writer you will work hard to construct a text: generating and discarding ideas, trying to figure out your point, sketching out alternative organizations, and then trying to signal that point and structure to your reader. But the story does not end there because your various readers have to work equally hard to construct a meaning based on your text. That is, readers want to construct *in their own minds* a coherent text, with a hierarchical organization (like an issue tree) based on key points. And they need to see a purpose for reading. Readers want to know *Why read this? What is the point?* and *How is all of this connected?*—and they want to find answers as quickly as possible.

Readers begin to predict the structure of a text and its meaning as soon as they face a page. They will look for cues that the text might offer about point, purpose, and structure, but if they do not find them, they will go ahead and construct their own version of the text *and assume that is what you intended.* It may be helpful to imagine your readers as needing to *write* your text for themselves. Thus, their own goals and interests will strongly influence what they look for in a text and the meaning they make out of it. Your goal, then, as a writer is to do the best job you can to make sure a reader's process through your text and his or her understanding matches or comes close to what you intended.

Once your ideas are down in a draft of some form, **revision** lets you anticipate how readers will respond and adjust the text to get the response you hope for. **Local revision** involves editing, correcting spelling and grammar, and making local improvements in wording or sentence style. **Global revision** involves looking at the big picture, but that does not mean throwing the draft out and starting again, and it may not even take a lot of time. Global revision, however, does mean looking at the text as a whole—thinking globally about the major rhetorical decisions and plans you made—now that the draft is complete. Global revisions alter the focus, organization, argument, or detail to improve the overall text.

Revision begins with a tricky reading process, a close reading of your own text. You know from your own experience in school that readers can read for different purposes: to skim a text to prepare for class discussion and then to read more carefully for an examination. In business, because writing usually augments a transaction between people, readers look for information but they also look for how well a piece of writing is adapted to their needs. The key to your success as a reviser is your ability to read your own writing critically for different features and purposes. At the simplest level, revision is *re-vision*—the process of stepping back from one's text and seeing it anew. . . .

STRATEGY 1: LOOK FOR WRITER-BASED PROSE

Why is it that even experienced writers typically choose to draft and revise rather than write a final text in one pass? Suppose you put time into planning your paper; you followed a good model; you thought about your reader as you chose what to say. Why should your first draft not do the job? One reason is because first drafts often contain large sections of **writer-based prose.** Writer-based prose appears when writers are essentially talking to themselves, talking through a problem, exploring their own knowledge, or trying to get their ideas out. In fact, producing writer-based prose is often a smart problem-solving strategy. Instead of getting blocked or spending hours staring at a blank page trying to write a perfect text the first time through, writers can literally walk through their own memory, talking out on paper what they know. Other concerns, including what the reader needs to hear, are put on temporary hold. Writer-based prose is, in fact, a very effective strategy for searching your memory and for dealing with difficult topics or lots of information. Writer-based prose may also be the best a writer can do in a new situation. A writer facing an unfamiliar task may produce writer-based prose, uncertain of what readers expect.

Whether it is a strategy or consequence, the downside of this strategy is that the text it produces is typically focused on the writer's thinking—not the readers' questions or needs—and it often comes out organized as a narrative or a river of connections. A number of studies of writers new to organizations demonstrate the reasoning behind writer-based prose and its ill-effect. New employees write consciously and unconsciously to report their discovery process or to survey all they know. And their readers impatiently wade through the river looking for the specific ideas and information they need.

Here [see Figure 1] is the first draft of a progress report written by four students in an organizational psychology course who were doing a consulting project with a local organization, the Oskaloosa Brewing Company. As a reader, put yourself first in the position of Professor Charns. He is reading the report to answer three questions: As analysts, what assumptions and decisions did these students make in setting up their study? Why did they make them? And where are they in the project now? Then take on the role of the client, the company vice president

---

Draft 1

Group Progress Report

(1) Work began on our project with the initial group decision to evaluate the Oskaloosa Brewing Company. Oskaloosa Brewing Company is a regionally located brewery manufacturing several different types of beer, notably River City and Brough Cream Ale. This beer is marketed under various names in Pennsylvania and other neighboring states. As a group, we decided to analyze this organization because two of our group members had had frequent customer contact with the sales department. Also, we were aware that Oskaloosa Brewing had been losing money for the past five years, and we felt we might be able to find some obvious problems in its organizational structure.

(2) Our first meeting, held February 17th, was with the head of the sales department, Jim Tucker. Generally, he gave us an outline of the organization, from president to worker, and discussed the various departments that we might ultimately decide to analyze. The two that seemed the most promising and more applicable to the project were the sales and production departments. After a few group meetings and discussions with the personnel manager, Susan Harris, and our advisor, Professor Charns, we felt it best suited our needs and Oskaloosa Brewing's needs to evaluate their bottling department.

(3) During the next week we had a discussion with the superintendent of production, Henry Holt, and made plans for interviewing the supervisors and line workers. Also, we had a tour of the bottling department that gave us a first-hand look at the production process. Before beginning our interviewing, our group met several times to formulate appropriate questions to use in interviewing, for both the supervisors and the workers. We also had a meeting with Professor Charns to discuss this matter.

(4) The next step was the actual interviewing process. During the weeks of March 14–18 and March 21–25, our group met several times at Oskaloosa Brewing and interviewed ten supervisors and twelve workers. Finally, during this past week, we have had several group meetings to discuss our findings and the potential problem areas within the bottling department. Also, we have spent time organizing the writing of our progress report.

1 of 2

FIGURE I The Oskaloosa Brewing Progress Memo, Draft I

(5)   The bottling and packaging division is located in a separate building, adjacent to the brewery, where the beer is actually manufactured. From the brewery the beer is piped into one of five lines (four bottling lines and one canning line) in the bottling house, where the bottles are filled, crowned, pasteurized, labeled, packaged in cases, and either shipped out or stored in the warehouse. The head of this operation, and others, is production manager Phil Smith. Next in line under him in direct control of the bottling house is the superintendent of bottling and packaging, Henry Holt. In addition, there are a total of ten supervisors who report directly to Henry Holt and who oversee the daily operations and coordinate and direct the twenty to thirty union workers who operate the lines.

(6)   During production, each supervisor fills out a data sheet to explain what was actually produced during each hour. This form also includes the exact time when a breakdown occurred, what it was caused by, and when production was resumed. Some supervisors' positions are production-staff-oriented. One takes care of supplying the raw material (bottles, caps, labels, and boxes) for production. Another is responsible for the union workers' assignments each day.

These workers are not all permanently assigned to a production-line position. Workers called "floaters" are used, filling in for a sick worker or helping out after a breakdown.

(7)   The union employees are generally older than 35, some in their late fifties. Most have been with the company many years and are accustomed to having more workers per a slower moving line. . . .

2 of 2

FIGURE I   *(Continued)*

who follows their progress and wants to know: O.K. What is the problem (i.e., how did they define it)? And what did they conclude? Would this draft answer these questions for either of its intended readers?

Put yourself in the shoes of the professor. What would you look for in a progress report? According to Charns, he used this report to evaluate the group's progress: Were they on schedule; were they on task; did he need to intervene? However, he didn't need a blow-by-blow story to do that. As an evaluator he wanted to see whether they knew how to analyze an organization: Were they making good decisions (that is, decisions they could justify in this report); had they made any discoveries about this company? His needs as a reader, then, reflected his dual role as a teacher (supervisor) and an evaluator.

When we showed this draft to a manager (with comparable experience and responsibility to the Oskaloosa VP), we got a very different response. Here was a reader looking quickly for the information she wanted and building an image of the writers' business savvy based on their text. Here is part of her response as she read and thought aloud (the student text she reads is underlined):

Work began on our project with the initial group decision to evaluate . . . *Ok. Our project What project is this? I must have a dozen "projects" I keep tabs on. And who is this group?* This beer is marketed . . . *blah, blah, I'm tempted to skim. This must be a student project. But why am I reading about the fact somebody bought a lot of beer for their frat? Maybe the next paragraph.*

   Our first meeting . . . *Ok, they saw Jim and Susan, . . . looked at bottling, . . . wrote their paper. And now they are telling me where my packaging division is located! This is like a shaggy plant tour story. They are just wasting my time. And I suppose I should say that I am also forming an image of them as rather naive, sort of bumbling around the plant, interrupting my staff with questions. I mean, what are they after? Do they have any idea of what they are doing?*

Now put yourself in the shoes of the professor. What would you look for in a progress report? How would you evaluate this group as decision makers? Have they learned anything about analyzing organizations? How would you evaluate their progress at this point? Have they made any discoveries, or are they just going through the steps?

   Fortunately, this is not the draft the writers turned in to their professor or the company manager. The revised draft [as shown in Figure 2] was written after a short conference with a writing instructor who instead of offering advice, asked the writers to predict what each of their readers would be looking for. It took ten minutes to step back from their draft and to rethink it from the perspective of their professor and the Oskaloosa manager. They found that they needed global revision—a revision that kept the substance of their writer-based first draft, but transformed it into reader-based text. As you compare these two drafts, notice the narrative and survey organization in the first draft and the "I did it" focus that are often a tip-off to writer-based prose, and how they improved the second draft.

   How would you characterize the differences between drafts one and two? One clear difference is the use of a conventional memo/report format to focus the reader's attention. But beyond the visual display of information, the writers moved away from narrative organization, an "I" focus, and a survey form or "textbook" pattern of organization. Watch for these three patterns as you revise.

Narrative Organization

The first four paragraphs of the first draft are organized as a narrative, starting with the phrase "Work began. . . ." We are given a story of the writers' discovery process. Notice how all of the facts are presented in terms of when they were discovered, not in terms of their implications or logical connections. The writers want to tell us what happened and when; the reader, on the other hand, wants to ask "why?" and "so what?"

Draft 2

MEMORANDUM

TO: Professor Martin Charns

FROM: Nancy Lowenberg, Todd Scott, Rosemary Nisson,
    Larry Vollen

DATE: March 31, 1987

RE: Progress Report: The Oskaloosa Brewing Company

Why Oskaloosa Brewing?

Oskaloosa Brewing Company is a regionally located brewery manufacturing several different types of beer, notably River City and Brough Cream Ale. As a group, we decided to analyze this organization because two of our group members have frequent contact with the sales department. Also, we were aware that Oskaloosa Brewing had been losing money for the past five years and we felt we might be able to find some obvious problems in its organizational structure.

Initial Steps: Where to Concentrate?

After several interviews with top management and a group discussion, we felt it best suited our needs, and Oskaloosa Brewing's needs, to evaluate the production department. Our first meeting, held February 17, was with the head of the sales department, Jim Tucker. He gave us an outline of the organization and described the two major departments, sales and production. He indicated that there were more obvious problems in the production department, a belief also suggested by Susan Harris, the personnel manager.

Next Step

The next step involved a familiarization with the plant and its employees. First, we toured the plant to gain an understanding of the brewing and bottling processes. Next, during the weeks of March 14–18 and March 21–25, we interviewed ten supervisors and twelve workers. Finally, during the past week we had group meetings to exchange information and discuss potential problems.

The Production Process

Knowledge of the actual production process is imperative in understanding the effects of various problems on efficient production. Therefore, we have included a brief summary of this process.

The bottling and packaging division is located in a separate building, adjacent to the brewery, where the beer is actually manufactured. From the brewery the beer is piped into one of five lines (four bottling lines and one canning line) in the bottling house, where the bottles are filled, crowned, pasteurized, labeled, packaged in cases, and either shipped out or stored in the warehouse.

Problems

Through extensive interviews with supervisors and union employees, we have recognized four apparent problems within the bottling house operations. The first is that the employees' goals do not match those of the company. . . . This is especially apparent in the union employees, whose loyalty lies with the union instead of the company. This attitude is well-founded, as the union ensures them of job security and benefits. . . .

FIGURE 2 The Oskaloosa Brewing Progress Memo, Draft 2

A narrative organization is tempting to write because it is a prefabricated order and easy to generate. All of us walk around with stories in our head, and chronology is a common rhetorical move. Instead of creating a **hierarchical** organization among ideas or worrying about a reader, the writer can simply remember his or her own discovery process and write a story. Remember that in a hierarchical structure, such as an issue tree, the ideas at the top of the structure work as the organizing concepts that include other ideas. The alternative is often a string of ideas simply linked by association or by the order in which the writer thought about them. Papers that start out, "In studying the reasons for the current decline in our return customers, ..." are often a dead give-away. They tell us we are going to watch the writer's mind at work and follow him or her through the process of thinking out conclusions. Following one's own associations makes the text easier to write. But another reason new employees are tempted to write narrative reports is that they were often rewarded for narratives at some point in their career *as students.* They fail to realize that in business the reader is someone who expects to *use* this text (not check off whether or not they did the assignment).

A narrative pattern, of course, has the virtue of any form of drama—it keeps you in suspense by withholding closure. But this drama is an effective strategy only if the audience is willing to wait that long for the point. Most professional and academic readers are impatient, and they tend to interpret such narrative, step-by-step structures either as wandering and confused (Is there a point?) or as a form of hedging. Narrative structures may be read as veiled attempts to hide what really happened or the writers' actual position. Although a progress report naturally involves narrative, how has Draft 2 been able to *use* the narrative to answer readers' questions?

The "I" Focus

The second feature of Draft 1 is that it is a discovery story starring the writers. Its drama, such as it is, is squarely focused on the writer: "I did/I thought/I felt. . . ." Of the fourteen sentences in the first three paragraphs, ten are grammatically focused on the writer's thoughts and actions rather than on the issues. For example: "Work began . . . ," "We decided . . . ," Also, "we were aware . . . and we felt. . . ." Generally speaking, the reader is more interested in issues and ideas than in the fact that the writer thought them.

In pointing out the "I" focus in Draft 1, we are not saying that writers cannot refer to themselves or begin a sentence with "I," as many learned in school. Sometimes a specific reference to oneself is exactly the information a reader needs, and a reader may respond to the honesty and directness. Use "I" or "we" to make a claim or when it is an important piece of information, not just as a convenient way to start a sentence. In Draft 2, the students are clearly present as people doing the research, but the focus is on the information the reader wants to hear.

Survey Form or Textbook Organization

In the fifth paragraph of Draft 1 the writers begin to organize their material in a new way. Instead of a narrative, we are given a survey of what the writers observed. Here, the raw facts of the bottling process dictated the organization of the paragraph. Yet the client-reader already knows this, and the professor probably does not care. In the language of computer science we could say the writers are performing a "memory dump": printing out information in the exact form in which they stored it in memory. Notice how in the revised version the writers try to use their observations to understand production problems.

The problem with a survey or "textbook" pattern is that it ignores the reader's need for a different organization of the information. Suppose, for example, you are writing to model airplane builders about wind resistance. The information you need comes out of a physics text, but that text is organized around the field of physics; it starts with subatomic particles and works up from there. To meet the needs of your reader, you have to adapt that knowledge, not lift it intact from the text. Sometimes writers can simply survey their knowledge, but generally the writer's main task is to use knowledge rather than reprint it.

To sum up, in Draft 2 of the Oskaloosa report, the writers made a real attempt to write for their readers. Among other things, the report is now organized around major questions readers might have, it uses headings to display the overall organization of the report, and it makes better use of topic sentences that tell the reader what each paragraph contains and why to read it. Most important, it focuses more on the crucial information the reader wants to obtain.

Obviously this version could still be improved. But it shows the writers attempting to transform writer-based prose into reader-based prose and change their narrative and survey pattern into a more issue-centered top-to-bottom organization.

STRATEGY 2: TEST YOUR TEXT FOR A READER-BASED STRUCTURE

Reading your text for writer-based prose lets you spot places where you were still exploring ideas or talking to yourself—places that probably call for some sort of global or structural revision to make this text a reader-based document. But that does not tell you how to revise. **Reader-based prose** foregrounds and makes explicit the information a reader needs or expects to find. Reader-based prose tries to anticipate and support an active reader, one who probably will use your writing for some specific end. What do you want your reader to see, think, or do? How will your reader respond? One important approach to global revision is to look at your text as a conversation with the reader in which you set up some initial agreements and expectations and then fulfill your promises. We offer three ways to test for a reader-based structure: *testing your drafts against your initial plans, using clues that reveal this plan to a reader, and keeping the promises you made in your writing.*

Does Your Text Reflect Your Plans?

Texts have a way of running off by themselves. The more text you produce, the more convinced you are that your prose is complete, readable, even entertaining. This is natural given the commitment it takes to write anything, but because texts often drift away from your intentions, you need a way to keep your text honest or to realize that you have come up with a better approach. So instead of reading your text "as written" and just going with the flow, start by setting up an image of your purpose and plans in your mind's eye, then test your text against that image. Can you find any evidence of your plans in the text?

To get a good image of your plan, return to . . . your planning notes to review your plans consciously. In your mind or on paper, restate the plans *to do* and *say* that you produced prior to your current draft, and find the exact places in your text where your writing satisfies (or departs from) your plans to reach a reader. You could set the mental exercise up as a checklist [Figure 3].

Holding your draft up against the backdrop of your own plans can help you notice how well the two fit together. Did you have important goals, or good points in your notes that have just not appeared in your text yet? Did you simply forget ideas? Or did you find—as many writers do—that the act of writing was itself an inventive, generative process? Your plans may have changed as a result of writing. If that happened, what should you do? At this point, inexperienced writers often abandon their old plans and follow wherever the text seems to be taking them. However, experienced writers make another move. They go back into planning and look for possible ways to consolidate their new ideas with other old plans. They try to build a new plan that makes use of the good parts of both ideas. They use this round of planning to guide their global revisions consciously.

My plans for writing, to do and say were...

The important Goals and Purposes I gave myself...
My Key Points...
How I wanted my Reader to respond (and what other responses I anticipated)...
My choice of Text Conventions...
How I planned to make all of this work together...

As a checklist, I could evaluate my writing this way:

Plans *to do* and *say*	Draft	Text Reference
Goals...	✓	------------------------
Key points...	✓	------------------------
Intended Response...	✓	---------------------
Text Conventions, Models...	✓	-------------------------
Making it all work together...	????	*oops, gotta work on this!!!*

FIGURE 3   Checking Your Text against Your Plans

This strategy is obviously one you will use more than once—a kind of in-process evaluation that lets you keep checking in with your goals for writing and checking your text against the big picture of what you want it to do.

Cues That Reveal Your Plans to Your Reader

Maybe you are satisfied that you have indeed <u>defined</u> a real and shared <u>problem</u> for your reader, and <u>compared</u> some <u>alternative</u> ways to respond to it, <u>supported</u> them with <u>examples</u>, while <u>proposing</u> your favored course of <u>action</u>. But will the reader recognize all of those rhetorical moves? It is possible that you "talked about" this information, but did not make your good rhetorical plan to define, compare, support and propose fully apparent or explicit.

In most pieces of writing there are two conversations going on. One is the information that the reader needs or expects to find: the recommendations that you want to make, the results of your study, the idea you are proposing, the specifics of a solution. As we shall see, this conversation is held together, usually, by a strong chain of topics and an appropriate rhetorical pattern. But the second conversation is explicitly between the writer and the reader, announcing what the reader will find, in what order, and reminding the reader of where he or she is in the text.

This second conversation is often called **metadiscourse.** *Meta-* is a Greek prefix that means "along with" or "among," and the basic strategy is to include explicit statements and cues to the reader that announce and reinforce your intentions along with your content. There are two main ways writers give such cues. One is to talk directly to the reader, inserting metacomments that preview what will come, remind, predict, or summarize. This lets the writer step back, make the plan of the text more visible, and direct traffic, by telling the reader, "In the next section I will argue that. . . ." The other kind of cues work more like traffic signals—they are the conventional words and phrases that signal transitions, or logic, or the structure of ideas. Readers often expect to find these metacomments and signalling cues in some standard places. Some common places to insert cues to the reader include:

Title, title page
Table of contents
Abstract
Introduction or first paragraphs
Headings
The beginning and end of paragraphs (i.e., topic sentences)
Entire paragraphs in between long sections

So review your text first to see if you have used enough cues to make your plan clear, and second to see if you have included cues in places readers expect to see some guidance on what to look for and how to read this document [Figure 4].

Cues that signal your plan and guide the reader:

Cues that lead the reader forward

To show addition: *To show time:*

Again	Moreover,	At length	And then
And	Nor,	Immediately thereafter,	Later,
And then,	Too,	Soon,	Previously,
Besides	Next,	After a few hours,	Formerly,
Equally important,	First, second, etc.	Afterwards,	First, second, etc.
Finally	Lastly,	Finally,	Next, etc.
Further,	What's more,	Then	
Furthermore,			

Cues that make the reader stop and compare

But	Notwithstanding,	Although
Yet,	On the other hand,	Although this is true,
And yet,	On the contrary,	While this is true,
However,	After all,	Conversely,
Still	For all that,	Simultaneously,
Nevertheless,	In contrast,	Meanwhile
Nonetheless,	At the same time,	In the meantime,

Cues that develop and summarize

			To signal in a relationship:
To give examples:	*To emphasize:*	*To repeat:*	
For instance,	Obviously,	In brief,	Finally
For example,	In fact,	In short,	Because
To demonstrate,	As a matter of fact,	As I have said,	Yet
To illustrate,	Indeed,	As I have noted,	For instance,
As an illustration,	In any case,	In other words,	
	In any event,	That is,	

To introduce conclusions: *To summarize:*

Hence,	In brief,
Therefore,	On the whole,
Accordingly,	Summing up,
Consequently,	To conclude,
Thus,	In conclusion,
As a result,	

FIGURE 4   Giving Cues to the Reader

Metacomment cues that announce and reinforce your intentions

To ask a question about your topic or the argument unfolding:

What series of events led to event? . . .
To answer that question . . .

To preview what will come:

In the next section, we will see how this formula applies . . .
The third paragraph will reveal how . . .

To summarize what has been said thus far:

In the preceding pages, I've described . . .
Thus far, I have argued . . .

To comment on your writing and thinking as it unfolds:

I haven't mentioned yet that . . .
I'm talking about . . .
My main point is . . .

FIGURE 4   *(Continued)*

There is an endless variety of sentences and phrases that can be invented and inserted to announce and reinforce your main ideas and the progression that you want your reader to follow. To show you the power of metadiscourse and how it plays out in a text, here is an excerpt from a shareowner's letter with the metadiscourse highlighted [as shown in Figure 5]. We offer the original paragraph numbers with the cues underlined and the rhetorical purpose of the cues as we read them. . . .

Did You Keep Your Promises?

Your text started out with the best of intentions—a strong rhetorical plan and cues that keep your reader on track. The next test is to see if you followed through on the promises that you made. Read your text as if you were outlining its key points and promises and then look back to see if you have delivered the necessary detail. For instance:

Chairman's Letter

¶ Rhetorical Purpose/Cues

¶		
1		Often the greatest opportunities.........................
4	ties to history	Our strong 1984 results speak for themselves......
5		Pacific Telesis Group earned $829 million............
6	forces question	What happened? How did the corporation that many observers predicted would be the biggest loser in the AT&T breakup turn out to be one of the biggest winners?
7	links date with solution	To answer that question, you have to go back to 1980, when we developed..........................
8	signals logic	Our employees mobilized to make it work. And work it did. More specifically:..............
8c	signals order & emphasis	Third, and very important, we've built a relationship with the California Public Utilities Commission..........
	previews, emphasizes	a subject I'll return to later on in this letter. ....

The Future of Your Investment

34	previews	A corporation, particularly one as new as ours, must operate with a clearly articulated and widely understood vision of its future....................
37	summarizes	In the preceding pages I've described to you our strategies for deploying technology, marketing technology, and diversifying into new lines ....
38	summarizes, emphasizes signals logic, emphasis uses authority	I've talked about our determination .... But there is another very significant factor I haven't mentioned. I'm talking about the people who work for the Pacific Telesis Group

FIGURE 5 Metadiscourse in the Chairman's Letter

- Your problem/purpose statement promises four main points and an extended example: Does each paragraph keep that promise? By referring back to your announced plan and delivering four main (i.e., well-developed) points in the same order that you promised?
- Your topic sentence in the seventh paragraph promises the two key instances that support a legal precedent: Does the paragraph deliver them in the order and detail necessary?

John S. Harris

The Project Worksheet for Efficient Writing Management

John S. Harris is Professor Emeritus of English at Brigham Young University. The author of several books on technical writing, he was the founding president of the Association of Teachers of Technical Writing.

When employees produce poor quality documents, the fault may lie not with their incompetence or lack of devotion to the job, but with inadequate and unclear assignments from their project managers or publications managers. The Project Worksheet is designed to help managers give initial writing assignments, effectively manage documents during production, and evaluate finished products. The following pages describe the Worksheet and suggest ways publications managers can use it effectively.

THE PROJECT WORKSHEET

The Project Worksheet consists of a series of questions that help writers and managers consider the factors involved in planning a document. I developed the Worksheet years ago, and faculty in the technical writing program at Brigham Young University have successfully used it to teach technical writing for some years. After graduation, many of our students continue to use it in industry to plan documents and make document assignments.

What the Writer Must Consider

The manager and writer must determine the document's intended readers, its purpose, scope, form, length, graphic aids, and sources of information.[1] They must know such things before the writing begins. Failure to consider them results in inefficient document production and poor-quality documents.

Experienced writers in an organization usually answer these planning questions intuitively or subconsciously. Or they get the answers by direct conference, by study of past documents, or through the grapevine. They then efficiently plan and write good documents.

But less-experienced writers learn the answers to the questions only by the inefficient and expensive trial-and-error method of submitting draft documents and then repeatedly amending them as management requires.

Managers—whether of engineering projects, scientific studies, or computer software documentation—can reduce the trial-and-error writing cycle by providing complete answers for such planning questions during a writing assignment conference. But managers cannot provide the answers to the planning questions without first knowing the questions and their consequences. The Project Worksheet in Figure 1 guides both writers and managers in considering the important factors in planning documents. It can, of course, be adapted for special situations. Managers may want to reformat it to allow more space for answers, and some may wish to put it into a word processor as a template. Let us first examine the Worksheet questions and then consider procedures for using the Worksheet for document management.

Housekeeping Headings

Headings such as assignment date, due date, writer, and assignment authority are self-explanatory. Still, listing them together permits the use of the Worksheet as a tracking document for projects. The *document subject* is the file subject. The *document title* may be the same, or it may be something less prosaic—especially if the document is going out of house. Thus a document subject could be *Dental Hygiene,* and the document title could be *Your Teeth: An Owner's Manual.*

Primary Reader

No factor is more important for the writer than full understanding of the nature and needs of the document's primary reader. Experienced writers know this, of course. But even they do not always know everything that management knows and believes about the readers of the projected document. Too frequently both managers and writers assume they understand who the readers will be, but they

---

[1]Although this . . . [essay] explains how managers can use the Project Worksheet for managing documents, the books listed under Selected Readings explain how writers can use the Worksheet and other methods to design technical documents for various audiences and purposes.

PROJECT WORKSHEET

Subject _____ Title _____

Assigned by _____ Approved _____ Assignment date _____

Writer _____ Agreed to _____ Due date _____

Primary Reader:

    Technical level (education, experience, etc.):
    Position (title or organizational relationship):
    Attitude toward subject:
    Other factors:

Secondary Readers (others who may read the document):

    Technical level:
    Position:
    Attitude toward subject:
    Other factors:

Reader's Purpose:

    What should the reader know after reading?
    What should the reader be able to do after reading?
    What attitude should the reader have after reading?
    How will the reader access the material?

Writer's Purpose:

    Intellectual purpose:
    Career or monetary purpose:

Logistics:

    Sources (lab reports, library research, etc.):
    Physical size limits (if any):
    Form or medium prescribed or desirable:
    Available aids (graphics, etc.):
    Means of production:
    Outline (Preliminary):

Distribution and Disposition:

FIGURE I   Project Worksheet

may have two quite different audiences in mind. Using the Worksheet, managers and writers can reach an understanding of this and other questions.

Reader's Technical Level

By carefully considering the technical level of the reader, the manager and writer can decide the proper level of explanation needed. How much information does the reader already have? Are definitions of *photo-grammetry* or *perihelion* or *debentures* needed? Considering such things can help avoid losing the reader in a maze of technical jargon, or alternatively avoid alienating a reader with overly simplistic explanations. In the trade-off between those two extremes, erring on the side of simplicity is nearly always better.

Reader's Organizational Relationship

The writer must also consider whether the document will go to the executive tower or to the shop floor. Since writers often prepare documents for someone else's signature, the manager—and the writer—must consider the organizational relationship of the reader *to the document's signer.* Thus, if the document will be signed by the boss, the writer must—for the task—carefully assume the persona of the boss. Is the signer addressing superiors, subordinates, or peers? Such situations raise delicate questions, and the manager must recognize the entailed problems and pass on an understanding of those problems to the writer.

Reader's Attitude

Similarly, the attitude of the reader must be considered. Is the reader hostile to the new manufacturing procedure because it will result in downsizing her department? Or is the reader skeptical about the efficacy of new cutting-edge technology because it has not yet been proved? Or is the reader torpid and apathetic and needing a stiff jolt to see the promise—or the threat—of the subject of the document?

Other Factors

Often even more subtle factors about the reader must be considered. Some of these may overlap the preceding questions, but they may also affect the kind of document needed. Is the reader a white, conservative Republican woman of sixty, raised on an Iowa farm, or a young Afro-American male from the South Bronx, or a highly educated but condescending academic? Is the reader an Ivy-League MBA obsessed with this year's bottom line, a high school dropout production-line worker who is fearful that increased automation will eliminate jobs, a displaced homemaker with an acute case of computer anxiety, or an immigrant with limited understanding of English? Such demographic factors may require delicate adjustments in the document.

Secondary Reader

Often a secondary reader must also be considered. The company annual report may be read by both stockholders and the CEO. An advertising brochure intended for customers may also be read by the competition. An environmental impact statement may be read by both the Douglas Fir Plywood Association and the Sierra Club. A report intended for Level 2 management may have to get past a Level 3 gatekeeper. Such situations require that the writer juggle the factors of two or more audiences at once, and this requires some skill. If the readers differ in attitude, the writer must watch out for red-flag statements that may cause a secondary reader to charge. If the differences are in technical level, the secondary reader's need for more elementary explanations or more technical information can often be taken care of in footnotes, glossaries, sidebars, or appendices.

The Reader's Purpose

Organizational documents have purposes as varied as securing approval for a new manufacturing process, justifying spending public funds for a reclamation project, instructing a software purchase on how to use a spreadsheet, or easing the worries of environmentalists about the effect of a new highway near a prime trout-fishing stream. Four basic questions about the reader's purposes need consideration:

- What should the reader know after the reading?
- What should the reader be able to do after reading?
- What attitude should the reader have after reading?
- How will the reader access the information?

What Should the Reader Know after Reading?

What factual information does the reader need to obtain from the document? Does the reader need to know the percentage of radial keratotomy patients who can pass a 20/40 eye examination without glasses after the operation? The percentage of impurities remaining in palladium after the macrocycle solvent extraction process? The change in numbers of predator kills of sheep after reintroduction of wolves to the grazing range? Often the specific information the reader is to gain from the document can be presented in lists, outlines, tables, or graphs.

What Should the Reader Be Able to Do after Reading?

What the reader can *do* after reading may be quite different from what the reader *knows*. Knowing the names for the stages of mitosis is different from being able to recognize telophase through a microscope. Knowing the stoichiometric ratio for combustion of a gasoline/air mixture is different from being able to adjust a fuel-injection system to achieve the ideal mixture.

What a reader can do after reading may depend on technical skills. It may also depend on the reader's capability to make executive decisions based on the information contained in the document. Thus the reader's needs probably extend beyond gaining general information to applying specific knowledge. A reader may use the information on color changes in a chemical solution to perform titration in the assay laboratory. Or a reader may use the information on the size of natural gas reserves of the overthrust belt to decide the feasibility of building a 24-inch pipeline from Wyoming to California.

What Attitude Should the Reader Have after Reading?

Earlier the Worksheet asked about the reader's attitude before reading. Here the Worksheet asks about the attitude the writer and manager want the reader to have *after* reading the document. Do they want the reader to believe and feel that the proposed SDI weapons system is an effective deterrent to foreign aggression? Or do they want the reader to believe that it is a horrendously expensive and impractical pork-barrel boondoggle? Though such attitudes in technical documents may be based on hard data, they are nonetheless attitudes—sometimes emotionally charged attitudes—and the writer must realize that shaping those attitudes is sometimes part of the job, even a moral responsibility. The technical writer uses different tools for shaping attitudes than the politician or advertiser does, and may have a higher regard for truth, but like the politician or advertiser, the technical writer may be an attitude shaper.

And one other attitude deserves consideration: the attitude that the reader should have toward the preparer of the document. The writer wants the reader to think that the document was prepared by a credible and meticulous professional who cares about the subject, the needs of the reader, and the needs of his or her employer. Though this point may seem obvious, writers do not always automatically consider it.

How Will the Reader Access the Information?

We read novels, murder mysteries, and perhaps newspaper editorials beginning-to-end, but almost everything else, we read in some other fashion. We do not read dictionaries, phone books, repair manuals, computer documentation, or encyclopedias beginning-to-end. We often read textbooks and journal articles in a nonlinear fashion too. Sometimes the text is well designed to help us get the critical information quickly. Or, exasperatingly, the text may bury critical information in the middle of a full-page paragraph. Whatever way the text is designed, readers scan, or skip around trying to find what they want without reading every blessed word.

Realizing what information the reader wants, the writer or manager can design documents so that the information is easy to find. Information accessibility relies on such devices as

- Headings
- Outlines
- Indexes

- Graphics
- Underlining
- Sidebars
- Glosses
- Varied typefaces
- Cross-references
- Bulleted lists (like this).

USA Today uses many of these devices. The *New York Times* uses fewer. *USA Today* lacks the substance of the *Times,* but it is easier to get the news from it quickly. A similar comparison could be made between *Popular Science* and *Scientific American*. The differences in format reflect how the writers and editors anticipate the publications will be read.

The writer and manager must carefully consider whether the reader will read beginning-to-end or access pieces of data individually. The basic consideration should be the importance of the information being conveyed. The more important the information, the more accessible it should be. The document manager must help the writer anticipate *all* the ways that *all the readers* will want to access the information and require the writer to provide the machinery that will allow that access. Most writers of technical and scientific documents should pay much more attention to accessibility. Few readers of technical documents read them beginning-to-end.

Writer's Purpose

The writer's and reader's purposes for a document differ just as the buyer's and seller's purposes differ. Both have their agendas. The writer—or the manager—may want to have a project approved. The target reader may want to know if the project is feasible.

The writer probably has potential salary, prestige, and promotion purposes for writing, but may also be writing because the task is challenging or intellectually interesting. The writer and manager should decide in advance whether the project is routine and can be done rather perfunctorily, or whether it will affect the safety of the user, the prosperity of the company, the security of the nation, the preservation of the environment—or the continued employment of the writer. These questions help the writer and manager determine the priorities of the task.

Logistics

The logistics are the nuts and bolts things—such as where the information comes from, the size of the projected document, and what form the document will take. These too must be considered in advance.

Sources of Information

Will the document be based on observation, personal opinion, government reports, a literature search, engineers' notes, customer interviews, public opinion polls, laboratory tests, compilations of vendors' brochures, or what? Frequently, lead time may be required, so early consideration of sources is wise to allow tests to be run, surveys to be conducted, or publications to be ordered.

Size of Document Expected

A two-page memo will not do if a twenty-page report is expected, and a twenty-page report will not do if a two-page memo is expected. Again, the question should be resolved between the writer and the manager before the writing begins.

Form or Medium Prescribed or Desirable

The form or medium of the eventual document should also be considered in advance. Should the document follow the company style sheet? Would a film, a wall chart, a wallet card, or a video be a more useful medium? *Such choices should be made according to subject, purpose, reader (user), and cost.*

Graphic Aids Available

Would photographs or art work be useful? Are facilities available to prepare them? Is there time? Are funds available? Again, these must be considered early.

Means of Document Production

The means of production of the document also need to be considered early. Will the document be laid out on a Macintosh with PageMaker software? Will it be photocopied or multilithed? How will the means of production affect the legibility and credibility of the document, and how much will it cost?

Outline

Even at the early planning stages, the writer and manager should consider the topics to be treated, their order, and their proportion of the expected document. Such an outline can, of course, be revised during preparation.

Distribution and Disposition

How many copies will be needed? Where do they go? What happens to them? The manager and writer should consider these questions—and their many consequences—in advance. Will the document be read by the competition? Will it come back to haunt everybody ten years later? Will it be subpoenaed in court? The wise consider such things.

PROJECT PLANNING CONFERENCE

The manager and the writer must come to an agreement about a document's audience, purpose, and scope, and the logistics of its development, production, and distribution. Such negotiation is most efficient in a planning conference. Usually the manager calls the conference, but the negotiation can take a variety of forms, depending on the situation and the experience of the writers.

With an inexperienced writer, the manager may say, "Willoughby, I want you to write a proposal for my signature recommending adoption of the hyperbaric procedure. It should be addressed to Allardyce, the plant manger. He is a chemical engineer with twenty years' experience on the job. But it will also be read by Sung. She is the company comptroller and has ultimate approval power on expenditures. She is pretty conservative, but her attention to cash-flow probably saved the company during the last recession."

With a more experienced writer, the manager can more inductively and democratically ask, "Kim, we need a brochure about our prefabricated forms for concrete. What ways do you see to make a brochure that will meet the needs of all of its readers?" Obviously, here the manager respects the writer's knowledge and expects useful suggestions.

Whatever the method, the manager and writer should agree upon the answers, write them down—Yes, and sign off—each keeping a copy. Such a signed-off Worksheet is then a kind of contract agreed to by both parties, and both can feel more secure with it. The manager is now more secure knowing that the assignment is clearly understood, and the writer is more secure knowing what is expected.

POSTCONFERENCE MANAGEMENT

Obviously other factors may arise during the writing. The manager may receive new data, or the writer may see new ways to deal with the problems. Depending on their impact on the document, such things may require additional conferences and a renegotiating of the specifications.

As needed, the manager may discreetly ask about the progress of the document, or may even ask to see sections in draft. However, if the initial assignment has been done carefully, a responsible and skilled writer should be able to produce a sound document with little intermediate prodding.

The writer properly attaches the Project Worksheet to the draft form of the document when submitting it. Then the manager checks to see if it fulfills the assignment. If it does, the manager approves it for production.

Quite often, however, problems appear. Of these, failures to match the assignment are the most obvious, and the manager can ask for a rewrite to make the document match the specifications. The most common problem seems to be a failure to consider adequately the technical levels of all the readers.

Or it may become clear that the specifications need changing. If for good reasons they do need changing, the manager can change them and ask that the new specifications be followed, but the manager who has agreed to the original specifications is less likely to make such changes capriciously or arbitrarily. If lessening the power to make arbitrary changes seems to take the fun out of being boss, then the manager should recognize that the resultant empowerment of the writer is not only enlightened and trendy, but also pragmatically effective.

OTHER APPLICATIONS

If the writer's manager does not make such detailed assignments, the writer obviously can still use the Project Worksheet for planning. Or the writer can use the Worksheet to initiate a conference and negotiate with the manager. We have found that managers are often impressed when writers request such conferences and demonstrate through their questions—the Worksheet questions—that they have a clear understanding of the situation.

The Worksheet is also effective in negotiating contracts in freelance writing assignments, or as an assignment device between editor and writer. In the university, a professor can make writing assignments following the Worksheet questions. Or a sharp student can use the questions to clarify a vaguely given assignment. And many problems of master's theses and doctoral dissertations would be avoided if candidates and supervising professors negotiated the specifications for theses and dissertations through the Worksheet.

In much technical writing, the most important work is the planning done before the first words of the document are written. If managers will use the Project Worksheet approach to document management, their writers will work more efficiently and produce better documents.

SELECTED READINGS

Anderson, P. V. (ed.), *Teaching Technical Writing: Teaching Audience Analysis and Adaptation,* ATTW Anthology No. 1, Association of Teachers of Technical Writing, St. Paul, Minnesota, 1980.

Caernarven-Smith, P. *Audience Analysis and Response,* Firman Technical Publications, Pembroke, Massachusetts, 1983.

Mathes, J. C., and D. Stevenson, *Designing Technical Reports: Writing for Audiences in Organizations* (2nd Edition), Macmillan, New York, 1991.

Pearsall, T., *Audience Analysis for Technical Writing,* Glencoe Press, Beverly Hills, California, 1969.

Souther, J. W., and M. L. White, *Technical Report Writing* (2nd Edition), John Wiley & Sons, New York, 1977.

Spilka, R., Orality and Literacy in the Workplace: Process- and Text-Based Strategies for Multiple-Audience Adaptation. *Journal of Business and Technical Communication, 4:1,* pp. 44–67, 1990.

Dorothy A. Winsor

Communication Failures Contributing to the Challenger Accident: An Example for Technical Communicators

Dorothy A. Winsor, Associate Professor at Iowa State University, conducts research on the writing of engineers. Her work has won the NCTE (National Council of Teachers of English) Outstanding Research Award on three separate occasions.

A technological failure such as the explosion of the space shuttle Challenger can be puzzling in retrospect. Investigation often reveals that various people in the organization involved knew that the failure was likely and knew how to prevent it, and yet that knowledge was not shared within the organization as a whole. How does it happen that such important knowledge is not communicated? In the case of the Challenger, why did those who knew of the problem with the shuttle's solid rocket boosters not convince those in power to stop the launch?

The answer to this question lies in a complex set of factors, the most important of which seem to be (1) managers and engineers viewing the same facts from different perspectives, and (2) the general difficulty of either sending or receiving bad news, particularly when it must be passed to superiors or outsiders. An analysis of the communication failures that contributed to the Challenger accident is potentially of great interest to engineers and their managers because a large part of an engineer's job is to communicate both good and bad news upward to management for decision making. The Challenger explosion was a horrifying public

COMMUNICATION FAILURES CONTRIBUTING TO THE CHALLENGER ACCIDENT: AN EXAMPLE FOR TECHNICAL COMMUNICATORS. By Dorothy A. Winsor. © 1988 IEEE (The Institute of Electrical and Electronics Engineers, Inc.). Reprinted with permission from *IEEE Transactions on Professional Communications* 31 (September 1988): 101–7. Also reprinted with the kind permission of Dorothy A. Winsor.

event, but it resulted from factors that are probably at work more quietly in many other organizations.

The first of these factors—managers and engineers viewing the same facts from different perspectives—suggests that knowledge is not simply seeing facts but rather interpreting them, and that interpretation varies depending upon one's vantage point. Communication, then, is not just shared information; it is shared interpretation. Achieving shared interpretation within an organization is relatively easy if the sender and receiver of communication share the same corporate role and hence the same concerns and values. If sender and receiver are from different corporate subcultures, however (as they often are in technical communication), then achieving shared interpretation is more difficult [1, 2]. In an appendix to the *Report of the Presidential Commission on the Space Shuttle Accident,* for instance, Commission Member R. P. Feynman notes the difference between probability estimates of flight failure and loss of life by managers—1 in 100,000—and engineers—1 in 100 [3, V. 1, p. F-1]. This difference occurred despite the fact that the managers were almost all engineers by training. Presumably, managers and working engineers had much the same background and many of the same facts at their disposal, but they interpreted the facts quite differently because they approached them from different points of view.

By the same token, communication about the solid rocket booster joint that failed was made more difficult because it was bad news. Research has repeatedly shown that bad news is often not passed upward in organizations [4, 5]. Moreover, even when bad news is sent, people are less likely to believe it than good news [6, 7]. In the shuttle disaster, bad news moved up only slowly from engineers to management within NASA; Marshall Space Center, where the shuttle program was headquartered; and Morton Thiokol International (MTI), the contractor responsible for the solid rocket boosters. It also moved slowly among the organizations because they were in a hierarchical relationship, with MTI dependent on Marshall for the contract and Marshall dependent on NASA for funds and career opportunities.

Additionally, the three organizations seemed to view one another as outsiders despite the fact that they were working jointly on the same project and, in the case of NASA and Marshall, were nominally part of the same agency. So the taboo against airing organizational dirty linen in public was added to the general difficulties of bad news transmission [8]. Communication about O-ring problems, then, had to overcome the barriers to moving bad news between engineering and management subcultures, up through organizational hierarchies, and out to other organizations. Under these circumstances, it is hardly surprising that the communication failed.

The following paragraphs explain, first, what failed physically on the shuttle system, and then what failed in the organizations' attempts to understand those physical problems and communicate about them in the two-year period before the January 28, 1986, launch. I hope that this presentation will give engineers and engineering managers some insight into how to minimize the occurrence of events like the Challenger accident.

PHYSICAL CAUSE OF THE ACCIDENT

The physical cause of the Challenger explosion was the failure of a rubber seal in the solid rocket booster. The shuttle system consisted of three parts: the orbiter, which contained crew and experimental equipment; a large tank of liquid fuel, which was used by the orbiter's engines during liftoff; and two solid rocket boosters, which assisted at liftoff and were jettisoned to be recovered and reused on later flights. These solid rocket boosters (SRBs) were made in segments, which were stacked together at the launch site. The joints between the segments were sealed with two O-rings, which were protected from the heat of combustion by putty. The joint was pressure sealed, meaning that during rocket firing, expanding gases from burning fuel pushed the putty into the air space in the joint; this compressed air, in turn, pushed the O-ring into place and held it there. The second O-ring in each joint, in theory, provided redundancy or backup for the primary ring. During the Challenger launch, the O-rings in one of the SRB joints failed to seal, allowing hot gases to escape from the side of the SRB and burn a hole into the nearby liquid fuel tank, which exploded approximately 73 seconds into the flight.

In hindsight, the failure of the O-rings should not have been unexpected. From early 1984 on, postflight evidence increasingly showed that the joint seals were failing to meet design expectations. After every shuttle flight, the SRBs were recovered, disassembled for inspection, and readied for reuse in future flights. The inspections looked for any anomaly indicating the O-rings had not functioned as they should, and specifically for anything such as charred or eroded surfaces, which would indicate that the rings had failed to seal or had come into close contact with the heat of combustion. Before February 1984, only one O-ring anomaly had been found on the first nine flights. Beginning with the tenth shuttle flight, however, launched two years before Challenger on February 3, 1984, anomalies occurred on more than 50 percent of the flights [3, V. 1, p. 155].

We now know that the increased number of O-ring anomalies after early 1984 was probably caused by the use of increased pressure in the leak check done on the SRB joints after assembly at the launch site. The leak check involved blowing air into the joint through a hole located between the two O-rings and testing to see if the joint pressurized or sealed. Flights one through seven were tested at a pressure of 50 psi, flights eight and nine at 100 psi. From flight ten on, tests were done at 200 psi. In May 1985, MTI experiments showed that the increased pressure was likely to have blown holes through the putty that shielded the primary O-ring from the hot blast of ignition [3, V. 1, p. 156]. The holes were particularly damaging, because they not only allowed the hot gases to penetrate but actually focused them, so that they would cause maximum O-ring erosion, which is the eating away of the edge of the O-ring by the hot gases rushing past before the ring seals. Ironically, the pressure at which the leak check was conducted was increased to 200 psi because of concern that the joints were not sealing. Some officials believed that the blow holes were necessary—unless they existed, the putty rather than the O-ring could seal the joints during the test, and a defective O-ring could escape detection [3, V. 1, p. 134].

EARLY RESPONSES TO BAD NEWS: DISBELIEF AND FAILURE TO SEND UPWARD

When O-ring anomalies first began appearing in early 1984, neither engineers nor management at MTI treated them as serious problems in their communications to Marshall. They did not send a grave interpretation of the data upward and, judging by internal documents, did not believe one themselves. Marshall's reactions are more ambiguous, for they treated the O-ring situation as serious when they communicated downward to MTI but as relatively minor when they communicated up to NASA headquarters.

After MTI engineers saw erosion on the February 3, 1984, flight, they filed a problem report, and O-rings were entered into formal problem tracking systems at both Marshall and MTI. In an action that illustrates the difficulty of accepting bad news, MTI claimed (in a subsequent briefing to Marshall) that the problem was not serious because even if the primary O-ring were damaged, the second ring would provide redundancy and seal [3, V. 1, p. 128]. Tests done more than a year earlier had shown the secondary O-rings to be unreliable because of a phenomenon called joint rotation. Joint rotation means that under the pressure of launch, the two sides of the O-ring joint bent apart, widening the gap the O-ring had to seal. Joint rotation was apparently especially hard on the secondary ring, making it likely to pull completely out of its groove and never seal at all.

As a result of these tests, Marshall had changed the joint's classification from criticality 1R (a critical system with backup) to criticality 1 (a critical system without such backup) in December 1982. However, those involved in shuttle design apparently found this change hard to accept. Marshall personnel, for instance, apparently believed that they had redundancy in all but exceptional cases [3, V, 1, p. 128]. At MTI, the difficulty of belief was even more marked. Some MTI engineers and officials told the Presidential Commission that they were not notified of the criticality rating change, although their names appear on distribution lists on MTI documents [3, V. 1, p. 128]. Did these people lie about being notified? It seems more likely that they literally could not keep the bad news in mind.

Marshall's response to MTI's briefing illustrates what would be a pattern in their reaction to the O-ring difficulties. Some people at Marshall were willing to say that there was a serious problem—as long as any failure was perceived as MTI's. On February 28, for instance, John Miller, chief of the solid rocket motor branch at Marshall, wrote to his superior, George Hardy, through project engineer Keith Coates, urging that tests be done to see if the leak checks were causing problems. In an unusual recognition of the seriousness of the matter, Miller said O-ring failure could be "catastrophic" [3, V. 1, p. 245]. The next day, Coates also wrote to Hardy saying that MTI's briefing had minimized the extent of joint rotation possible and thus was too optimistic [3, V. 1, p. 128].

When the O-ring problem had to be claimed as Marshall's, however, and NASA had to be informed, Marshall, too, became optimistic. On March 8, 1984, a Flight Readiness Review for the eleventh shuttle flight was held at Marshall.

These reviews were held at four levels, with each level resolving what problems it could and passing unresolved issues on to the next higher one. The March 8 meeting was Level III, meaning that it was a meeting between Marshall and its contractors and that the highest officials present were the Marshall project managers. At the meeting, MTI reported that maximum erosion on the O-rings would be 0.09 inch and that tests had shown that the rings would function with 0.095 inch of erosion. The 0.005-inch difference appears to be an extremely small safety margin. Rather than report a serious problem to a Level II meeting at Johnson Space Center, however, Marshall apparently accepted the margin, because this same information was entered in the Marshall problem assessment report with a note that future flights need therefore not be delayed. The Marshall problem tracking record reads: "Remedial action—none required" [3, V. 1, p. 128].

The 0.005-inch safety margin was also used as a rationale to justify no flight interruptions at the Level I briefing of top NASA personnel on March 27. NASA accepted Marshall's recommendation but wrote to Lawrence Mulloy, SRB project manager at Marshall, asking for further study of the O-rings. Mulloy had Marshall engineer Lawrence Wear ask MTI to identify the cause of erosion, determine its seriousness, and define any necessary changes.

Internal MTI documents show that the contractor was examining the problem but with little sense of urgency, again evidencing the tendency to see the problem in the best light possible. MTI analyzed the erosion history and test data and, on May 4, presented Marshall with a plan for studying the O-rings to produce the information NASA had asked for. The information was not actually produced, however, until a briefing on August 19, 1985—16 months later.

Despite its optimism to NASA, Marshall was apparently uncomfortable with this pace and pressed for prompter action than MTI was giving. On July 2, L. H. Sayers, MTI's Director of Engineering Design, suggested by phone to Marshall's Ben Powers that tests done to date by MTI were sufficient [3, V. 1, p. 134]. This again implies that MTI did not perceive the matter to be crucial or dangerous.

Early signs of serious O-ring problems, then, were generally not believed at MTI, were accepted at Marshall only when it was possible to see the problem as MTI's, and were not sent upward to NASA headquarters.

CONTINUED BAD NEWS REJECTION DESPITE CONTRADICTORY EVIDENCE

The optimistic view of the O-rings persisted at both MTI and Marshall over the 1984–85 period despite mounting evidence that the rings were not functioning well. This evidence had to do with the effect of cold on the rings and the amount of erosion that could occur in an O-ring.

On January 24, 1985, the fifteenth shuttle flight was launched at a temperature of 53° F, the lowest up to that time. It showed much greater O-ring erosion than any previous flight. Tests have since shown that cold reduces O-ring resiliency and increases the time to seal, thereby exacerbating the problems the joint was already experiencing. The damage on the fifteenth flight was severe enough to bring concern about the rings to the fore again. On January 31, Mulloy wrote to Wear asking him to get MTI to prepare information on O-ring erosion for a Level III Flight Readiness Review scheduled for February 8. At that review, MTI personnel mentioned the cold as a factor in the damage but labeled the risk "acceptable," mostly because they assumed the secondary ring would seal if the first one failed.

An optimistic interpretation of the data on cold was held by both managers and engineers at MTI, and one of the primary advocates of continued launch was MTI engineer Roger Boisjoly, who would later be one of the primary opponents to the launch of Challenger. The split between managerial and engineering interpretation of the data did not develop for four or five more months. MTI's claim that the secondary O-rings would seal is a further example of retaining previous theories in the face of contradictory evidence because, as noted above, the redundancy of the secondary ring had been in question since late 1982.

As they had done earlier, Marshall management accepted MTI's rationale, at least in what it told NASA. A Level I Review was held on February 21. At this meeting, the influence of temperature was not mentioned, and only a single reference was made to O-ring erosion, saying redundancy made the risk acceptable [3, V. 1, p. 136]. Although each of the next four flights experienced joint seal problems, neither MTI nor Marshall seemed unduly concerned. Perhaps the very frequency of the problem added to its acceptability because the damage kept occurring with no serious consequences.

On June 25, however, one of the joints from a flight that had been launched April 29 was examined and found to have severe erosion of not only the primary ring, but even the secondary ring, calling its redundancy into question once again. As mentioned above, MTI had predicted maximum erosion of 0.09 inch for the primary seal. The primary O-ring on this joint was eroded 0.171 inch, almost double the predicted maximum and far beyond the 0.095 inch MTI had claimed to know was safe. This was bad news indeed. It could not be ignored, since an engineer from NASA headquarters was present when the damage was discovered. The engineer wrote to Michael Weeks at NASA on June 28, reporting the damage. The joint affected was a nozzle joint—that is, the joint linking the SRB to the flared section at its base. The NASA engineer blamed the damage on the fact that, in contrast to the rest of the SRB joints, which were now pressure-tested at 200 psi, nozzle joints were still being tested at only 100 psi, which might have permitted a defective ring to escape notice [3, V. 1, pp. 137–38].

In July 1985, Mulloy placed a launch constraint on the nozzle joints. This, in theory, meant that no other flights would take place until O-ring erosion at the nozzle joint had been fixed or shown not to be a problem. By including only the nozzle joints in the constraint, Mulloy was taking the most optimistic view possible of the problem. He reasoned that the nozzle joint had failed, not because of defective design, but because of a defective ring that had escaped notice in the nozzle joint's less rigorous leak test. Thus he believed that the leak test, and not the joint, was problematic. As a consequence, although he had assigned the launch constraint, Mulloy waived it for every subsequent flight, including Challenger, believing he was justified because Marshall increased to 200 psi the leak check pressure on the nozzle joint.

The launch constraint was treated as bad news by both MTI and Marshall. MTI officials testifying before the Commission all said they did not know about it, although subsequent MTI documents refer by document number to the report imposing the constraint [3, V. 1, p. 137]. The officials apparently did not take the news in. NASA officials, on the other hand, seem genuinely not to have been informed of the constraint, although regulations required that Level II be told. Marshall seems to have kept the news to itself rather than pass it out and up to NASA [3, V. 1, p. 138].

INTERNAL VERSUS EXTERNAL COMMUNICATION OF CONCERN FROM MTI ENGINEERS.

Despite MTI's ignorance of the launch constraint, the damage discovered in late June seems to have galvanized its engineers into action. Among them, at least, there seems to have been increased recognition of the problem's existence and its seriousness. MTI engineer Roger Boisjoly, for instance, became increasingly insistent about the potential danger from the O-rings. On July 22, his activity report predicted loss of the contract or flight failure if no solution was found. On July 31, he sent the following memo to R. K. Lund, MTI's Vice President of Engineering [3, V. 1, pp. 249–50]:

SUBJECT: *SRM O-ring Erosion/Potential Failure Criticality.*

This letter is written to insure that management is fully aware of the seriousness of the current O-ring erosion problem in the SRM joints from an engineering standpoint. The mistakenly accepted position on the joint problem was to fly without fear of failure and to run a series of design evaluations which would ultimately lead to a solution or at least a significant reduction of the erosion problem. This position is now drastically changed as a result of the SRM 16A nozzle joint erosion which eroded a secondary O-ring with the primary O-ring never sealing.

If the same scenario should occur in a field joint (and it could), then it is a jump ball as to the success or failure of the joint because the secondary O-ring cannot respond to the clevis opening rate and may not be capable of pressurization. The result would be a catastrophe of the highest order—loss of human life.

An unofficial team [a memo defining the team and its purpose was never published] with leader was formed on 19 July 1985 and was tasked with solving the problem for both the short and long term. This unofficial team is essentially nonexistent at this time. In my opinion, the team must be officially given the responsibility and the authority to execute the work that needs to be done on a non-interference basis (full time assignment until completed).

It is my honest and very real fear that if we do not take immediate action to dedicate a team to solve the problem with the field joint having the number one priority, then we stand in jeopardy of losing a flight along with all the launch pad facilities.

It is evident from this memo that Boisjoly's interpretation of the data had changed and that he was trying to communicate his new interpretation to his management. This memo does not give much space to new or old factual information, but rather concentrates on what the facts mean. Boisjoly's concern is evidenced both by the way he faults his own company (particularly his own management) and by his use of emotional language unusual in engineering documents. The memo implies, for instance, that MTI management may have an inaccurate understanding of the situation. The company's previous position was "mistakenly accepted." The company had planned to solve the problem but had not actually gotten to work to do so. The situation had changed "drastically," and a "catastrophe" could result. Boisjoly's memo and subsequent similar ones evidently had some effect on their receiver, because on the night before Challenger flew, Lund did at least begin by recommending against launch. Boisjoly's concern, however, was kept within MTI. He marked his memo *COMPANY PRIVATE* at both top and bottom. Although he was sufficiently alarmed to try to reach his superiors, he still attempted to keep bad news from the prying eyes of outsiders.

Bad news went to Marshall only in response to specific questions from them. On August 9, MTI engineer Brian Russell wrote to Marshall's Jim Thomas about results of tests investigating the effect of cold on the O-rings. The tests had been initiated after the January 24 flight showed such severe damage. The tone of Russell's letter makes an instructive contrast to Boisjoly's [3, V. 5, pp. 1568–69]:

SUBJECT: *Actions Pertaining to SRM Field Joint Secondary Seal*

Per your request, this letter contains the answers to the two questions you asked at the July Problem Review Board telecon.

1. *Question:* If the field joint secondary seal lifts off the metal mating surfaces during motor pressurization, how soon will it return to a position where contact is re-established?

   Answer: Bench test data indicate that the o-ring resiliency (its capability to follow the metal) is a function of temperature and the rate of case expansion. MTI measured the force of the o-ring against Instron plattens, which simulated the nominal squeeze on the o-ring and approximated the case expansion distance and rate.

   At 100° F the o-ring maintained contact. At 75° F the o-ring lost contact for 2.4 seconds. At 50° F the o-ring did not re-establish contact in ten minutes at which time the test was terminated.

   The conclusion is that secondary sealing capability in the SRM field joint cannot be guaranteed.

2. *Question:* If the primary o-ring does not seal, will the secondary seal seat in sufficient time to prevent joint leakage?

   Answer: MTI has no reason to suspect that the primary seal would ever fail after pressure equilibrium is reached, i.e., after the ignition transient. If the primary o-ring were to fail from 0 to 170 milliseconds, there is a very high probability that the secondary o-ring would hold pressure since the case has not expanded appreciably at this point. If the primary seal were to fail from 170 to 330 milliseconds, the probability of the secondary seal holding is reduced. From 330 to 660 milliseconds the chance of the secondary seal holding is small. This is a direct result of the o-ring's slow response compared to the metal case segments as the joint rotates.

Please call me or Mr. Roger Boisjoly if you have additional questions concerning this issue.

Russell and Boisjoly were actually of much the same opinion on the dangers of the joint, and Boisjoly helped Russell write the memo shown here, but Russell is speaking to outsiders and this affects the memo. The memo is not in any way untruthful, but it is not very communicative. In a sense, it is the opposite of Boisjoly's memo. It gives just the facts, providing little interpretation. Its tone is adamantly objective, in contrast to Boisjoly's more emotional one. In retrospect, some of the facts Russell gives should have been frightening. Note, for instance, that if there was joint rotation, the secondary O-ring never sealed when it was tested at 50° F. The conclusion reached from this information was "that secondary sealing . . . cannot be guaranteed." This is a negative wording of the finding, which was that at 50° F or below, MTI could pretty well guarantee no secondary seal.

That this memo did not communicate its intent is shown by the fact that the people who read it were uncertain about what it meant. Thomas copied the memo to be sent to NASA headquarters, but when the memo went through Mulloy's office for his signature, Mulloy returned it to Thomas saying it sounded like old news. The NASA official to whom Thomas was sending it has since said that even had he received the memo, he might not have understood it. "I don't know if anybody at that time understood the joint well enough to realize that the data was crucial," he said. When Mulloy was asked why he had not treated the temperature data as more important, he said he had not realized its significance, adding, "There were a whole lot of people who weren't smart enough to look behind the veil and say, 'Gee, I wonder what this means' " [9]. As can be seen, the urgency in Boisjoly's memo had not been conveyed to Marshall. Marshall officials had been given the facts about the effects of cold on the O-rings. They did not, however, interpret those facts in the same alarmed manner Boisjoly and Russell did, and Russell's memo did not attempt to communicate the more pessimistic interpretation.

Thus, MTI engineers concluded that the O-ring problems were serious before their management did. However, in their written communication, they varied the extent to which they voiced that seriousness, depending on whether their audience was internal or external.

THE SPLIT BETWEEN MANAGERS AND ENGINEERS

As is suggested by Boisjoly's memo above, MTI managers and engineers were beginning to disagree over the seriousness of the O-ring problem, and engineers had a difficult time communicating their view upward. Support from the miniculture of the task force probably made it easier than it had been previously for engineers to recognize the problem and speak up about it. On October 1, MTI engineers Roger Ebeling and S. R. Stein complained in separate internal MTI memos to management that the O-ring task force was being slowed by administrative delays and lack of cooperation [3, V. 1, pp. 252–53]. Both men complained that, although the task force members regarded their work as urgent, administrators required that all testing and design be done according to routines established for more leisurely long-term development.

On October 3, the team met with Joe Kilminster, MTI's Vice President of Space Programs, to discuss these administrative difficulties, but apparently the members were not successful in convincing him of the gravity of the situation. On October 4, Roger Boisjoly's activity report complained bitterly that "upper management apparently feels" MTI has the SRB contract "for sure and the customer be damned" [3, V. 1, p. 255]. In December, Ebeling actually told fellow task force members that MTI should not ship any more SRBs to Marshall until the problem was solved. However, consistent with patterns of bad news transmittal, he did not tell this to any of his superiors [3, V. 1, p. 142].

In January 1986, final preparation for Challenger's flight began. A launch scheduled for January 27 was cancelled and rescheduled for the next day. The temperature at launch time was 36° F, 17° colder than it had been for any previous launch. When MTI engineers, including Ebeling, Russell, and Boisjoly, heard of the predicted low temperatures, they became alarmed enough to convince Lund, their Vice President of Engineering, to recommend that the launch be delayed until the temperature of the joints reached 53° F, the previous lowest launch temperature. In their argument, they cited the information from Russell's memo and the severe erosion from the 53° F launch the previous January.

In a teleconference involving numerous MTI and Marshall managers and engineers, Lund, MTI's Vice President of Engineering, did recommend delaying launch. Marshall was apparently surprised by MTI's action and, refusing to accept the bad news, they resisted the recommendation. George Hardy, who headed engineering at Marshall, said he was "appalled" [3, V. 1, p. 94]. He later testified to the Presidential Commission that he meant he was appalled at MTI's data, but MTI personnel believed he said he was appalled at their recommendation. Mulloy apparently asked if they expected him to wait until April to launch.

In general, Marshall challenged, not MTI's facts, but the conclusions drawn from them. Mulloy and Hardy conceded that the primary ring might be slower to seal in the cold and that the increased time to seal would allow more erosion to take place. But they argued that the ring would seal eventually and that the joint could sustain three times the worst erosion they had seen to date and still have a good seal. In the time before the primary ring sealed, joint rotation would not yet have taken place; therefore, the secondary ring would still be good. Moreover, they believed that temperature could not be the deciding factor because, although they had had severe erosion at 53° F, they had also had it at 75° F. (It is true that erosion occurred at various temperatures. However, the Commission later pointed out that 15 percent of the flights launched above 65° F had O-ring anomalies, while 100 percent of those launched below that temperature had them.) In the face of Marshall's opposition, Joe Kilminster, MTI's Vice-President of Space Programs, asked that MTI have a private caucus off the phone line.

During the caucus, it became obvious that MTI was split along role lines. The engineers continued to argue against launch. Boisjoly says that, in the caucus, "there was never one comment in favor . . . of launching by any engineer or other nonmanagement person in the room" [3, V. 1, p. 93]. And in words that describe his listeners shutting out bad news, he says he himself argued until it became clear "that no one wanted to hear what I had to say" [3, V. 4, p. 697].

At this point, Jerald Mason, MTI's Senior Vice President, said it was obvious that all present would not reach agreement and that a management decision would have to be made. He polled the other three vice presidents in the room, first asking Lund, who had presented the recommendation not to launch, to take off his "engineering hat" and put on his "management hat" [3, V. 1, p. 94]. When Lund changed his role, he changed his position, and the four managers voted unanimously to launch. Engineer Brian Russell describes an atmosphere in which it was difficult to maintain opposition to the launch, wondering "whether I would have the courage, if asked, . . . to stand up and say no" [3, V. 4, p. 822]. As it happened, no engineer was asked to vote, and MTI went back on the conference line and reversed its earlier warning.

During the time MTI was caucusing, Marshall engineer Ben Powers also told his immediate supervisors that he agreed with MTI's recommendation not to launch, but his supervisors did not pass his view upward to Hardy and Mulloy. Similarly, Hardy and Mulloy did not pass on to NASA officials the fact that MTI engineers were opposed to the launch. At 11:38 the next morning, Challenger took off.

CONCLUSION

Looking at prelaunch miscommunications, then, several factors are apparent. First, no one at MTI or Marshall wanted to believe the growing evidence of O-ring problems. Second, even when MTI engineers came to believe that a problem existed, they had a difficult time convincing their management, with its different perspective on operations, to interpret the facts in the same light. In turn, on the night before launch, MTI personnel were unable to convince Marshall of the situation's gravity, even though they looked at the same facts, because Marshall, too, saw things differently. Finally, both engineers and managers at MTI were especially reluctant to communicate bad news to those outside the company.

All of this suggests a number of precautions engineers and their managers might take in the face of the same kind of pressure-induced miscommunication. From a manager's point of view, one of the most important precautions is to establish an atmosphere in which engineers feel free to communicate bad news as well as good. A number of writers have offered advice on how this can be done [8, 10, 11, 12, 13]. Establishing an open atmosphere takes time and a concerted effort from the whole organization. It cannot be done on short notice when emergencies arise.

In addition, pressures for holding back bad news should be anticipated to be especially strong when contractors are involved. Encouraging bad news transmittal is difficult when the bearer of bad tidings is afraid of losing a contract. Contracts can, perhaps, be designed to lessen this fear, but those issuing the contracts should be alert for any sign of problems, since full disclosure of bad news is unlikely in this situation.

Lastly, managers and engineers alike should anticipate that they are probably erring on the side of optimism in interpreting data bearing on already established designs and programs. Failure to believe bad news is probably caused by a number of factors, including reluctance to admit that one was wrong, fear of practical consequences such as expensive redesign, and a kind of intellectual inertia that makes it easier to persist in an already established belief than to change it. Such optimism can, however, have disastrous consequences, especially when coupled with other forces, as the Challenger accident demonstrates.

REFERENCES

1. Gregory, K. L., "Native-View Paradigms: Multiple Cultures and Culture Conflicts in Organizations," *Administrative Science Quarterly 28* (1983), 359–76.
2. Riley, P., "A Structurationist Account of Political Culture," *Administrative Science Quarterly 28* (1983), 414–37.
3. Presidential Commission on the Space Shuttle Challenger Accident, *Report of the Presidential Commission on the Space Shuttle Accident,* 5 vols, Washington: GPO, 1986.
4. Housel, T. J., and Davis, W. E., "The Reduction of Upward Communication Distortion," *Journal of Business Communication 14* (1977), 49–65.
5. Frank, A. D., "Can Water Flow Uphill?" *Training and Development Journal 38* (1984), 118–28.
6. Rasberry, R. W., and Lemoine, L. F., *Effective Managerial Communication,* Boston: Kent, 1986, pp. 63–97.
7. Applebaum, R. L., and Anatol, K. W. E., *Effective Oral Communication for Business and the Professions,* Chicago: SRA, 1982, pp. 259–262.
8. Steele, F., *The Open Organization: The Impact of Secrecy and Disclosure on People and Organizations,* Reading, MA: Addison-Wesley, 1975.
9. Zaldivar, R. A., "Rocket Chief Was Warned and Failed to Act," *Detroit Free Press,* May 23, 1986.
10. Vance, C. C., "How to Encourage Upward Communication from Your Employees," *Association Management 28* (1976), 56–59.
11. Driver, R. W., "Opening the Channels of Upward Communication," *Supervisory Management 25* (1980), 24–29.
12. Macleod, J. S., "How to Unmuzzle Employees," *Employment Relations Today 11* (1984), 49–54.
13. Thomas, P., "Plugging the Communication Channel—How Managers Stop Upward Communication," *Supervisory Management 30* (1985), 7–10.

Part 2

Problems with Language

In Lewis Carroll's *Through the Looking Glass,* Humpty Dumpty and Alice have the following exchange:

> "When *I* use a word," Humpty Dumpty said, in a rather scornful tone, "it means just what I choose it to mean—nothing more or less."
>
> "The question is," said Alice, "whether you *can* make words mean so many different things."

After reading some examples of business and technical writing, it is easy to suspect that Humpty Dumpty has been promoted to project manager—or even company president.

It is not always clear why business and technical writers get themselves into the problems with language that they do. At times, they mistakenly assume their readers will know what they are talking about, or that their goal is to impress rather than inform their readers. At other times, they simply imitate their superiors or model their documents after those in the files, as if the filing cabinet were a repository for sacred writings handed down from on high. (For further discussion of the dangers that can lie lurking in the file cabinet, see the essay by Lee Clark Johns in part three of *Strategies.*)

To solve problems with language, business and technical writers must remember their readers and then be guided by two principles:

- they should write everything as clearly, straightforwardly, and simply as possible; and
- they should write in a manner or style with which they would be comfortable speaking.

These two principles will help writers avoid problems with jargon, gobbledygook, legalese, and sexist and biased language. They will also serve as a guide in selecting the appropriate style for different writing situations.

Jargon is simply technical language unique to a profession or occupation. It only becomes a problem when writers use it in writing to an audience without the necessary background or training to understand it. A systems engineer writing to another systems engineer could rightly assume that the following passage would make sense to his or her reader:

> If you are using a CONFIG.SYS file to modify version 2.0 of the Personal Computer DOS, you may want to create a CONFIG.SYS file to use with the program instead of the AUTOEXEC.BAT file.

Unfortunately, this passage comes from an introductory computer manual.

Jargon in the extreme becomes gobbledygook, mindless gibberish akin to double-talk and characterized by pretentiousness. For example, the host of a popular television game show tells contestants who are losing that they are in "a deficit situation"; ordinary No. 2 pencils become, on the requisition forms of some bureaucrats, "writing implements, standard issue"; politicians wary of angering voters call tax increases "revenue enhancements" and "misspeak themselves" rather than lie when they are caught making false claims or leveling inaccurate charges against their opponents; "turn out the lights when you leave" becomes "illumination is required to be extinguished upon vacation of the premises"; the threat that "anyone caught smoking on the premises will be dealt with accordingly" may at first cause someone to think twice before lighting up, but, in the final analysis, we may well wonder just what the threat is really supposed to mean.

Gobbledygook also includes words and phrases that writers mistakenly think sound the way business and technical writing should sound. They may be in vogue or out of date, but either way, they have unfortunately come to be viewed by some writers as examples of good professional writing practice. People no longer meet for discussions—they "interface"—and every issue about which they interface has "parameters," "viable" and otherwise. Some of these same people see nothing wrong with beginning their letters as follows:

> Pursuant to yours of the twelfth, enclosed herewith please find our check in the amount of $24.95.

Subsequently in the same letter, they "are pleased to advise that," and they go on to refer to data "hereunder discussed" or "above referenced."

Legalese is an overreliance on legal terminology—or legal *sounding* terminology—when Plain English will serve the reader and the writer as well.

Sexist or biased language is linguistic discrimination usually, although not exclusively, against women and groups that are perceived to be minorities. Just as all managers are not men, so all secretaries are not women. Common sense, sensitivity to audience and language, and fairness can help writers avoid sexism and bias.

The five essays in this section of *Strategies* address further the issues raised in these introductory comments. In a now classic piece, Stuart Chase sends up gobbledygook, the linguistic overkill that is all too common in the writing of government officials, lawyers, and, sadly, even some academics. William Zinsser follows with an essay that suggests that it is possible to write for the world of work and still sound like a human being. Alan Siegel follows Zinsser with a brief history of the Plain English movement in his account of the work his company has done in helping clients reduce legalese in the documents they produce. Guidelines developed at the University of Wisconsin offer ways to avoid language that stereotypes or denigrates both men and women. Rosalie Maggio concludes this section of *Strategies* with her suggestions on how to recognize, evaluate, and avoid various kinds of linguistic bias.

Stuart Chase

Gobbledygook

Stuart Chase worked for many years as a consultant to various government agencies; his other books include The Tyranny of Words *(1938) and* Democracy Under Pressure *(1945).*

Said Franklin Roosevelt, in one of his early presidential speeches: "I see one-third of a nation ill-housed, ill-clad, ill-nourished." Translated into standard bureaucratic prose his statement would read:

> It is evident that a substantial number of persons within the Continental boundaries of the United States have inadequate financial resources with which to purchase the products of agricultural communities and industrial establishments. It would appear that for a considerable segment of the population, possibly as much as 33.3333* of the total, there are inadequate housing facilities, and an equally significant proportion is deprived of the proper types of clothing and nutriment.
> *Not carried beyond four places.

This rousing satire on gobbledygook—or talk among the bureaucrats—is adapted from a report[1] prepared by the Federal Security Agency in an attempt to break out of the verbal squirrel cage. "Gobbledygook" was coined by an exasperated Congressman, Maury Maverick of Texas, and means using two, or three, or ten words in the place of one, or using a five-syllable word where a single syllable would suffice. Maverick was censuring the forbidding prose of executive departments in Washington, but the term has now spread to windy and pretentious language in general.

"Gobbledygook" itself is a good example of the way a language grows. There was no word for the event before Maverick's invention; one had to say: "You

[1]This and succeeding quotations from FSA report by special permission of the author, Milton Hall.

know, that terrible, involved, polysyllabic language those government people use down in Washington." Now one word takes the place of a dozen.

A British member of Parliament, A. P. Herbert, also exasperated with bureaucratic jargon, translated Nelson's immortal phrase, "England expects every man to do his duty":

> England anticipates that, as regards the current emergency, personnel will face up to the issues, and exercise appropriately the functions allocated to their respective occupational groups.

A New Zealand official made the following report after surveying a plot of ground for an athletic field:[2]

> It is obvious from the difference in elevation with relation to the short depth of the property that the contour is such as to preclude any reasonable developmental potential for active recreation.

Seems the plot was too steep.

An office manager sent this memo to his chief:

> Verbal contact with Mr. Blank regarding the attached notification of promotion has elicited the attached representation intimating that he prefers to decline the assignment.

Seems Mr. Blank didn't want the job.

> A doctor testified at an English trial that one of the parties was suffering from "circumorbital haematoma."

Seems the party had a black eye.

> In August 1952 the U.S. Department of Agriculture put out a pamphlet entitled: "Cultural and Pathogenic Variability in Single-Condial and Hyphaltip Isolates of Hemlin-Thosporium Turcicum Pass."

Seems it was about corn leaf disease.

On reaching the top of the Finsteraarhorn in 1845, M. Dollfus-Ausset, when he got his breath, exclaimed:

> The soul communes in the infinite with those icy peaks which seem to have their roots in the bowels of eternity.

Seems he enjoyed the view.

A government department announced:

> Voucherable expenditures necessary to provide adequate dental treatment required as adjunct to medical treatment being rendered a pay patient in in-patient status may be incurred as required at the expense of the Public Health Service.

Seems you can charge your dentist bill to the Public Health Service. Or can you?

---

[2]This item and the next two are from the piece on gobbledygook by W. E. Farbstein, *New York Times,* March 29, 1953.

LEGAL TALK

Gobbledygook not only flourishes in government bureaus but grows wild and lush in the law, the universities, and sometimes among the literati. Mr. Micawber was a master of gobbledygook, which he hoped would improve his fortunes. It is almost always found in offices too big for face-to-face talk. Gobbledygook can be defined as squandering words, packing a message with excess baggage and so introducing semantic "noise." Or it can be scrambling words in a message so that meaning does not come through. The directions on cans, bottles, and packages for putting the contents to use are often a good illustration. Gobbledygook must not be confused with double talk, however, for the intentions of the sender are usually honest.

I offer you a round fruit and say, "Have an orange." Not so an expert in legal phraseology, as parodied by editors of *Labor:*

> I hereby give and convey to you, all and singular, my estate and interests, right, title, claim and advantages of and in said orange, together with all rind, juice, pulp and pits, and all rights and advantages therein . . . anything hereinbefore or hereinafter or in any other deed or deeds, instrument or instruments of whatever nature or kind whatsoever, to the contrary, in any wise, notwithstanding.

The state of Ohio, after five years of work, has redrafted its legal code in modern English, eliminating 4,500 sections and doubtless a blizzard of "whereases" and "hereinafters." Legal terms of necessity must be closely tied to their referents, but the early solons tried to do this the hard way, by adding synonyms. They hoped to trap the physical event in a net of words, but instead they created a mumbo-jumbo beyond the power of the layman, and even many a lawyer, to translate. Legal talk is studded with tautologies, such as "cease and desist," "give and convey," "irrelevant, incompetent, and immaterial." Furthermore, legal jargon is a dead language; it is not spoken and it is not growing. An official of one of the big insurance companies calls their branch of it "bafflegab." Here is a sample from his collection.[3]

> One-half to his mother, if living, if not to his father, and one-half to his mother-in-law, if living, if not to his mother, if living, if not to his father. Thereafter payment is to be made in a single sum to his brothers. On the one-half payable to his mother, if living, if not to his father, he does not bring in his mother-in-law as the next payee to receive, although on the one-half to his mother-in-law, he does bring in the mother or father.

You apply for an insurance policy, pass the tests, and instead of a straightforward "here is your policy," you receive something like this:

> This policy is issued in consideration of the application therefor, copy of which application is attached hereto and made part hereof, and of the payment for said insurance on the life of the above-named insured.

---

[3]Interview with Clifford B. Reeves by Sylvia F. Porter, *New York Evening Post,* March 14, 1952.

ACADEMIC TALK

The pedagogues may be less repetitious than the lawyers, but many use even longer words. It is a symbol of their calling to prefer Greek and Latin derivatives to Anglo-Saxon. Thus instead of saying: "I like short clear words," many a professor would think it more seemly to say: "I prefer an abbreviated phraseology, distinguished for its lucidity." Your professor is sometimes right, the longer word may carry the meaning better—but not because it is long. Allen Upward in his book *The New Word* warmly advocates Anglo-Saxon English as against what he calls "Mediterranean" English, with its polysyllables built up like a skyscraper.

Professional pedagogy, still alternating between the Middle Ages and modern science, can produce what Henshaw Ward once called the most repellent prose known to man. It takes an iron will to read as much as a page of it. Here is a sample of what is known in some quarters as "pedageese":

> Realization has grown that the curriculum or the experiences of learners change and improve only as those who are most directly involved examine their goals, improve their understandings and increase their skill in performing the tasks necessary to reach newly defined goals. This places the focus upon teacher, lay citizen and learner as partners in curricular improvement and as the individuals who must change, if there is to be curriculum change.

I think there is an idea concealed here somewhere. I think it means: "If we are going to change the curriculum, teacher, parent, and student must all help." The reader is invited to get out his semantic decoder and check on my translation. Observe there is no technical language in this gem of pedageese, beyond possibly the word "curriculum." It is just a simple idea heavily oververbalized.

In another kind of academic talk the author may display his learning to conceal a lack of ideas. A bright instructor, for instance, in need of prestige may select a common sense proposition for the subject of a learned monograph—say, "Modern cities are hard to live in"—and adorn it with imposing polysyllables: "Urban existence in the perpendicular declivities of megalopolis . . ." et cetera. He coins some new terms to transfix the reader—"mega-decibel" or "strato-cosmopolis"—and works them vigorously. He is careful to add a page or two of differential equations to show the "scatter." And then he publishes, with 147 footnotes and a bibliography to knock your eye out. If the authorities are dozing, it can be worth an associate professorship.

While we are on the campus, however, we must not forget that the technical language of the natural sciences and some terms in the social sciences, forbidding as they may sound to the layman, are quite necessary. Without them, specialists could not communicate what they find. Trouble arises when experts expect the uninitiated to understand the words; when they tell the jury, for instance, that the defendant is suffering from "circumorbital haematoma."

Here are two authentic quotations. Which was written by a distinguished modern author, and which by a patient in a mental hospital? You will find the answer at the end of the [selection].

1. Have just been to supper. Did not knowing what the woodchuck sent me here. How when the blue blue blue on the said anyone can do it that tries. Such is the presidential candidate.

2. No history of a family to close with those and close. Never shall he be alone to be alone to be alone to be alone to be alone to lend a hand and leave it left and wasted.

REDUCING THE GOBBLE

As government and business offices grow larger, the need for doing something about gobbledygook increases. Fortunately the biggest office in the world is working hard to reduce it. The Federal Security Agency in Washington,[4] with nearly 100 million clients on its books, began analyzing its communication lines some years ago, with gratifying results. Surveys find trouble in three main areas: correspondence with clients about their social security problems, office memos, official reports.

Clarity and brevity, as well as common humanity, are urgently needed in this vast establishment which deals with disability, old age, and unemployment. The surveys found instead many cases of long-windedness, foggy meanings, clichés, and singsong phrases, and gross neglect of the reader's point of view. Rather than talking to a real person, the writer was talking to himself. "We often write like a man walking on stilts."

Here is a typical case of long-windedness:

> *Gobbledygook as found:* "We are wondering if sufficient time has passed so that you are in a position to indicate whether favorable action may now be taken on our recommendation for the reclassification of Mrs. Blank, junior clerk-stenographer, CAF 2, to assistant clerk-stenographer, CAF 3?"
>
> *Suggested improvement:* "Have you yet been able to act on our recommendation to reclassify Mrs. Blank?"

Another case:

> Although the Central Efficiency Rating Committee recognizes that there are many desirable changes that could be made in the present efficiency rating system in order to make it more realistic and more workable than it now is, this committee is of the opinion that no further change should be made in the present system during the current year. Because of conditions prevailing throughout the country and the resultant turnover in personnel, and difficulty in administering the Federal programs, further mechanical improvement in the present rating system would require staff retraining and other administrative expense which would seem best withheld until the official termination of hostilities, and until restoration of regular operations.

---

[4]Now the Department of Health and Human Services.

The FSA invites us to squeeze the gobbledygook out of this statement. Here is my attempt:

> The Central Efficiency Rating Committee recognizes that desirable changes could be made in the present system. We believe, however, that no change should be attempted until the war is over.

This cuts the statement from 111 to 30 words, about one-quarter of the original, but perhaps the reader can do still better. What of importance have I left out?

Sometimes in a book which I am reading for information—not for literary pleasure—I run a pencil through the surplus words. Often I can cut a section to half its length with an improvement in clarity. Magazines like *The Reader's Digest* have reduced this process to an art. Are long-windedness and obscurity a cultural lag from the days when writing was reserved for priests and cloistered scholars? The more words and the deeper the mystery, the greater their prestige and the firmer the hold on their jobs. And the better the candidates's chance today to have his doctoral thesis accepted.

The FSA surveys found that a great deal of writing was obscure although not necessarily prolix. Here is a letter sent to more than 100,000 inquirers, a classic example of murky prose. To clarify it, one needs to *add* words, not cut them:

> In order to be fully insured, an individual must have earned $50 or more in covered employment for as many quarters of coverage as half the calendar quarters elapsing between 1936 and the quarter in which he reaches age 65 or dies, whichever first occurs.

Probably no one without the technical jargon of the office could translate this: nevertheless, it was sent out to drive clients mad for seven years. One poor fellow wrote back: "I am no longer in covered employment. I have an outside job now."

Many words and phrases in officialese seem to come out automatically, as if from lower centers of the brain. In this standardized prose people never *get* jobs, they "secure employment"; *before* and *after* become "prior to" and "subsequent to"; one does not *do*, one "performs"; nobody *knows* a thing, he is "fully cognizant"; one never *says*, he "indicates." A great favorite at present is "implement."

Some charming boners occur in this talking-in-one's-sleep. For instance:

> The problem of extending coverage to all employees, regardless of size, is not as simple as surface appearances indicate.
> Though the proportions of all males and females in ages 16–45 are essentially the same . . .
> Dairy cattle, usually and commonly embraced in dairying . . .

In its manual to employees, the FSA suggests the following:

Instead of	Use
give consideration to	consider
make inquiry regarding	inquire
is of the opinion	believes
comes into conflict with	conflicts
information which is of a confidential nature	confidential information

Professional or office gobbledygook often arises from using the passive rather than the active voice. Instead of looking you in the eye, as it were, and writing "This act requires . . . ," the office worker looks out of the window and writes: "It is required by this statute that . . ." When the bureau chief says, "We expect Congress to cut your budget," the message is only too clear; but usually he says, "It is expected that the departmental budget estimates will be reduced by Congress."

> GOBBLED: "All letters prepared for the signature of the Administrator will be single spaced."
> UNGOBBLED: "Single space all letters for the Administrator." (Thus cutting 13 words to 7.)

Only People Can Read

The FSA surveys pick up the point . . . that human communication involves a listener as well as a speaker. Only people can read, though a lot of writing seems to be addressed to beings in outer space. To whom are you talking? The sender of the officialese message often forgets the chap on the other end of the line.

A woman with two small children wrote the FSA asking what she should do about payments, as her husband had lost his memory. "If he never gets able to work," she said, "and stays in an institution would I be able to draw any benefits? . . . I don't know how I am going to live and raise my children since he is disable to work. Please give me some information. . . ."

To this human appeal, she received a shattering blast of gobbledygook, beginning, "State unemployment compensation laws do not provide any benefits for sick or disabled individuals . . . in order to qualify an individual must have a certain number of quarters of coverage . . ." et cetera, et cetera. Certainly if the writer had been thinking about the poor woman he would not have dragged in unessential material about old-age insurance. If he had pictured a mother without means to care for her children, he would have told her where she might get help—from the local office which handles aid to dependent children, for instance.

Gobbledygook of this kind would largely evaporate if we thought of our messages as two way—in the above case, if we pictured ourselves talking on the doorstep of a shabby house to a woman with two children tugging at her skirts, who in her distress does not know which way to turn.

Results of the Survey

The FSA survey showed that office documents could be cut 20 to 50 percent, with an improvement in clarity and a great saving to taxpayers in paper and payrolls.

A handbook was prepared and distributed to key officials.[5] They read it, thought about it, and presently began calling section meetings to discuss gob-

---

[5] By Milton Hall.

bledygook. More booklets were ordered, and the local output of documents began to improve. A Correspondence Review Section was established as a kind of laboratory to test murky messages. A supervisor could send up samples for analysis and suggestions. The handbook is now used for training new members; and many employees keep it on their desks along with the dictionary. Outside the Bureau some 25,000 copies have been sold (at 20 cents each) to individuals, governments, business firms, all over the world. It is now used officially in the Veterans Administration and in the Department of Agriculture.

The handbook makes clear the enormous amount of gobbledygook which automatically spreads in any large office, together with ways and means to keep it under control. I would guess that at least half of all the words circulating around the bureaus of the world are "irrelevant, incompetent, and immaterial"— to use a favorite legalism; or are just plain "unnecessary"—to ungobble it.

My favorite story of removing the gobble from gobbledygook concerns the Bureau of Standards at Washington. I have told it before but perhaps the reader will forgive the repetition. A New York plumber wrote the Bureau that he had found hydrochloric acid fine for cleaning drains, and was it harmless? Washington replied: "The efficacy of hydrochloric acid is indisputable, but the chlorine residue is incompatible with metallic permanence."

The plumber wrote back that he was mighty glad the Bureau agreed with him. The Bureau replied with a note of alarm: "We cannot assume responsibility for the production of toxic and noxious residues with hydrochloric acid, and suggest that you use an alternate procedure." The plumber was happy to learn that the Bureau still agreed with him.

Whereupon Washington exploded: "Don't use hydrochloric acid; it eats hell out of the pipes!"

Note: The second quotation on page 66 comes from Gertrude Stein's *Lucy Church Amiably.*

William Zinsser

Writing in Your Job

William Zinsser is a noted author, columnist, editor, teacher of writing, and critic.

The memo, the business letter, the administrative report, the financial analysis, the marketing proposal, the note to the boss, the fax, the Post-it—all the pieces of paper that circulate through your office every day are forms of writing. Take them seriously. Countless careers rise or fall on the ability or inability of employees to state a set of facts, summarize a meeting or present an idea coherently.

Most people work for institutions: businesses, banks, insurance firms, law firms, government agencies, school systems, nonprofit organizations and other entities. Many of them are managers whose writing goes out to the public: the president addressing the stockholders, the banker explaining a change in procedure, the school principal writing a newsletter to parents. Whoever they are, they are often so afraid of writing that their sentences lack all humanity—and so do their institutions. It's hard to imagine that these are real places where real men and women come to work every morning.

But just because people work for an institution they don't have to write like one. Institutions can be warmed up. Administrators can be turned into human beings. Information can be imparted clearly and without pomposity. It's a question of remembering that readers identify with people, not with abstractions like "profitability," or with Latinate nouns like "utilization" and "implementation," or with inert constructions in which nobody can be visualized doing something ("pre-feasibility studies are in the paperwork stage").

Nobody has made the point better than George Orwell in his translation into modern bureaucratic fuzz of this famous verse from Ecclesiastes:

I returned and saw under the sun that the race is not to the swift, nor the battle to the strong, neither yet bread to the wise, nor yet riches to men of understanding, nor yet favor to men of skill; but time and chance happeneth to them all.

Orwell's version goes:

Objective consideration of contemporary phenomena compels the conclusion that success or failure in competitive activities exhibits no tendency to be commensurate with innate capacity, but that a considerable element of the unpredictable must invariably be taken into account.

First notice how the two passages look. The one at the top invites us to read it. The words are short and have air around them; they convey the rhythms of human speech. The second one is clotted with long words. It tells us instantly that a ponderous mind is at work. We don't want to go anywhere with a mind that expresses itself in such suffocating language. We don't even start to read.

Also notice what the two passages say. Gone from the second one are the short words and vivid images from everyday life—the race and the battle, the bread and the riches—and in their place have waddled the long and flabby nouns of generalized meaning. Gone is any sense of what one person did ("I returned") or what he realized ("saw") about one of life's central mysteries: the capriciousness of fate.

Let me illustrate how this disease infects the writing that most people do in their jobs. I'll use school principals as my first example, not because they are the worst offenders (they aren't) but because I happen to have such an example. My points are intended for all the men and women who work in all the organizations where language has lost its humanity and nobody knows what the people in charge are talking about.

My encounter with the principals began when I got a call from Ernest B. Fleishman, superintendent of schools in Greenwich, Connecticut. "We'd like you to come and 'dejargonize' us," he said. "We don't think we can teach students to write unless all of us at the top of the school system clean up our own writing." He said he would send me some typical materials that had originated within the system. His idea was for me to analyze the writing and then conduct a workshop.

What appealed to me was the willingness of Dr. Fleishman and his colleagues to make themselves vulnerable. Vulnerability has a strength of its own. We decided on a date, and soon a fat envelope arrived. It contained various internal memos and mimeographed newsletters that had been mailed to parents from the 16 elementary, junior, and senior high schools.

The newsletters had a cheery and informal look. Obviously the system was making an effort to communicate warmly with its families. But even at first glance certain chilly phrases caught my eye—"prioritized evaluative procedures," "modified departmentalized schedule"—and one principal promised that his school would provide "enhanced positive learning environments." Just as obviously the system wasn't communicating as warmly as it thought it was.

I studied the principals' material and divided it into good and bad examples. On the appointed morning in Greenwich I found 40 principals and curriculum coordinators assembled and eager to learn. I told them I could only applaud them for submitting to a process that so threatened their identity. In the national clamor over why Johnny can't write, Dr. Fleishman was the first adult in my experience who admitted that youth has no monopoly on verbal sludge and that the problem must also be attacked at the top.

I told the principals that we want to think of the men and women who run our children's schools as people not unlike ourselves. We are suspicious of pretentiousness, of all the fad words that the social scientists have coined to avoid making themselves clear to ordinary mortals. I urged them to be natural. How we write and how we talk is how we define ourselves.

I asked them to listen to how they were defining themselves to the community. I had made copies of certain bad examples, changing the names of the schools and the principals. I explained that I would read some of the examples aloud. Later we would see if they could turn what they had written into plain English. This was my first example:

> Dear Parent:
>
> We have established a special phone communication system to provide additional opportunities for parent input. During this year we will give added emphasis to the goal of communication and utilize a variety of means to accomplish this goal. Your inputs, from the unique position as a parent, will help us to plan and implement an educational plan that meets the needs of your child. An open dialogue, feedback and sharing of information between parents and teachers will enable us to work with your child in the most effective manner.
>
> Dr. George B. Jones
> Principal

That's the kind of communication I don't want to receive, unique though my parent inputs might be. I'd like to be told that the school is going to make it easier for me to telephone the teachers and that they hope I'll call often to discuss how my children are getting along. Instead the parent gets junk: "special phone communication system," "added emphasis to the goal of communication," "plan and implement an educational plan." As for "open dialogue, feedback and sharing of information," they are three ways of saying the same thing.

Dr. Jones is clearly a man who means well, and his plan is one we all want: a chance to pick up the phone and tell the principal what a great kid Johnny is despite that unfortunate incident in the playground last Tuesday. But Dr. Jones doesn't sound like a person I want to call. In fact, he doesn't sound like a person. His message could have been tapped out by a computer. He is squandering a rich resource: himself.

Another example that I chose was a "Principal's Greeting" sent to parents at the start of the school year. It consisted of two paragraphs that were very different:

> Fundamentally, Foster is a good school. Pupils who require help in certain subjects or study skills areas are receiving special attention. In the school year ahead we seek to provide enhanced positive learning environments. Children, and staff, must work in an atmosphere that is conducive to learning. Wide varieties of instructional materials are needed. Careful attention to individual abilities and learning styles is required. Coopera-

tion between school and home is extremely important to the learning process. All of us should be aware of desired educational objectives for every child.

Keep informed about what is planned for our children this year and let us know about your own questions and about any special needs your child may have. I have met many of you in the first few weeks. Please continue to stop in to introduce yourself or to talk about Foster. I look forward to a very productive year for all of us.

Dr. Ray B. Dawson
Principal

In the second paragraph I'm being greeted by a person; in the first I'm hearing from an educator. I like the real Dr. Dawson of Paragraph 2. He talks in warm and comfortable phrases: "Keep informed," "let us know," "I have met," "Please continue," "I look forward."

By contrast, Educator Dawson of Paragraph 1 never uses "I" or even suggests a sense of "I." He falls back on the jargon of his profession, where he feels safe, not stopping to notice that he really isn't telling the parent anything. What are "study skills areas" and how do they differ from "subjects"? What are "enhanced positive learning environments" and how do they differ from "an atmosphere that is conducive to learning"? What are "wide varieties of instructional materials": pencils, textbooks, filmstrips? What exactly are "learning styles"? What "educational objectives" are "desired," and who desires them?

The second paragraph, in short, is warm and personal; the other is pedantic and vague. This was a pattern I found repeatedly. Whenever the principals wrote to notify the parents of some human detail, they wrote with humanity:

> It seems that traffic is beginning to pile up again in front of the school. If you can possibly do so, please come to the rear of the school for your child at the end of the day.

> I would appreciate it if you would speak with your children about their behavior in the cafeteria. Many of you would be totally dismayed if you could observe the manners of your children while they are eating. Check occasionally to see if they owe money for lunch. Sometimes children are very slow in repaying.

But when the educators wrote to explain how they proposed to do their educating, they vanished without a trace:

> In this document you will find the program goals and objectives that have been identified and prioritized. Evaluative procedures for the objectives were also established based on acceptable criteria.

> Prior to the implementation of the above practice, students were given very little exposure to multiple choice questions. It is felt that the use of practice questions correlated to the unit that a student is presently studying has had an extremely positive effect as the test scores confirm.

After I had read various good and bad examples, the principals began to hear the difference between their true selves and their educator selves. The problem was how to close the gap. I recited my four articles of faith: clarity, simplicity, brevity and humanity. I explained about using active verbs and avoiding windy "concept nouns." I told them not to use the private vocabulary of education as a crutch; almost any subject can be made accessible in good English.

These were all basic tenets, but the principals wrote them down as if they had never heard them before—and maybe they hadn't, or at least not for many years. Perhaps that's why bureaucratic prose becomes so turgid, whatever the bureaucracy. Once an administrator rises to a certain level, nobody ever points out to him again the beauty of a simple declarative sentence or shows him how his writing has become swollen with pompous generalizations.

Finally our workshop got down to work. I distributed my copies and asked the principals to rewrite the more knotty sentences. It was a grim moment. They had met the enemy for the first time. They scribbled on their pads and scratched out what they had scribbled. Some didn't write anything. Some crumpled their paper. They began to look like writers. An awful silence hung over the room, broken only by the crossing out of sentences and the crumpling of paper. They began to sound like writers.

As the day went on, they slowly relaxed. They began to write in the first person and to use active verbs. For a while they still couldn't loosen their grip on long words and vague nouns ("parent communication response"). But gradually their sentences became human. When I asked them to tackle "Evaluative procedures for the objectives were also established based on acceptable criteria," one of them wrote: "At the end of the year we will evaluate our progress." Another wrote: "We will see how well we have succeeded."

That's the kind of plain talk a parent wants. It's also what stockholders want from their corporation, what customers want from their bank, what the widow wants from the agency that's handling her social security. There is a deep yearning for human contact and a resentment of bombast. Recently I got a "Dear Customer" letter from the company that supplies my computer needs. It began: "Effective March 30 we will be migrating our end user order entry and supplies referral processing to a new telemarketing center." I finally figured out that they had a new 800 number and that the end user was me. Any institution that won't take the trouble in its writing to be both clear and personal will lose friends, customers, and money. Let me put it another way for business executives: a shortfall will be experienced in anticipated profitability.

Here's an example of how organizations throw away their humanity with pretentious language. It's a "customer bulletin" distributed by a major corporation. The sole purpose of a customer bulletin is to give helpful information to a customer. This one begins: "Companies are increasingly turning to capacity planning techniques to determine when future processing loads will exceed processing capabilities." That sentence is no favor to the customer; it's congealed with Orwellian nouns like "capacity" and "capabilities" that convey no procedures that a customer can picture. What *are* capacity planning techniques? Whose capacity is being planned? By whom? The second sentence says: "Capacity planning adds objectivity to the decision-making process." More dead nouns. The third sentence says: "Management is given enhanced decision participation in key areas of information system resources."

The customer has to stop after every sentence and translate it. The bulletin might as well be in French. He starts with the first sentence—the one about capacity planning techniques. Translated, that means "It helps to know when you're giving your computer more than it can handle." The second sentence— "Capacity planning adds objectivity to the decision-making process"—means you should know the facts before you decide. The third sentence—the one about enhanced decision participation—means "The more you know about your system, the better it will work." It could also mean several other things.

But the customer isn't going to keep translating. Soon he's going to start looking for another company. He thinks, "If these guys are so smart, why can't they tell me what they do? Maybe they're *not* so smart." The bulletin goes on to say that "for future cost avoidance, productivity has been enhanced." That seems to mean that the product will be free: all costs have been avoided. Next the bulletin assures the customer that "the system is delivered with functionality." That means it works. I would hope so.

Finally, at the end, we get a glimmer of humanity. The writer of the bulletin asks a satisfied customer why he chose this system. The man says he chose it because of the company's reputation for service. He says: "A computer is like a sophisticated pencil. You don't care how it works, but if it breaks you want someone there to fix it." Notice how refreshing that sentence is after all the garbage that preceded it: in its language (comfortable words), in its details that we can visualize (the pencil), and above all in its humanity. The writer has taken the coldness out of a technical process by relating it to an experience we're all familiar with—waiting for the repairman when something breaks. I'm reminded of a sign I saw in the New York subway which proves that even a huge municipal bureaucracy can talk to its constituents humanely: "If you ride the subway regularly you may have seen signs directing you to trains you've never heard of before. These are only new names for very familiar trains."

Still, plain talk will not be easily achieved in corporate America. Too much vanity is on the line. Managers at every level are prisoners of the notion that a simple style reflects a simple mind. Actually a simple style is the result of hard work and hard thinking; a muddled style reflects a muddled thinker or a person too arrogant or too dumb or too lazy to organize his thoughts. Remember that what you write is often the only chance you'll get to present yourself to someone whose business or money or goodwill you want. If what you write is ornate or pompous or fuzzy, that's how you'll be perceived. The reader has no other choice.

I learned about corporate America by venturing out into it, after Greenwich, to conduct workshops for some major corporations, which also asked to be dejargonized. "We don't even understand our own memos anymore," they told me. I worked with the men and women who write the vast amounts of material these companies generate for internal and external consumption. The internal material consists of house organs and newsletters whose purpose is to tell employees what's happening at their "facility" and to give them a sense of belonging. The external material includes the glossy magazines and annual reports that go to stockholders, the speeches delivered by high executives, the releases sent to the press, and the consumer manuals that explain how the product works. I found almost all of it lacking in human juices and much of it impenetrable.

Typical of the sentences in the newsletters was this one:

> Announced concurrently with the above enhancements were changes to the System Support Program, a program product which operates in conjunction with the NCP. Among the additional functional enhancements are dynamic reconfiguration and inter-systems communications.

There's no joy for the writer in such work, and certainly none for the reader. It's language out of *Star Trek,* and if I were an employee I wouldn't be cheered—or informed—by these efforts to raise my morale. I would soon stop reading them. I told the corporate writers that they had to find the people behind the fine achievements being described. "Go to the engineer who conceived the new system," I said, "or to the designer who designed it, or to the technician who assembled it, and get them to tell you in their own words how the idea came to them, or how they put it together, or how it will be used by real people out in the real world." The way to warm up any institution is to locate the missing "I." Remember: "I" is the most interesting element in any story.

The writers explained that they often did interview the engineer but couldn't get him to talk English. They showed me some typical quotes. The engineers spoke in an arcane language studded with acronyms ("Sub-system support is available only with VSAG or TNA"). I said that the writers had to keep going back to the engineer until he finally made himself intelligible. They said the engineer didn't *want* to be made intelligible: if he spoke too simply he would look like a jerk to his peers. I said that their responsibility was to the facts and to the reader, not to the vanity of the engineer. I urged them to believe in themselves as writers and not to relinquish control. They replied that this was easier said than done in hierarchical corporations, where approval of written reports is required at a succession of higher levels. I sensed an undercurrent of fear: Do things the company way and don't risk your job trying to make the company human.

High executives were equally victimized by wanting to sound important. One corporation had a monthly newsletter to enable "management" to share its concerns with middle managers and lower employees. Prominent in every issue was a message of exhortation from the division vice-president, whom I'll call Thomas Bell. Judging by his monthly message, he was a pompous ass, saying nothing and saying it in inflated verbiage.

When I mentioned this, the writers said that Thomas Bell was actually a diffident man and a good executive. They pointed out that he doesn't write the message himself; it's written for him. I said that Mr. Bell was being done a disservice—that the writers should go to him every month (with a tape recorder, if necessary) and stay there until he talked about his concerns in the same language he would use when he got home and talked to Mrs. Bell.

What I realized was that most executives in America don't write what appears over their signature or what they say in their speeches. They have surrendered the qualities that make them unique. If they and their institutions seem cold, it's because they acquiesce in the process of being pumped up and dried out. Preoccupied with their high technology, they forget that some of the most powerful tools they possess—for good and for bad—are words.

If you work for an institution, whatever your job, whatever your level, be yourself when you write. You will stand out as a real person among the robots, and your example might even persuade Thomas Bell to write his own stuff.

Alan Siegel

The Plain English Revolution

Alan Siegel, a leading consultant in the Plain English movement, is the Chairman and Chief Executive Officer of Siegel & Gale, a New York firm specializing in language simplification.

> *LOUIS XVI: C'est une grande révolte?*
>
> *DUC DE LA ROCHEFOUCAULD— LIANCOURT: Non, Sire, c'est une grande révolution. (When the news arrived at Versailles of the Fall of the Bastille, 1789)*

Three-and-a-half years ago I wrote an article . . . on the movement away from legalese toward forms and documents that can be understood by ordinary citizens.* At that time, the movement consisted of isolated revolts against legalistic gobbledygook and bureaucratese. A handful of business documents—mainly insurance policies and loan notes—had been simplified by forward-looking companies. In government, the scattered advocates of clear communication shared the lament of Alfred Kahn, then head of the Civil Aeronautics Board: "There's nothing I can do but cry. I feel so lonely and futile."

Now the plain language movement is becoming a revolution. . . . But the Bastille has not fallen, yet. Part of the legal community remains resistant to change. Some supposed practitioners of Plain English have confused simplicity with simplemindedness. And the greatest obstacle has yet to be fully overcome—a misunderstanding of what Plain English is all about.

---

*"To lift the curse of legalese—Simplify, Simplify," *Across the Board*, June 1977.

THE PLAIN ENGLISH REVOLUTION. By Alan Siegel. From *Across the Board* (February 1981) reprinted by permission of *Across the Board* and its publisher, The Conference Board, 845 Third Avenue, New York, New York 10022.

That misunderstanding was dramatized when New Jersey's Plain English Bill, passed by the legislature, expired in February 1980, through a pocket veto by Gov. Brendan Byrne, who said: " . . . although regulating the protection of consumers is a worthy goal, legislating the style of a society's prose is another thing." *

But the movement has nothing to do with "legislating the style of a society's prose." We are not trying to turn English into one-syllable words, or to translate Saul Bellow into baby-talk. We are not even trying to do away with professional jargon, though that is a tempting target. Let lawyers talk to lawyers, or accountants to accountants, as they please. So long as they understand one another, that is just fine.

Well, what is the Plain English movement all about? *Its aim is to make functional documents function, whether they are put out by business or by government. If a consumer is expected to abide by a formal document—an insurance policy, a mortgage, a lease, a warranty, a tax form—then the consumer should be able to understand the document.* Carl Felsenfeld, vice president and counsel for Citicorp N.A., expresses the concept in legal terms: "There is growing dissatisfaction with contracts where consumers merely 'sign here' and can't, under any reputable system of contract law, be deemed to have agreed to all the printed verbiage." In social terms, I would say that people are learning a new right: the right to understand.

It's been argued that the fault lies with consumers, or with the educational system that does not train them to cope with legalese, or accounting, or insurance terminology. But that argument is neither reasonable nor realistic. In fact, documents meant for consumers have been made much more difficult than they really need to be. Redressing the balance—making sure that the documents consumers are expected to understand are made understandable—is a matter of simple fairness, and simple efficiency. In the long run it's in the interests of business as well as the consumer.

Five years of experience have proved that consumer contracts and forms of all kinds can be made much more understandable without sacrificing legal effectiveness. The very first plain language loan note, which I helped write for Citibank of New York, was introduced in 1975. Since then, Citicorp counsel Felsenfeld notes, "We've lost no money and there has been no litigation as a result of simplification."

That loan note remains a good example of just what plain language means. Among other things:

- A personal tone was used throughout—"I" and "me" rather than "the undersigned" or "Borrower," "you" and "yours" rather than "the Bank."
- Language was radically simplified. For example, "To repay my loan, I promise to pay you . . ." instead of "For value received, the undersigned (jointly and severally) hereby promise(s) to pay . . ."
- When unfamiliar terms couldn't be eliminated, we added explanatory phrases. For instance, ". . . if this loan is refinanced—that is, replaced by a new note—you

---

*Rewritten and reintroduced in June, the bill was passed again. This time Gov. Bryne signed it (on October 10th), out of deference to its "unusually large" number of sponsors—65 of the Assembly's 80 members.

[the bank] will refund the unearned finance charge, figured by the rule of 78—a commonly used formula for figuring rebates or installment loans."

- We shortened the sentences wherever possible and even used contractions ("I'll pay this sum . . ."). To enhance clarity, we chose active instead of passive verb forms where possible.
- Improvements in design included the use of larger (12-point) type printed in green on light brown stock. Compared with the previously intimidating format, this visually appealing approach suggests immediately that the document is supposed to be read.

The Citibank note taught us a fundamental lesson about simplifying language. Consumer contracts have traditionally been adapted from mercantile contracts, which feature verbose protective clauses accumulated over the years in an effort to cover all possible contingencies. Many of these provisions have no practical value in the consumer marketplace. So the first task in simplifying a document is not rewriting it in Plain English, but identifying clauses taken from commercial contracts that can be eliminated from the consumer contract without jeopardizing its validity. The secret to doing this lies in analyzing actual business experience to see which provisions are really being used.

In the case of the Citibank note, the provisions describing the lender's protections in case of default proved to be the biggest challenge. The traditional note listed a string of contingencies more than 180 words long:

Before

In the event of default in the payment of this or any other Obligation or the performance or observance of any term or covenant contained herein or in any note or other contract or agreement evidencing or relating to any Obligation or any Collateral on the Borrower's part to be performed or observed; or the undesigned Borrower shall die; or any of the undersigned become insolvent or make an assignment for the benefit of creditors; or a petition shall be filed by or against any of the undersigned under any provision of the Bankruptcy Act; or any money, securities or property of the undersigned now or hereafter on deposit with or in the possession or under the control of the Bank shall be attached or become subject to distraint proceedings or any order of process of any court; or the Bank shall deem itself to be insecure, then and in any such event, the Bank shall have the right (at its option), without demand or notice of any kind, to declare all or any part of the Obligations to be immediately due and payable. . . .

But analysis of Citibank's business experience disclosed that the typical consumer loan transaction needs only one event of default—failure to pay. With one additional protection added, the following replaced all the fine print above:

After

Default   I'll be in default—
1.   If I don't pay an installment on time, or
2.   If any other creditor tries by legal process to take any money of mine in your possession.

A number of insurance companies, whose product, after all, is words on paper, continue to get good marks for simplification. The Massachusetts Savings Bank Life Insurance (SBLI) Whole Life Policy used to begin like this:

> IN CONSIDERATION OF THE APPLICATION for this policy (copy attached hereto) which is the basis of and a part of this contract and of the payment of an annual premium as hereinafter specified for the basic policy as of the Date of Issue as specified herein and on the anniversary of such date in each year during the continuance of this contract until premiums have been paid for the number of years indicated in the POLICY SPECIFICATIONS or until the prior death of the Insured . . .

By contrast, the cover of their simplified policy begins with this message:

> Please take the time to read your SBLI policy carefully. Your SBLI representative will be glad to answer any questions. You may return this policy within ten days after receiving it. Deliver it to any SBLI agency. We'll promptly refund all premiums paid for it.

The insurer's promise to pay the consumer used to be phrased in these forbidding terms:

> The Bank Hereby Agrees upon Surrender of this Policy to Pay the Face Amount Specified Above, less any indebtedness on or secured hereunder . . . upon receipt of due proof of the Insured's death to the beneficiary named in the Application herefor or to such other beneficiary as may be entitled thereto under the provisions hereof, or if no such beneficiary survives the Insured, then to the Owner or to the estate of the Owner.

In Plain English, policyholders can understand what they are buying:

> We will pay the face amount when we receive proof of the Insured's death. We will pay the named Beneficiary. If no Beneficiary survives the Insured, we'll pay the Owner of this policy, or the Owner's estate.
>
> Any amount owed to us under this policy will be deducted. We'll refund any premiums paid beyond the month of death.

Perhaps the most ambitious simplification program so far has been that undertaken by the St. Paul Fire and Marine Insurance Company, of St. Paul, Minnesota. In 1975 they became one of the first large insurers to begin simplifying their policies. As a pilot project, the company produced an easy-to-read "personal liability catastrophe policy" that used personalized examples in colloquial English:

Before

a.　Automobile and Watercraft Liability:

1.　any Relative with respect to (i) an Automobile owned by the Named Insured or a Relative, or (ii) a Non-owned Automobile, provided his actual operation or (if he is not operating) the other actual use thereof is with the permission of the owner and is within the scope of such permission, or

2.　any person while using an Automobile or Watercraft, owned by, loaned or hired for use in behalf of the Named Insured and any person or organization legally responsible for the use thereof is within the scope of such permission.

After

We'll also cover any person or organization legally responsible for the use of a car, if it's used by you or with your permission. But again, the use has to be for the intended purpose.

You loan your station wagon to a teacher to drive a group of children to the zoo. She and the school are covered by this policy if she actually drives to the zoo, but not if she lets the children off at the zoo and drives to her parents' farm 30 miles away.

By 1978, St. Paul had reduced the number of policy forms in its commercial business package from 366 to 150. Plans now call for most of its commercial insurance policies to be rewritten and reprinted in a more readable format by 1982. The company has appointed a Manager of Forms Simplification, with his own staff and a detailed style manual, to oversee the effort.

The financial field is a particularly promising new area for Plain English efforts. A case in point: Sanford C. Bernstein & Co. is a New York broker-dealer advising clients on investments as well as trading for them. It is subject to regulation under the Federal Securities Exchange Act of 1934 and the Investment Advisers Act of 1940, as well as regulation by the New York Stock Exchange, the National Association of Securities Dealers, the Board of Governors of the Federal Reserve System, and state securities laws in various jurisdictions in which it does business or has clients. These various regulations extend to every aspect of the firm's business, from the content of advertising and soliciting materials to client reports, statements, and the consents that must be obtained before the firm can effect certain types of transaction on clients' behalf. Nonetheless, the firm was able to replace jargon with Plain English, while satisfying legal and technical requirements.

Before

It is the express intention of the undersigned to create an estate or account as joint tenants with rights of survivorship and not as tenants in common. In the event of the death of either of the undersigned, the entire interest in the joint account shall be vested in the survivor or survivors on the same terms and conditions as theretofore held, without in any manner releasing the decedent's estate from the liability provided for in the next preceding paragraph.

After

Other signers share your interest equally. If one of you dies, the account will continue and the other people who've signed the agreement will own the entire interest in it.

Where obscure terms could not be eliminated, the new forms explain them. The meaning of the term "clearing agents," for instance, is described in this way:

Clearing agents. We can use other broker/dealers as clearing agents to execute transactions, to hold securities in custody on your behalf and to perform other routine procedures in connection with your account. These clearing agents will act at our direction, and they won't have any part in investment decisions.

Especially complex ideas are not only explained but accompanied by illustrative examples:

Buying on margin. In managing your account, we may buy securities on margin. This means that we'll loan you part of the cost of the securities and charge you interest on the loan. By buying on margin, we can purchase more securities on your behalf than we could

Plain Language Scorecard

- Thirty-four states have laws or regulations setting standards for clear language in insurance policies. In 1978, New York became the first state to require that business contracts "primarily for personal, family, or household purposes" be plainly written in everyday language. Since then, New Jersey, Maine, Connecticut, and Hawaii have passed similar laws requiring consumer contracts to be understandable to the public, and a score of other states are considering such laws.

- Hundreds of corporations across the country are simplifying their documents—employee benefit manuals, brokerage account agreements, trusts, the notes to financial statements, customer correspondence, billing statements, and internal communications, ranging from corporate policy manuals to the humble memo.

- President Carter, following up a promise for "regulations in plain English for a change," issued two unprecedented Executive Orders to Federal agencies telling them to simplify paperwork and eliminate gobbledy-gook from regulations. Agencies that have started to do so include the Environmental Protection Agency and the Department of Health and Human Services (formerly HEW). The Civil Aeronautics Board, where Alfred Kahn started a push for Plain English, is rewriting the notices it requires to be posted at airline counters and printed on the backs of tickets. The Federal Acquisition Regulation Project (FARP), in the Office of Federal Procurement Policy (Department of Management and Budget), offers great potential; FARP is rewriting and recodifying the massive Federal regulations on procurement of good and services.

- Major law firms, such as Shearman & Sterling in New York, have launched programs to train their young lawyers in clear legal drafting.

- The Internal Revenue Service committed over one million dollars in an all-out effort to simplify the Federal income tax forms. Simplified prototypes were presented to the Congress in November.

- The 1978 Amendments to the Constitution of the State of Hawaii include a provision that, "Insofar as practicable, all governmental writing meant for the public . . . should be plainly worded, avoiding the use of technical terms."

- On April 15, 1980, Gov. Hugh Carey of New York issued an Executive Order directing all state agencies to write their forms and regulations in plain language.

- Last fall, New York City's Department of Consumer Affairs proposed simplified language for more than 40 of its regulations covering deceptive and unconscionable business practices.

- The prestigious Practising Law Institute held a program and published a course book on "Drafting Documents in Plain Language." Another such program is planned for 1981.

- A lobbying group for the movement, Plain Talk, Inc., has been established in Washington, D.C.

- The National Institute for Education, a Federal research agency, has funded the Document Design Project, which is being run by the American Institutes of Research in conjunction with Siegel & Gale and Carnegie-Mellon University. The project's purpose is to study and encourage the use of simplified public documents.

—A.S.

using only the cash and securities actually available in your account at the time. Because it increases the amount we can invest on your behalf, buying on margin can increase the return on your investment. However, it can also increase the amount of loss on your account.

Example: Assume that the value of A Manufacturing Corp. stock in your portfolio has increased from $5,000 to $10,000. Because of changed market conditions, we decide to sell it and buy Z Industries common stock for your portfolio. If, for instance, the margin requirement is 50%, in theory we could purchase $20,000 worth of Z Industries stock for your account by using the $10,000 proceeds from the sale of your A Manufacturing stock and loaning you the additional $10,000.

In government, the promise—and difficulty—of the Plain English movement is illustrated by the case of the Food Stamp program. That program, begun in 1961, was substantially revised by the Food Stamp Act of 1977. As a result, the forms that had been used by the states were made obsolete. There was a bewildering variety of these forms to begin with. Wisconsin's, for instance, was 37 pages long. To help the states design simple, client-oriented forms reflecting the changes in the law, the Food and Nutrition Service (FNS) of the Department of Agriculture commissioned my firm to come up with new forms to offer to the states as models.

Our approach is illustrated by the new titles we gave our forms:

Old	**New**
Notice of Expiration	Continuing Your Food Stamps
Notice of Eligibility, Denial or Pending Status	Action Taken on Your Food Stamp Case
Tax Dependency Form	Student Tax Report

Significantly, the new forms ran into a special kind of opposition. First, *we* discovered that not all state governments agreed that the forms should *help* people to use the program, as the law intended. Food Stamp recipients are seen as one measure of a state's poverty, so some officials were reluctant to "encourage" their use by simplifying the paperwork. Some states had actually sought to discourage applicants by using forms that needlessly demanded embarrassing information, such as whether the applicant was a drug user or had a criminal record. Some state officials also felt that the language of the new forms, which included words such as "please," "thank you," and "sincerely," was too nice to people wanting money from the government.

Second, we found that state caseworkers themselves often appeared threatened by new, simplified forms. Some felt that their authority was being undermined. For example, people whose applications for food stamps are turned down or who are cut off are allowed by law to request a fair hearing. We embodied this feature of the law in the forms themselves: letters bearing bad news to Food Stamp clients included a perforated tear-off portion that could be sent back to appeal the decision. Some caseworkers objected heatedly: they felt the tear-offs

---

Plain Research Is Needed

The writing of Plain English documents is still a developing field. Researchers have discovered that the reader's comprehension has as much to do with what is inside his or her head as it has to do with what is on the printed page. Diagnostic research can help clarify what information must be explained at greater length.

A study conducted by law professor Jeffrey Davis, described in *Virginia Law Review* 63: No. 6, 1977, provides an excellent example. Davis prepared a simplified installment purchase contract for a refrigerator, showed it to shoppers from various backgrounds, and then asked them questions about its contents. The contract included this definition of default: "I will be in default if I fail to pay an installment on time or if I sell or fail to take proper care of the collateral." Yet most of the readers—even the well educated—failed to understand that they would be in default if they carelessly damaged the refrigerator. They persisted in the common but mistaken belief that they could not be in default if they made all their payments on time.

—A.S.

---

would cast doubts on their competence and would generate "unnecessary" hearings, increasing their already heavy workload.

After more than two years, the objections seem to have been overcome. Our model forms are used now by about half the states, and one FNS official commented earlier this year: "I've never heard a critical comment about the forms themselves." In fact, as in other simplifications projects, researchers got a strong impression that staff objections had as much to do with the fact the forms were new and unfamiliar as with their style or content.

Earlier in this piece I mentioned that some Plain English efforts have confused simplicity with simplemindedness. A major source of this confusion is the quick-and-dirty "readability formula" that equates clarity with short words and sentences. Here is a rider to a health insurance policy obviously written to such a formula. It is both uncommunicative and insulting to the reader's intelligence.

THE DRUG SPECIAL ENDORSEMENT

The DRUG SPECIAL is a small Endorsement. It goes with your [name of company] Contract. It depends on that other Contract to say who's covered. It is added coverage, with big extra benefits.

The DRUG SPECIAL gives you special help in paying for DRUGS . . .

Some attempts at simplification produce a kind of black humor. One life insurance company offers a "simplified" policy requiring that, "The Insured must die while this policy is in force." In a grisly parody that might be called "Dick and Jane Become Underwriters," the policy continues:

The Insured's death may be caused by accidental Bodily Injury. If so the Beneficiary may be paid an additional amount. It will be equal to the Face Amount. Three things must all happen. (1) Death must have been caused by Accidental Bodily Injury. Not by sickness. Not by anything else. . . .

Another misuse of the movement is the way some lawyers are cynically using "simplified" language, notably in apartment leases, to misinform consumers and to mislead them by playing on their ignorance. The nation's first plain language law went into effect in New York State in November 1978. One year later, the State Consumer Protection Board found that most revised lease forms "force tenants to surrender nearly every right they have under law." Rosemary S. Pooler, the board's executive director, referring to the lease prepared by the New York City Bar Association, commented: "It is ironic that tenants were in some respects better off with the 'legalese' of the 1965 lease since the . . . 'plain language' lease seems to have taken almost every opportunity to resolve legal issues in favor of landlords, sometimes at the expense of existing statutory and decisional law."

To illustrate her point, Mrs. Pooler gave, among others, the following examples:

RENT PAYMENT PROVISION

Pre-plain language version (1965):
"The tenant will pay the rent as herein provided."

Plain language version (1978):
"Tenant will pay the rent without any deductions, even if permitted by law."

The new document flies in the face of New York State's Multiple Dwelling Law (Sec. 302-a); Real Property Law (Sec. 235-2); Real Property Actions and Proceedings (Sec. 756 and Article 7-A); and several recent court decisions, all of which give tenants the right to reduce or withhold rent under certain circumstances—for instance, when these are serious building code violations.

ASSIGNMENT AND SUBLETTING

Pre-plain language version (1965):
"(Tenant) will not assign this lease or underlet the leased premises or any part thereof without the landlord's written consent, which landlord agrees not to withhold unreasonably."

Plain language version (1978):
"Tenant shall not assign this lease or enter into a sublease unless it is allowed by a law of the State of New York."

Since 1975, New York's Real Property Law (Sec. 226-b) has given tenants the right to request their landlord's permission to sublet or assign their lease. If the landlord withholds permission unreasonably, he must let the tenant break the lease. The 1965 version is a better statement of existing law than the plain language version, which implies that subletting is allowed by state law only in very special situations.

INSTALLATION OF LOCKS, CHAINS, ETC.

Pre-plain language version (1965):
None.

Plain language version (1978): "Tenant will not, without landlord's written approval: . . . . 3. Put in any locks or chain guards or change any lock-cylinders on the doors of the Apartment."

But New York State's Multiple Dwelling Law (Sec. 51-C) gives tenants of multiple dwellings the right to install additional locks without the landlord's permission, provided that the landlord who requests a key is given one.

Though many lawyers have seen the light, some of them—and their corporate clients—misguidedly try to protect themselves by insisting on too-precise standards for compliance. Such traditionalists do not like the Plain English laws that have been passed in New York, Maine, and Hawaii. Those laws simply require that each document for a residential lease or for a loan, property or services of less than $50,000 for personal, family or household purposes be:

1. Written in a clear and coherent manner using words with common everyday meanings;
2. Appropriately divided and captioned by its various sections.

Ironically, some lawyers and executives who are quick to decry regulatory minutiae in other areas object to this general approach to plain language legislation. They yearn for the false security of the simplistic "readability formula" approach, which ignores less easily quantified elements such as grammar, logic, and organization. They fear that without an exact definition of "clear, coherent, everyday language," compliance will elude them. Lawyers such as Wilbur H. Friedman, chairman of the New York County Lawyers' Association Special Committee on Consumer Agreements, direfully predicted that New York's law would create "upheaval" among businesses and an "absolutely staggering" burden on the courts.

Safe English

The *New York Times* (Aug. 31, 1980) reported the first case under the Plain English law on contracts:

"The New York State Attorney General was puzzled by this sentence.

'The liability of the bank is expressly limited to the exercise of ordinary diligence and care to prevent the opening of the within-mentioned safe deposit box during the within-mentioned term, or any extension or renewal thereof, by any person other than the lessee or his duly authorized representative and failure to exercise such diligence or care shall not be inferable from any alleged loss, absence or disappearance of any of its contents, nor shall the bank be liable for permitting a colessee or an attorney in fact of the lessee to have access to and remove the contents of said safe deposit box after the lessee's death or disability and before the bank has written knowledge of such death or disability.'

"Saying, 'I defy anyone, lawyer or lay person, to understand or explain what that means,' Attorney General Robert Abrams sued the Lincoln Savings Bank in New York City . . . demanding that it simplify a customer agreement on safe-deposit boxes.

"The case is settled . . . The former 121-word sentence now says: 'Our liability with respect to property deposited in the box is limited to ordinary care by our employees in the performance of their duties in preventing the opening of the box during the term of the lease by anyone other than you, persons authorized by you or persons authorized by the law.' "

But these fears were unfounded: there has been no flood of lawsuits. The arguments of the technically minded still have their appeal, however, as demonstrated by Connecticut's plain language law. It lists two alternate sets of criteria for plain language, with nine and eleven different tests respectively, including these:

1. The average number of words per sentence is less than twenty-two; and
2. No sentence in the contract exceeds fifty words; and
3. The average number of words per paragraph is less than seventy-five; and
4. No paragraph in the contract exceeds one hundred fifty words; and
5. The average number of syllables per word is less than 1.55; and . . .

Cookbook detail like this only guarantees that the spirit of plain language legislation will be lost in attempts to follow it to the letter: the clarity of a sentence becomes less important than seeing that it has "less than 1.55" syllables per word. And what would happen to interstate commerce if one of Connecticut's neighbors were to require less than 20 words per sentence instead of 22, or 160 words per paragraph instead of 150?

Keeping plain language laws simple, like the law in New York State, is essential. The courts should be left free to judge particular cases according to a general standard, as they do now with the concept of the "reasonable man." Lawyers who hang back, in favor of the traditional approach to legal language, should remember the traditional result: as Harold Laski put it, in every revolution the lawyers lead the way to the guillotine.

University of Wisconsin–Extension Equal Opportunities Program Office and Department of Agricultural Journalism

Guide to Nonsexist Language

With a little thought, you can use accurate, lively, figurative language in your classrooms, publications, columns, newsletters, workshops, broadcasts and telecommunications—and still represent people fairly. Breaking away from sexist language and traditional patterns can refresh your style.

You can follow two abbreviated rules to check material for bias: Would you say the same thing about a person of the opposite sex? Would you like it said about you? That's the bottom line. Use your own good sense on whether a joke, comment or image is funny—or whether it unfairly exploits people and perpetuates stereotypes.

Most fairness rules improve communication. Use explicit, active words; give concrete examples, specifics and anecdotes to demonstrate facts; present your message in context of the "big picture"; draw your reader in; use parallel forms; present a balanced view; and avoid clichés and generalizations that limit communication. Balanced language rules also guide you to choose or create balanced visual images.

Good communication respects individual worth, dignity, integrity and capacity. It treats people equally despite their sex, race, age, disability, socioeconomic background or creed. And it expresses fairness and balance. To communicate effectively, use real people, describe their unique characteristics and offer specific information. Using stereotypes or composites stifles communication and neglects human potential.

Routinely using male nouns and pronouns to refer to all people excludes more than half the population. There have been many studies that show that when the generic "he" is used, people in fact think it refers to men, rather than men *and* women. Making nouns plural to ensure plural pronouns can help you avoid using the singular "generic" male pronoun.

GUIDE TO NONSEXIST LANGUAGE. By the University of Wisconsin–Extension Equal Opportunities Program Office and Department of Agricultural Journalism. Reprinted by kind permission of the University of Wisconsin–Extension.

Many professional titles and workplace terms exclude women and unfairly link men with their earning capacity, while others patronize and subordinate women. Such nongeneric titles reinforce assumptions restricting women and men to stereotypical roles, inaccurately identify people, and give false images of people and how they live and work.

As a communicator, teacher or illustrator, you can help correct and eliminate irrelevant and inaccurate concepts about what it means to be male or female, black or white, young or old, rich or poor, healthy or disabled, or to hold a particular belief.

Editing, publishing and style manuals recognize the need for creating accurate, quality messages without slighting anyone and are beginning to prescribe standards for writing and evaluating manuscripts that represent people without stereotyping them.

Colleges and universities may want to develop their own booklets on nonsexist communication. For example, Franklin and Marshall College (PA) and Michigan State University have developed their own materials on bias-free language.

PRONOUNS: Each Person, to the Best of Her or His Ability

1. *Address Your Reader*

   No. If he studies hard, a student can make the honor roll.

   Yes. If you study hard, you can make the honor roll.

2. *Eliminate the Pronoun*

   No. Each nurse determines the best way she can treat a patient.

   Yes. Each nurse determines the best way to treat a patient.

3. *Replace Pronouns with Articles*

   No. A careful secretary consults her dictionary often.

   Yes. A careful secretary consults a dictionary often.

4. *Use Plural Nouns and Pronouns*

   No. Teach the child to walk by himself.

   Yes. Teach children to walk by themselves.

   He is expanding his operation.

   They are expanding their operation.

   Everyone needs his own space.

   All people need their own space.

5. *Alternate Male and Female Pronouns Throughout Text*

   No. The baby tries to put everything he finds in his mouth.

   Yes. The baby tries to put everything she finds in her mouth.

6. *Use Both Pronouns and Vary Their Order*

   No. A worker with minor children should make sure his will is up to date.

   Yes. A worker with minor children should make sure her or his will is up to date.

7. *Use Specific, Genderless Nouns*

   No. The average man on the street speaks his mind on the issues.

   Yes. The average voter speaks out on political issues.

8. *Substitute Job Titles or Descriptions*
   No. He gave a test on Monday.

   Yes. The professor gave a test on Monday.

9. *Repeat the Noun or Use a Synonym*
   No. The professor who gets published frequently will have a better chance when he goes before the tenure board.

   Yes. The professor who gets published frequently will have a better change when faculty tenure is granted.

(*Note:* We don't recommend using "their" to refer to a singular noun.)

(*Note:* Nations, battleships, gas tanks and other objects have no gender.)

Titles: People Working
Replace Language Stereotyping Men

No	Yes
Businessman/men	Business person/people, people in business, executive, merchant, industrialist, entrepreneur, manager
Cameraman	Camera operator, photographer
Chairman	Chairperson, chair, moderator, group leader, department head, presiding officer
Congressmen	Members of Congress, Representatives, congressmen and congresswomen
Craftsman	Craftsperson, artisan
Deliveryman/boy	Delivery driver/clerk, porter, deliverer, courier, messenger
Draftsman	Drafter
Fireman	Fire fighter
Foreman	Supervisor
Guys	Men, people
Headmaster	Principal
Kingpin	Key person, leader
Lumberman	Wood chopper, tree/lumber cutter
Male nurse	Nurse
Manhole/cover	Sewer hole, utility access/cover

Man-hours	Labor, staff/work hours, time
Man-made	Manufactured, handbuilt, hand made, synthetic, simulated, machine-made
Night watchman	Night guard, night watch
Policeman	Police officer, detective
Pressman	Press operator
Repairman, handyman	Repairer (Better: plumber, electrician, carpenter, steam fitter's apprentice)
Salesman/men	Salespeople, salesperson(s), sales agent(s), sales associate(s), sales representative(s), sales force
Spokesman	Representative, spokesperson, advocate, proponent
Sportsman	Sports/outdoor enthusiast (Better: hunter, fisher, canoer)
Sportsmanship	Fair play
Statesman	Political leader, public servant, diplomat
Statesmanship	Diplomacy
Steward/stewardess	Flight attendant
Weatherman	Weather reporter, meteorologist
Workmen	Workers

Replace Titles Stereotyping Women

No	Yes
Authoress	Author
Aviatrix	Pilot, aviator
Career girl/woman	Professor, engineer, mathematician, administrative assistant
Coed	Student
Gal, Girl, Girl Friday	Woman, secretary, assistant, aide (Better: full name)
Housewife, lady of the house	Homemaker, consumer, customer, shopper, parent

Lady/female doctor, lawyer	Doctor, lawyer
Little lady, better half	Spouse, partner
Maid, cleaning lady	Houseworker, housekeeper, custodian
Poetess	Poet
Sculptress	Sculptor
Usherette	Usher
Waitress	Wait person, waiter
Working wife/mother	Worker

Replace Stereotypical Adjectives and Expressions

No	Yes
Act like a lady and think like a man	Act and think sensitively and clearly
Act like a gentleman/man	Be polite, brave, keep your chin up
Dear Sir	Dear Madam or Sir, Dear Personnel Officer/Director, Dear Executive/Manager (Better: name)
Fatherland	Homeland, native land
Founding fathers	Pioneers, colonists, patriots, forebears, founders
Gentleman's agreement	Informal agreement, your word, oral contract, handshake
Ladylike, girlish, sissy, effeminate	Tender, cooperative, polite, neat, fearful, weak, illogical, inactive (Both male and female characteristics)
Lady luck	Luck
Layman, layman's terms	Lay, common, ordinary, informal, nontechnical
Maiden name	Birth name
Maiden voyage	First/premiere voyage
Male chauvinist	Chauvinist
Male ego	Ego
Man-sized	Husky, sizable, big, large, voracious

Man-to-man defense/talk	Player-to-player, person-to-person, face-to-face, one-to-one
Manly, tomboy	Courageous, strong, vigorous, adventurous, spirited, direct, competitive, physical, mechanical, logical, rude, active, messy, self-confident (Both female and male characteristics)
Mother doing dishes, father reading the paper	Men and women doing dishes, women and men reading the paper (Note: Also applies to visual images)
Mother Nature, Father Time	Nature, time
Mothering, fathering	Parenting, child-rearing
Motherly	Protective, supportive, kind
Unwed mother	Mother
Woman did well for a woman/ as well as a man	Woman did well, woman performed competently
Woman's/man's work	Avoid (Too broad; use specifics)
Women's page	Lifestyle, living section

Rosalie Maggio

Bias-Free Language: Some Guidelines

Rosalie Maggio is the author of The Nonsexist Word Finder *(1987),* How to Say It: Words, Phrases, Sentences, and Paragraphs for Every Situation *(1990), and* The Music Box Christmas *(1990). She has also won several literary honors and awards for her children's fiction and research on women's issues.*

Language both reflects and shapes society. The textbook on American government that consistently uses male pronouns for the president, even when not referring to a specific individual (e.g., "a president may cast his veto"), reflects the fact that all our presidents have so far been men. But it also shapes a society in which the idea of a female president somehow "doesn't sound right."

Culture shapes language and then language shapes culture. "Contrary to the assumption that language merely reflects social patterns such as sex-role stereotypes, research in linguistics and social psychology has shown that these are in fact facilitated and reinforced by language" (Marlis Hellinger, in *Language and Power,* ed., Cheris Kramarae et al.).

Biased language can also, says Sanford Berman, "powerfully harm people, as amply demonstrated by bigots' and tyrants' deliberate attempts to linguistically dehumanize and demean groups they intend to exploit, oppress, or exterminate. Calling Asians 'gooks' made it easier to kill them. Calling blacks 'niggers' made it simpler to enslave and brutalize them. Calling Native Americans 'primitives' and 'savages' made it okay to conquer and despoil them. And to talk of 'fishermen,' 'councilmen,' and 'longshoremen' is to clearly exclude and discourage women from those pursuits, to diminish and degrade them."

The question is asked: Isn't it silly to get upset about language when there are so many more important issues that need our attention?

First, it's to be hoped that there are enough of us working on issues large and small that the work will all get done—someday. Second, the interconnections between the way we think, speak, and act are beyond dispute. Language goes hand-in-hand with social change—both shaping it and reflecting it. Sexual harassment was not a term anyone used twenty years ago; today we have laws against it. How could we have the law without the language; how could we have the language without the law? In fact, the judicial system is a good argument for the importance of "mere words"; the legal profession devotes great energy to the precise interpretation of words—often with far-reaching and significant consequences.

On August 21, 1990, in the midst of the Iraqi offensive, front-age headlines told the big story: President Bush had used the word *hostages* for the first time. Up to that time, *detainee* had been used. The difference between two very similar words was of possible life-and-death proportions. In another situation—also said to be life-and-death by some people—the difference between *fetal tissue* and *unborn baby* (in referring to the very same thing) is arguably the most debated issue in the country. So, yes, words have power and deserve our attention.

Some people are like George Crabbe's friend: "Habit with him was all the test of truth,/it must be right: I've done it from my youth." They have come of age using *handicapped, black-and-white, leper, mankind,* and pseudogeneric *he;* these terms must therefore be correct. And yet if there's one thing consistent about language it is that language is constantly changing; when the *Random House Dictionary of the English Language: 2nd Edition* was published in 1988, it contained 50,000 new entries, most of them words that had come into use since 1966. There were also 75,000 new definitions. (Incidentally, *RHD-II* asks its readers to "use gender-neutral terms wherever possible" and it never uses *mankind* in definitions where *people* is meant, nor does it ever refer to anyone of unknown gender as *he*.) However, few supporters of bias-free language are asking for changes; it is rather a matter of choice—which of the many acceptable words available to us will we use?

A high school student who felt that nonsexist language did demand some changes said, "But you don't understand! You're trying to change the English language, which has been around a lot longer than women have!"

One reviewer of the first edition commented, "There's no fun in limiting how you say a thing." Perhaps not. Yet few people complain about looking up a point of grammar or usage or checking the dictionary for a correct spelling. Most writers are very fussy about finding the precise best word, the exact rhythmic vehicle for their ideas. Whether or not these limits "spoil their fun" is an individual judgment. However, most of us accept that saying or writing the first thing that comes to mind is not often the way we want to be remembered. So if we have to think a little, if we have to search for the unbiased word, the inclusive phrase, it is not any more effort than we expend on proper grammar, spelling, and style.

Other people fear "losing" words, as though there weren't more where those came from. We are limited only by our imaginations; vague, inaccurate, and disrespectful words can be thrown overboard with no loss to society and no impoverishment of the language.

Others are tired of having to "watch what they say." But what they perhaps mean is that they're tired of being sensitive to others' requests. From childhood onward, we all learn to "watch what we say": we don't swear around our parents; we don't bring up certain topics around certain people; we speak differently to friend, boss, cleric, English teacher, lover, radio interviewer, child. Most of us are actually quite skilled at picking and choosing appropriate words; it seems odd that we are too "tired" to call people what they want to be called.

The greatest objection to bias-free language is that it will lead us to absurdities. Critics have posited something utterly ridiculous, cleverly demonstrated how silly it is, and then accounted themselves victorious in the battle against linguistic massacre. For example: "So I suppose now we're going to say: He/she ain't heavy, Father/Sister; he/she's my brother/sister." "I suppose next it will be 'ottoperson'." Cases have been built up against the mythic "woperson," "personipulate," and "personhole cover" (none of which has ever been advocated by any reputable sociolinguist). No grist appears too ridiculous for these mills. And, yes, they grind exceedingly small. Using a particular to condemn a universal is a fault in logic. But then ridicule, it is said, is the first and last argument of fools.

One of the most rewarding—and, for many people, the most unexpected—side effects of breaking away from traditional, biased language is a dramatic improvement in writing style. By replacing fuzzy, overgeneralized, cliché-ridden words with explicit, active words and by giving concrete examples and anecdotes instead of one-word-fits-all descriptions you can express yourself more dynamically, convincingly, and memorably.

"If those who have studied the art of writing are in accord on any one point, it is on this: the surest way to arouse and hold the attention of the reader is by being specific, definite, and concrete" (Strunk and White, *The Elements of Style*). Writers who talk about *brotherhood* or *spinsters* or *right-hand men* miss a chance to spark their writing with fresh descriptions; they leave their readers as uninspired as they are. Unthinking writing is also less informative. Why use the unrevealing *adman* when we could choose instead a precise, descriptive, inclusive word like *advertising executive, copywriter, account executive, ad writer,* or *media buyer?*

The word *man-made*, which seems so indispensable to us, doesn't actually say very much. Does it mean artificial? handmade? synthetic? fabricated? machine-made? custom-made? simulated? plastic? imitation? contrived?

Communication is—or ought to be—a two-way street. A speaker who uses *man* to mean *human being* while the audience hears it as *adult male* is an example of communication gone awry.

Bias-free language is logical, accurate, and realistic. Biased language is not. How logical is it to speak of the "discovery" of America, a land already inhabited by millions of people? Where is the accuracy in writing "Dear Sir" to a woman? Where is the realism in the full-page automobile advertisement that says in bold letters, "A good driver is a product of his environment," when more women than men influence car-buying decisions? Or how successful is the ad for a dot-matrix printer that says, "In 3,000 years, man's need to present his ideas hasn't changed.

But his tools have," when many of these printers are bought and used by women, who also have ideas they need to present? And when we use stereotypes to talk about people ("isn't that just like a welfare mother/Indian/girl/old man"), our speech and writing will be inaccurate and unrealistic most of the time.

DEFINITION OF TERMS

Bias/Bias-Free

Biased language communicates inaccurately about what it means to be male or female; black or white; young or old; straight, gay, or bi; rich or poor; from one ethnic group or another; disabled or temporarily able-bodied; or to hold a particular belief system. It reflects the same bias found in racism, sexism, ageism, handicappism, classism, ethnocentrism, anti-Semitism, homophobia, and other forms of discrimination.

Bias occurs in the language in several ways.

1. Leaving out individuals or groups. "Employees are welcome to bring their wives and children" leaves out those employees who might want to bring husbands, friends, or same-sex partners. "We are all immigrants in this country" leaves out Native Americans, who were here well before the first immigrants.
2. Making unwarranted assumptions. To address a sales letter about a new diaper to the mother assumes that the father won't be diapering the baby. To write "Anyone can use this fire safety ladder" assumes that all members of the household are able-bodied.
3. Calling individuals and groups by names or labels that they do not choose for themselves (e.g., *Gypsy, office girl, Eskimo, pygmy, Bushman, the elderly, colored man*) or terms that are derogatory (*fairy, libber, savage, bum, old goat*).
4. Stereotypical treatment that implies that all lesbians/Chinese/women/people with disabilities/teenagers are alike.
5. Unequal treatment of various groups in the same material.
6. Unnecessary mention of membership in a particular group. In a land of supposedly equal opportunity, of what importance is a person's race, sex, age, sexual orientation, disability, or creed? As soon as we mention one of these characteristics—without a good reason for doing so—we enter an area mined by potential linguistic disasters. Although there may be instances in which a person's sex, for example, is germane ("A recent study showed that female patients do not object to being cared for by male nurses"), most of the time it is not. Nor is mentioning a person's race, sexual orientation, disability, age, or belief system usually germane.

Bias can be overt or subtle. Jean Gaddy Wilson (in Brooks and Pinson, *Working With Words*) says, "Following one simple rule of writing or speaking will eliminate most biases. Ask yourself: Would you say the same thing about an affluent, white man?"

Inclusive/Exclusive

Inclusive language includes everyone; exclusive language excludes some people. The following quotation is inclusive: "The greatest revolution of our generation is the discovery that human beings, by changing the inner attitudes of their minds, can change the outer aspects of their lives" (William James). It is clear that James is speaking of all of us.

Examples of sex-exclusive writing fill most quotation books: "Man is the measure of all things" (Protagoras). "The People, though we think of a great entity when we use the word, means nothing more than so many millions of individual men" (James Bryce). "Man is nature's sole mistake" (W. S. Gilbert).

Sexist/Nonsexist

Sexist language promotes and maintains attitudes that stereotype people according to gender while assuming that the male is the norm—the significant gender. Nonsexist language treats all people equally and either does not refer to a person's sex at all when it is irrelevant or refers to men and women in symmetrical ways.

"A society in which women are taught anything but the management of a family, the care of men, and the creation of the future generation is a society which is on the way out" (L. Ron Hubbard). "Behind every successful man is a woman—with nothing to wear" (L. Grant Glickman). "Nothing makes a man and wife feel closer, these days, than a joint tax return" (Gil Stern). These quotations display various characteristics of sexist writing: (1) stereotyping an entire sex by what might be appropriate for some of it; (2) assuming male superiority; (3) using unparallel terms (*man and wife* should be either *wife and husband/husband and wife* or *woman and man/man and woman*).

The following quotations clearly refer to all people: "It's really hard to be roommates with people if your suitcases are much better than theirs" (J. D. Salinger). "If people don't want to come out to the ball park, nobody's going to stop them" (Yogi Berra). "If men and women of capacity refuse to take part in politics and government, they condemn themselves, as well as the people, to the punishment of living under bad government" (Senator Sam J. Ervin). "I studied the lives of great men and famous women, and I found that the men and women who got to the top were those who did the jobs they had in hand, with everything they had of energy and enthusiasm and hard work" (Harry S. Truman).

Gender-Free/Gender-Fair/Gender-Specific

Gender-free terms do not indicate sex and can be used for either women/girls or men/boys (e.g., *teacher, bureaucrat, employee, hiker, operations manager, child, clerk, sales rep, hospital patient, student, grandparent, chief executive officer*).

Writing or speech that is gender-fair involves the symmetrical use of gender-specific words (e.g., *Ms. Leinwohl/Mr. Kelly, councilwoman/councilman, young man/young woman*) and promotes fairness to both sexes in the larger context. To ensure gender-fairness, ask yourself often: Would I write the same thing in the same way about a person of the opposite sex? Would I mind if this were said of me?

If you are describing the behavior of children on the playground, to be gender-fair you will refer to girls and boys an approximately equal number of times, and you will carefully observe what the children do, and not just assume that only the boys will climb to the top of the jungle gym and that only the girls will play quiet games.

Researchers studying the same baby described its cries as "anger" when they were told it was a boy and as "fear" when they were told it was a girl (cited in Cheris Kramarae, *The Voices and Words of Women and Men*). We are all victims of our unconscious and most deeply held biases.

Gender-specific words (for example, *alderwoman, businessman, altar girl*) are neither good nor bad in themselves. However, they need to be used gender-fairly; terms for women and terms for men should be used an approximately equal number of times in contexts that do not discriminate against either of them. One problem with gender-specific words is that they identify and even emphasize a person's sex when it is not necessary (and is sometimes even objectionable) to do so. Another problem is that they are so seldom used gender-fairly.

Although gender-free terms are generally preferable, sometimes gender-neutral language obscures the reality of women's or men's oppression. *Battered spouse* implies that men and women are equally battered; this is far from true. *Parent* is too often taken to mean *mother* and obscures the fact that more and more fathers are very much involved in parenting; it is better here to use the gender-specific *fathers and mothers* or *mothers and fathers* than the gender-neutral *parents*.

Generic/Pseudogeneric

A generic is an all-purpose word that includes everybody (e.g., *workers, people, voters, civilians, elementary school students*). Generic pronouns include: *we, you, they*.

A pseudogeneric is a word that is used as though it included all people, but that in reality does not. *Mankind, forefathers, brotherhood,* and *alumni* are not generic because they leave out women. When used about Americans, *immigrants* leaves out all those who were here long before the first immigrants. "What a christian thing to do!" uses *christian* as a pseudogeneric for *kind* or *good-hearted* and leaves out all kind, good-hearted people who are not Christians.

Although some speakers and writers say that when they use *man* or *mankind* they mean everybody, their listeners and readers do not perceive the word that way and these terms are thus pseudogenerics. The pronoun *he* when used to mean *he and she* is another pseudogeneric.

Certain generic nouns are often assumed to refer only to men, for example, *politicians, physicians, lawyers, voters, legislators, clergy, farmers, colonists, immigrants, slaves, pioneers, settlers, members of the armed forces, judges, taxpayers*. References to "settlers, their wives, and children," or "those clergy permitted to have wives" are pseudogeneric.

In historical context it is particularly damaging for young people to read about settlers and explorers and pioneers as though they were all white men. Our language should describe the accomplishments of the human race in terms of all those who contributed to them.

SEX AND GENDER

An understanding of the difference between sex and gender is critical to the use of bias-free language.

Sex is biological: people with male genitals are male, and people with female genitals are female.

Gender is cultural: our notions of "masculine" tell us how we expect men to behave and our notions of "feminine" tell us how we expect women to behave. Words like *womanly/manly, tomboy/sissy, unfeminine/unmasculine* have nothing to do with the person's sex; they are culturally acquired, subjective concepts about character traits and expected behaviors that vary from one place to another, from one individual to another.

It is biologically impossible for a woman to be a sperm donor. It may be culturally unusual for a man to be a secretary, but it is not biologically impossible. To say "the secretary . . . she" assumes all secretaries are women and is sexist because the issue is gender, not sex. Gender describes an individual's personal, legal, and social status without reference to genetic sex; gender is a subjective cultural attitude. Sex is an objective biological fact. Gender varies according to the culture. Sex is a constant.

The difference between sex and gender is important because much sexist language arises from cultural determinations of what a woman or man "ought" to be. Once a society decides, for example, that to be a man means to hide one's emotions, bring home a paycheck, and be able to discuss football standings while to be a woman means to be soft-spoken, love shopping, babies, and recipes, and "never have anything to wear," much of the population becomes a contradiction in terms—unmanly men and unwomanly women. Crying, nagging, gossiping, and shrieking are assumed to be women's lot; rough-housing, drinking beer, telling dirty jokes, and being unable to find one's socks and keys are laid at men's collective door. Lists of stereotypes appear silly because very few people fit them. The best way to ensure unbiased writing and speaking is to describe people as individuals, not as members of a set.

Gender Role Words

Certain sex-linked words depend for their meanings on cultural stereotypes: *feminine/masculine, manly/womanly, boyish/girlish, husbandly/wifely, fatherly/motherly, unfeminine/unmasculine, unmanly/unwomanly*, etc. What a person understands by these words will vary from culture to culture and even within a culture. Because the words depend for their meanings on interpretations of stereotypical behavior or characteristics, they may be grossly inaccurate when applied to individuals. Somewhere, sometime, men and women have said, thought, or done everything the other sex has said, thought, or done except for a very few sex-linked biological activities (e.g., only women can give birth or nurse a baby, only a man can donate sperm or impregnate a woman). To describe a woman as unwomanly is a contradiction in terms; if a woman is doing it, saying it, wearing it, thinking it, it must be—by definition—womanly.

F. Scott Fitzgerald did not use "feminine" to describe the unforgettable Daisy in *The Great Gatsby*. He wrote instead, "She laughed again, as if she said something very witty, and held my hand for a moment, looking up into my face, promising that there was no one in the world she so much wanted to see. That was a way she had." Daisy's charm did not belong to Woman; it was uniquely hers. Replacing vague sex-linked descriptors with thoughtful words that describe an individual instead of a member of a set can lead to language that touches people's minds and hearts.

NAMING

Naming is power, which is why the issue of naming is one of the most important in bias-free language.

Self-Definition

People decide what they want to be called. The correct names for individuals and groups are always those by which they refer to themselves. This "tradition" is not always unchallenged. Haig Bosmajian (*The Language of Oppression*) says, "It isn't strange that those persons who insist on defining themselves, who insist on this elemental privilege of self-naming, self-definition, and self-identity encounter vigorous resistance. Predictably, the resistance usually comes from the oppressor or would-be oppressor and is a result of the fact that he or she does not want to relinquish the power which comes from the ability to define others."

Dr. Ian Hancock uses the term *exonym* for a name applied to a group by outsiders. For example, Romani peoples object to being called by the exonym *Gypsies.* They do not call themselves Gypsies. Among the many other exonyms are: the elderly, colored people, homosexuals, pagans, adolescents, Eskimos, pygmies, savages. The test for an exonym is whether people describe themselves as "redmen," "illegal aliens," "holy rollers," etc., or whether only outsiders describe them that way.

There is a very small but visible element today demanding that gay men "give back" the word *gay*—a good example of denying people the right to name themselves. A late-night radio caller said several times that gay men had "stolen" this word from "our" language. It was not clear what language gay men spoke.

A woman nicknamed "Betty" early in life had always preferred her full name, "Elizabeth." On her fortieth birthday, she reverted to Elizabeth. An acquaintance who heard about the change said sharply, "I'll call her Betty if I like!"

We can call them Betty if we like, but it's arrogant, insensitive, and uninformed: the only rule we have in this area says we call people what they want to be called.

"Insider/Outsider" Rule

A related rule says that insiders may describe themselves in ways that outsiders may not. "Crip" appears in *The Disability Rag;* this does not mean that the word is available to anyone who wants to use it. "Big Fag" is printed on a gay man's T-shirt. He may use that expression; a non-gay may not so label him. One junior-high student yells to another, "Hey, nigger!" This would be highly offensive and inflammatory if the speaker were not African American. A group of women talk about "going out with the girls," but a co-worker should not refer to them as "girls." When questioned about just such a situation, Miss Manners replied that "people are allowed more leeway in what they call themselves than in what they call others."

"People First" Rule

Haim Ginott taught us that labels are disabling; intuitively most of us recognize this and resist being labeled. The disability movement originated the "people first" rule, which says we don't call someone a "diabetic" but rather "a person with diabetes." Saying someone is "an AIDS victim" reduces the person to a disease, a label, a statistic; use instead "a person with/who has/living with AIDS." The 1990 Americans with Disabilities Act is a good example of correct wording. Name the person as a person first, and let qualifiers (age, sex, disability, race) follow, but (and this is crucial) only if they are relevant. Readers of a magazine aimed at an older audience were asked what they wanted to be called (elderly? senior citizens? seniors? golden agers?). They rejected all the terms; one said, "How about 'people'?" When high school students rejected labels like kids, teens, teenagers, youth, adolescents, and juveniles, and were asked in exasperation just what they would like to be called, they said, "Could we just be people?"

Women as Separate People

One of the most sexist maneuvers in the language has been the identification of women by their connections to husband, son, or father—often even after he is dead. Women are commonly identified as someone's widow while men are never referred to as anyone's widower. Marie Marvingt, a French-woman who lived around the turn of the century, was an inventor, adventurer, stunt woman, superathlete, aviator, and all-around scholar. She chose to be affianced to neither man (as a wife) nor God (as a religious), but it was not long before an uneasy male press found her a fit partner. She is still known today by the revealing label "the Fiancée of Danger." If a connection is relevant, make it mutual. Instead of "Frieda, his wife of seventeen years," write "Frieda and Eric, married for seventeen years."

It is difficult for some people to watch women doing unconventional things with their names. For years the etiquette books were able to tell us precisely how to address a single woman, a married woman, a divorced woman, or a widowed woman (there was no similar etiquette on men because they have always been just men and we have never had a code to signal their marital status). But now some women are Ms. and some are Mrs., some are married but keeping their birth names, others are hyphenating their last name with their husband's, and still others have constructed new names for themselves. Some women—including African American women who were denied this right earlier in our history—take great pride in using their husband's name. All these forms are correct. The same rule of self-definition applies here: call the woman what she wants to be called.

Part 3

Business and Technical Correspondence

Traditionally, business and technical correspondence has taken two forms: the letter and the memo. More recently, it has assumed a third electronic form, e-mail. This electronic innovation has brought with it a myriad of issues, most of which unfortunately remain unresolved. There has been little discussion of, let alone agreement about, the etiquette of e-mail.

In the past, business and technical correspondence *seemed* easier on some levels: letters went to people outside of the company; memos, to people inside the company. Letters were advertisements for writers and their companies. Memos provided a record of decisions made and actions taken within a department of a company.

Letters and memos announce or reaffirm policies, confirm decisions and conversations, and send or request information. Some letters and memos are routine; others concern pressing issues. In either case, letters and memos require careful writing.

Given the volume of correspondence many companies produce, it is easy for business and technical writers to be tempted to use shortcuts when writing. Looking for a shortcut can, however, court danger if the shortcut involves relying on form responses or copies of previous—though not necessarily good examples of—similar correspondence.

There is nothing wrong with form letters or memos, as long as such correspondence is appropriate to the given writing situation. A letter reminding customers of bills past due or a memo transmitting attached data is often an effective and time-efficient way to communicate—provided such letters and memos are appropriate to the situations and audiences with which they are used.

The danger with form letters and memos is that they can become a crutch that writers depend on even in situations in which they are inappropriate. Such misuse of form letters and memos can create new problems for writers. Writers may have to send a follow-up letter or memo to fix what they tried to do in the first piece of correspondence. Or worse, they might have to send *several* follow-up messages.

A process approach to business and technical correspondence will help determine when form letters and memos are appropriate and when something more original is required. The careful use of the writing process can also make follow-ups unnecessary.

The advent of electronic correspondence would seem to offer a boon to business and technical writers, but every silver lining has its cloud. The ease with which electronic correspondence can be transmitted has led some writers to become more casual—if not careless and sloppy—in their electronic correspondence than they would be in producing correspondence in the more traditional forms of letters and memos.

Electronic correspondence can be sent within and outside companies to multiple audiences across the globe. A misstatement, mistake, or unintended slight can suddenly be transmitted to a very large audience, making the writer look foolish or worse to a large body of readers. As I indicated earlier, there is yet no established or generally adopted etiquette for electronic correspondence. In the absence of such etiquette, writers would be wise to be conservative in such correspondence, adhering closely to the rules and principles that work so well for letters and memos. Whether the product is a letter, a memo, or an electronic message, a process approach is the safest approach to take.

The selections in this section of *Strategies* follow a decidedly process-oriented approach to business and technical correspondence. The late Malcolm Forbes writes from the dual perspective of a harried reader and a writer. Forbes, like everyone else in the world of work, was a busy person who was impatient with correspondence that wasted his time. David V. Lewis expands on Forbes's advice on how to write to people effectively by offering suggestions on how writers can use correspondence to sell themselves to readers.

Writers for the Royal Bank of Canada agree with Lewis, arguing that *all* letters can be letters that sell. They suggest writers use formulas for their correspondence to achieve this goal. Of course, when the purpose is to say "no," writers face special difficulties. As Allan A. Glatthorn suggests, writers can say "no" in their letters and still keep or make friends.

Harold K. Mintz and Marvin H. Swift show how a process approach can inform the writing of memos. Mintz offers a formula for writing better memos, while Swift shows how the writing process can help a manager rework and rethink a routine memo to produce a clear and effective message.

Lee Clark Johns takes a decidedly different approach—and tone—in her discussion of the pros and cons of relying on the file cabinet as a ready source for writing models. John S. Fielden and Ronald E. Dulek, on the other hand, suggest a model for efficient writing that helps not hinders the writing process. Finally, M. Jimmie Killingsworth discusses the challenges and opportunities created by electronic correspondence.

Malcolm Forbes

How to Write a Business Letter

Malcolm Forbes was President and Editor-in-Chief of Forbes Magazine.

A good business letter can get you a job interview.

Get you off the hook.

Or get you money.

It's totally asinine to blow your chances of getting *whatever* you want—with a business letter that turns people off instead of turning them on.

The best place to learn to write is in school. If you're still there, pick your teachers' brains.

If not, big deal. I learned to ride a motorcycle at 50 and fly balloons at 52. It's never too late to learn

Over 10,000 business letters come across my desk every year. They seem to fall into three categories: stultifying if not stupid, mundane (most of them), and first rate (rare). Here's the approach I've found that separates the winners from the losers (most of it's just good common sense)—it starts *before* you write your letter.

KNOW WHAT YOU WANT

If you don't, write it down—in one sentence. "I want to get an interview within the next two weeks." That simple. List the major points you want to get across—it'll keep you on course.

If you're *answering* a letter, check the points that need answering and keep the letter in front of you while you write. This way you won't forget anything—*that* would cause another round of letters.

HOW TO WRITE A BUSINESS LETTER. By Malcolm Forbes. Reprinted by permission of International Paper.

And for goodness' sake, answer promptly if you're going to answer at all. Don't sit on a letter—*that* invites the person on the other end to sit on whatever you want from *him*.

PLUNGE RIGHT IN

Call him by name—not "Dear Sir, Madam, or Ms." "Dear Mr. Chrisanthopoulos"—and be sure to spell it right. That'll get him (thus, you) off to a good start.

(Usually, you can get his name just by phoning his company—or from a business directory in your nearest library.)

Tell what your letter is about in the first paragraph. One or two sentences. Don't keep your reader guessing or he might file your letter away—even before he finishes it.

In the round file.

If you're answering a letter, refer to the date it was written. So the reader won't waste time hunting for it.

People who read business letters are as human as thee and me. Reading a letter shouldn't be a chore—*reward* the reader for the time he gives you.

WRITE SO HE'LL ENJOY IT

Write the entire letter from his point of view—what's in it for *him*? Beat him to the draw—surprise him by answering the questions and objections he might have.

Be positive—he'll be more receptive to what you have to say.

Be nice. Contrary to the cliché, genuinely nice guys most often finish first or very near it. I admit it's not easy when you've got a gripe. To be agreeable while disagreeing—that's an art.

Be natural—write the way you talk. Imagine him sitting in front of you— what would you *say* to him?

Business jargon too often is cold, stiff, unnatural.

Suppose I came up to you and said, "I acknowledge receipt of your letter and I beg to thank you." You'd think, "Huh? You're putting me on."

The acid test—read your letter out loud when you're done. You might get a shock—but you'll know for sure if it sounds natural.

Don't be cute or flippant. The reader won't take you seriously. This doesn't mean you've got to be dull. You prefer your letter to knock 'em dead rather than bore 'em to death.

Three points to remember:

Have a sense of humor. That's refreshing *anywhere*—a nice surprise in a business letter.

Be specific. If I tell you there's a new fuel that could save gasoline, you might not believe me. But suppose I tell you this:

"Gasohol"—10% alcohol, 90% gasoline—works as well as straight gasoline. Since you can make alcohol from grain or corn stalks, wood or wood waste, coal—even garbage, it's worth some real follow-through.

Now you've got something to sink your teeth into.

<u>Lean heavier on nouns and verbs, lighter on adjectives. Use the active voice instead of the passive.</u> Your writing will have more guts.

Which of these is stronger? Active voice: "I kicked out my money manager." Or, passive voice: "My money manager was kicked out by me." (By the way, neither is true. My son, Malcolm Jr., manages most Forbes money—he's a brilliant moneyman.)

GIVE IT THE BEST YOU'VE GOT

When you don't want something enough to make *the* effort, making *an* effort is a waste.

<u>Make your letter look appetizing</u>—or you'll strike out before you even get to bat. Type it—on good-quality 8 1/2" × 11" stationery. Keep it neat. And use paragraphing that makes it easier to read.

<u>Keep your letter short</u>—to one page, if possible. Keep your paragraphs short. After all, who's going to benefit if your letter is quick and easy to read?

You.

For emphasis, <u>underline</u> important words. And sometimes indent sentences as well as paragraphs.

Like this. See how well it works? (But save it for something special.)

<u>Make it perfect.</u> No typos, no misspellings, no factual errors. If you're sloppy and let mistakes slip by, the person reading your letter will think you don't know better or don't care. Do you?

<u>Be crystal clear.</u> You won't get what you're after if your reader doesn't get the message.

<u>Use good English.</u> If you're still in school, take all the English and writing courses you can. The way you write and speak can really help—or *hurt*.

If you're not in school (even if you are), get the little 71-page gem by Strunk & White, *Elements of Style*. It's in paperback. It's fun to read and loaded with tips on good English and good writing.

<u>Don't put on airs.</u> Pretense invariably impresses only the pretender.

<u>Don't exaggerate.</u> Even once. Your reader will suspect everything else you write.

<u>Distinguish opinions from facts.</u> Your opinions may be the best in the world. But they're not gospel. You owe it to your reader to let him know which is which. He'll appreciate it and he'll admire you. The dumbest people I know are those who Know It All.

<u>Be honest.</u> It'll get you farther in the long run. If you're not, you won't rest easy until you're found out. (The latter, not speaking from experience.)

Edit ruthlessly. Somebody ~~has~~ said that words are ~~a lot~~ like inflated money— the more ~~of them that~~ you use, the less each one ~~of them~~ is worth. ~~Right on.~~ Go through your entire letter ~~just~~ as many times as it takes. ~~Search out and~~ **A**nnihilate all unnecessary words ~~and~~ sentences—even ~~entire~~ *paragraphs.*

SUM IT UP AND GET OUT

The last paragraph should tell the reader exactly what you want *him* to do—or what *you're* going to do. Short and sweet. "May I have an appointment? Next Monday, the 16th, I'll call your secretary to see when it'll be most convenient for you."

Close with something simple like, "Sincerely." And for heaven's sake sign legibly. The biggest ego trip I know is a completely illegible signature.

Good luck.

I hope you get what you're after.

Sincerely,

Malcolm S. Forbes

David V. Lewis

Making Your Correspondence Get Results

David V. Lewis is an in-house consultant with Western Company of North America in Fort Worth, Texas, where he is involved in sales and management training.

If you turn out an average of five letters a day, you'll produce nearly twice as many words during the year as the typical professional writer.

These letters reflect *your* attitude—and obviously your organization's—toward customers. Realizing this, many progressive organizations train key people in the art of writing readable, results-getting letters. For example, the New York Life Insurance Company produces more than a million letters a year and uses its correspondence as a public relations tool.

"Anyone who writes a letter for New York Life holds a key position in our organization," says Nathan Kelne, vice president of the company. "By the letters our people write, they help determine how the public feels toward our company, and toward the life insurance business as a whole. Since they are instrumental in shaping the personality of the company, they are in a very real sense public relations writers."

For example, before New York Life launched its companywide letter-writing courses, here's how one of its executives tried to explain to a beneficiary the way the death claim was to be paid on a $3,000 policy:

> The monthly income per $1,000 under option 10 years certain is $7.93 per $1,000 total face amount of insurance, which total is $3,000.

Fortunately, the beneficiary was able to cut through the jargon. He replied:

MAKING YOUR CORRESPONDENCE GET RESULTS. By David V. Lewis. From *Secrets of Successful Writing, Speaking, and Listening* by David V. Lewis (AMACOM a division of the American Management Association, New York, 1982). Reprinted with the kind permission of the author.

As I understand your letter, you seem to be saying that I should receive a monthly check for the amount of $23.79 for at least 10 years.

But another policyholder's response to a similarly bewildering letter is more typical:

Please tell me what you want me to do, and I'll be glad to do it.

Some universally proven principles that can help you sell yourself and your company to the public will now be examined.

WRITE FOR HIM, NOT TO HIM

General Foods is an organization that believes strongly in creating a favorable image through its correspondence. The public relations people came across this letter signed by a marketing executive and ready for mailing:

```
Dear Sir:
    Enclosed please find a questionnaire which we are sending to
all our retail contacts in this state.
    Will you please answer this as soon as possible? It's very
important that we have an immediate reply. We're delaying final
plans for our retail sales program in this area until we get
answers to the questionnaire.
    With thanks in advance, we are,

                                   Gratefully yours, . . .
```

The letter is clear and to the point; it *does* communicate readily. But there's a major flaw. It points out benefits the company will receive instead of suggesting how the program will help the recipient. The letter is writer- rather than reader-oriented.

Psychologists say that each of us is basically interested in himself or herself. We want to know "what's in it for me?" Once you routinely approach letter writing from this point of view, you'll find yourself telling your readers in specific terms how the letter will benefit them.

The questionnaire letter was rewritten this way:

```
Dear Sir:
    Enclosed is a questionnaire on our proposed retail sales pro-
gram. Your advance opinion on how this new plan would help you
and others will be very useful in our evaluation of the program.
    If you will fill out and return this questionnaire as soon as
possible, we can let you and other retail contacts in this area
know promptly of changes in the present program. Thank you.

                                   Very truly yours, . . .
```

The revised version asks for your advance opinion, suggests how quick action might help you and *others,* and promises to let you know promptly of changes.

There's also another improvement. The tired and outdated "With thanks in advance" has been replaced by a modern "Thank you."

The best way to persuade your readers to your point of view is to show them that it will be worth their while to do so. This rule holds true for just about every successful letter, sermon, sales presentation, or advertisement. A good sales letter, for example, is much like a good ad in that it attempts to dramatize benefits to the reader.

What does it take to get—and hold—a reader's attention? The most powerful letters appeal to basic needs and emotions rather than to purely logical reasons. Mainly, management people want to know how to save time, money, and effort. Show them how your product or service can help them do one or more of these things, and you're likely to get their attention.

With effort, almost any letter can be oriented to the reader, even the usually hard-to-write collection letter. Here's such a specimen, written almost entirely from the writer's point of view:

```
Dear Sir:
    Our records show that you are three months delinquent in pay-
ment of your bill for $37.50.
    Perhaps this is an oversight on your part. Otherwise we can-
not understand why you have not taken care of this obligation.

                                  Very truly yours, . . .
```

Keeping the reader's self-esteem in mind, the letter could have been rewritten this way:

```
Dear Sir:
    We know you'll want to take care of your small past-due
account for $37.50. This will help you to maintain the fine
credit you have built up with us over the years.

                                      Sincerely, . . .
```

To help develop the "you" attitude, put yourself in your reader's shoes, then write from his or her point of view. Once you've developed this attitude, you'll automatically start telling your readers what's in it for them. Instead of saying, "I wish to thank you," you'll write, "Thank you." Instead of writing, "We'd like to have your business," you'll write, "You'll find our service can help your business in many ways."

PERSONALIZE YOUR LETTERS (THERE'S POWER IN PRONOUNS)

Some years ago, many would have considered this letter to have been perfectly proper and very effective:

Gentlemen:
   Enclosed herewith are the subject documents which were requested
in yours of the 10th. The documents will be duly reviewed and an
opinion rendered as to their relevancy in the involved litigation.

                              Very truly yours, . . .

The letter *was* all right—back in the horse-and-buggy days! Phrases like
"enclosed herewith," "duly received," and "involved litigation" would have
marked the writer as learned. But the executive who makes a habit of writing like
that today is generally regarded as an anachronism.

Current usage calls for clear, to-the-point letters, written mostly in conversational
language. Like good conversation, your letters should generally be friendly, filled with
personal references, and almost always informal in tone and language. Here's how
the horse-and-buggy letter might have been rewritten by the modern executive:

Dear Sam:
   Here are the documents you asked for. I'll look them over and
let you know if we can use them in our lawsuit.

                              Cordially, . . .

When experts write a letter, even a form letter, they generally try to make it
sound as personal as possible. Almost always, it contains a sprinkling of personal
pronouns. In orienting your letter to the reader, fill it with "you," "your," and
"yours." Use "I" and "me" sparingly.

Here's a case in point, a letter sent out by a mortgage company (emphasis
added by the author):

Dear Mr. Jones:
   We want to thank you for your query about *our* new mortgage
insurance plan.
   We are enclosing a pamphlet which outlines benefits of *our* new
policy and gives testimonials from some of *our* policyholders.
   We would like very much to enroll you within the next 30 days,
since we are offering a special low premium rate as *our* introductory offer.

                              Very truly yours, . . .

The repeated use of *we* and *our* clearly shows the letter is written with the mortgage company's interests at heart. But notice how substituting *you* and *yours* orients
the letter to the reader's interest. (Emphasis has again been added by the author.)

Dear Mr. Jones:
   Thank *you* for *your* recent inquiry about our new mortgage insurance plan.
   You'll find that the enclosed pamphlet outlines benefits of
the policy in detail. *You'll* also probably be interested in the
testimonials furnished by some policyholders.

If *you* sign up within the next 30 days, *you'll* be able to take
advantage of special premiums we're offering on an introductory
basis. Thank you.

Cordially, . . .

The word is out now in enlightened business circles. Companies are telling
their executives to regard every letter as a personal contact: to write your own way,
to develop your own style (within bounds), to make your letter distinctively *you*.

MASTERING TONE (YOUR PERSONALITY IN PRINT)

Writing a "rejection letter" that leaves the recipient's self-esteem intact and preserves
his or her goodwill is a difficult task. It requires tact, diplomacy, and empathy—
all of which must be effected through appropriate *tone*. For example, here's the way
one banker turned down a builder, a long-time customer:

Dear Mr. Jones:
We regret to inform you that your request for an additional
loan in the amount of $250,000 must be rejected. It was the judg-
ment of the loan committee that, with your present commitment,
such a loan would present too much of a risk for us.
Naturally, we look forward to doing business with you on your
existing commitment.

Sincerely, . . .

What's wrong: Tone, mainly—the *way* the rejection is phrased. It deflates the
reader's ego and makes it virtually certain the builder will do his banking elsewhere in the future.

True, the rejection was "justified." But why not let the reader down more
gently and leave the door open for future business? For example:

Dear Mr. Jones:
One of the most distasteful tasks we have is to turn down a
loan application, particularly when it is from a regular and
respected customer like you.
We know you'll be disappointed, and so are we. We share your
excitement about your plans to expand the Roseland Project, and
hope that circumstances will later warrant our working with you
on this and other projects.
However, the current economic outlook, combined with your
delinquent status on your existing loan, makes your new loan
application a questionable venture for us at the present time.
We will be glad to work with you in any way we can to resolve
your current financial problems, and we look forward to helping
you meet your future financial commitments.

Sincerely, . . .

The two letters say the same thing, but in vastly different ways. Suffice it to say that tone is as important to a letter as good muscle tone is to an athlete. A negatively phrased letter can have the same effect as an abrasive personality. As one New York Life executive put it:

> By its very nature, the "no" letter is a turndown and leaves the reader unsatisfied. Yet there are ways to soften the blow, and one of the best is simple candor. So, if you must say "no," state the reason first: "Currently there are no vacancies in that department. Therefore, I cannot offer you a position with the company at this time."
>
> A "no" response often requires positive alternatives if they are appropriate (for example, "Although, for the reasons mentioned, I cannot do as you ask, may I suggest that . . ."). You have to be more diplomatic and more sensitive if you're to have any chance of salvaging your reader's goodwill or ego.
>
> And that really is your goal in an effective "no" letter—to tell your readers something they don't want to hear in a way that compels understanding and, ideally, acceptance.

Unfortunately, tone requirements vary not only from person to person, but from one situation to another. For example, you would ordinarily use fast-paced, persuasive prose for a sales-promotion letter. In writing to a highly regarded law firm, you might use a slightly more "dignified" approach. If you're writing for a service-type organization, you might use a middle-of-the-road approach, as the government does.

Government letter writers are told to strive for a tone of "simple dignity," to make letters brief and to the point, and to avoid gobbledygook. "Don't act as if you're the only game in town," they're told. "On the other hand, don't bow and scrape just because you're a service organization."

Poor letters start with poor thinking on the part of the writer. Negative thoughts lead to negative words—and before you know it, there goes the old ball game. Here's a letter from a manager of a department store to a customer who complained that an appliance she had recently bought didn't work. (Negative words have been italicized by the author.)

```
Dear Sir:
     I am in receipt of your letter in which you state that the hair
dryer you purchased from us recently failed to meet the warranty
requirements.
     You claim that the dryer failed to do the things you say our
salesman promised it would do.
     Possibly you misunderstood the salesman's presentation. Or per-
haps you failed to follow instructions properly. We positively
know of no other customer who has made a similar complaint about
the dryer. The feeling is that it will do all that is stated if
properly used.
     However, we are willing to make some concessions for the alleged
faulty part. We will allow you to return it; however, we cannot
do so until you sign the enclosed card and return it to us.

                                        Very truly yours, . . .
```

The tone is unmistakably negative. "State" and "failed to meet" in the opening paragraph imply there's some doubt that the claim is valid. In the second paragraph, "you say" suggests the salesman didn't make the statement at all.

Such phrases as "you misunderstood," "you failed to follow instructions," and "if properly used" tell you in so many words that you're not too bright. And to wrap it up, the writer uses the negative "we cannot . . . until" instead of the positive "we will . . . as soon as."

Most letters aren't this negative, of course. But it doesn't take much to offend. Any of these negative words or phrases, in themselves, could have spoiled the tone of an otherwise effective letter.

If you've been guilty of taking a negative approach, study the following examples. In each case, the negative thought (emphasis added by the author) has been converted into a positive one.

Negative	Since you *failed* to say what size you wanted, we cannot send you the shirts.
Positive	You'll receive the shirts within two or three days after you send us your size on the enclosed form.
Negative	We *cannot* pay this bill in one lump sum as you requested.
Positive	We can clear up the balance in six months by paying you in monthly installments of $20.
Negative	We're sorry we *cannot* offer you billboard space for $200.
Positive	We can offer you excellent billboard space for $300.
Negative	We are *not* open on Saturday.
Positive	We are open from 8 A.M. to 8 P.M. daily, except Saturday and Sunday.

Negative words aren't the only cause of poor letter tone. Many letters are made more or less "neutral" by mechanical, impersonal, or discouraging language. Here's an example of each fault with a preferred alternative:

Interpersonal	Many new names are being added to our list of customers. It is always a pleasure to welcome our new friends.
Personal	It's a pleasure to welcome you as our customer, Mr. Jones. We will make every effort to serve you well.
Mechanical	This will acknowledge yours of the 10th requesting a copy of our company's annual report. A copy is enclosed herewith.
Friendly	Thanks for requesting a copy of our annual report, which is enclosed. We hope you will find it helpful.
Discouraging	Since we have a shortage of personnel at this time, we won't be able to process your order until the end of the month.
Encouraging	We should have more help shortly, which will enable us to get to your order by the end of the month.

Ideally, the tone of your letter should reflect the same ease in conver-sation that you enjoy when talking about your favorite hobby, business, or pastime.

HOW TO WRITE (MORE) THE WAY YOU TALK

This letter was sent out by a large department store. Imagine yourself as the recipient.

```
Dear Sir:
    We are in receipt of yours of July 10 and contents have been
duly noted.
    As per your request, we are forwarding herewith copies of our
new fall brochure. Thanking you in advance for any business you
will be so gracious as to do with us,

                            Yours truly, . . .
```

Conversational? Of course not. Who uses such language as "We are in receipt of . . . ," "duly noted . . . ," and "as per your request"? Practically no one. People just don't talk that way. Most of these phrases went out with the Model-T—or should have.

Face to face, the writer probably would have said, "Here's the fall brochure you requested. Let me know if we can serve you." It says the same thing, in less space, and without all the fuss.

Some business and professional people still feel it isn't quite proper to write conversationally, mainly because they have seen so much stilted business writing. But many progressive companies are telling their people to communicate in plain English, using only those technical terms that are absolutely necessary. As one executive of a major company said, "The best letters are more than just stand-ins for personal contact. They bridge any distance by the friendly way they have of talking things over person to person." And this from a U.S. Navy bulletin: "At best, writing is a poor substitute for talking. But the closer our writing comes to conversation, the better our exchange of ideas will be."

The consensus clearly is that informal, natural business writing is *in;* stilted business writing is *out.*

One word of caution about writing (more) the way you talk. Since World War II, readability experts have urged business people to "write more the way they talk" or "write as they talk."

Detractors have soft-pedaled the idea, claiming that most conversation is rambling, often incoherent, and frequently a bit too earthy. These are valid objections, but they miss the point. You're not being asked to write *exactly* the way you talk. Rather, you're being asked to bridge the gap between the spoken and the written word—to narrow the difference in the way you would *give* an order verbally and the way you would *write* that same order in a memo. Naturally, writing requires more restraint than speech, and the writer must normally use fewer words. But you can do these things and still capture the tone and cadence of spoken English.

The first step in making your correspondence more conversational is to rid your vocabulary of worn-out business phrases. Here are some of the more flagrant offenders. They were stylish once, but they've done their duty and need to be honorably discharged.

Old Hat	Conversational
At a later date	Later
If this should prove to be the case	If this is the case
This will acknowledge receipt of	Thank you for
Attached herewith please find	Here is; Enclosed is
We shall advise you accordingly	We'll let you know
Due to the fact that	Because
With regard to	About
Please notify the writer as to	Please let me know
Enclosed please find a stamped envelope	I've enclosed a stamped envelope
We are submitting herewith a duplicate copy	Here is a copy
In compliance with your request	Here is
We are submitting herewith our check in the amount of $75	Here is our check for $75
We beg to advise (acknowledge) that	[Begging is unnecessary]
The information will be duly recorded	We'll record the information
The subject typewriter	This typewriter
In compliance with your request	As you requested
We will ascertain the facts and advise accordingly	We'll let you know
The writer wishes to state	[Just say it]

This list is far from complete but you get the idea. Once you're mentally geared to writing more conversationally, you'll detect many other clichés. Try to eliminate them.

Contractions will also make your writing more conversational. They play a part—a very large part—in almost everyone's everyday conversation. Even your most learned associate doesn't say, "We shall endeavor to be there at eight o'clock"; he's more likely to say, "We'll try to be there at eight o'clock." Instead of saying, "I am going to the ball game," he'll probably say, "I'm going to the ball game." It takes less effort to say, "You needn't bother to call," than to say, "You need not bother to call." Contractions tend to make the spoken words flow more smoothly. That's why they make writing appear more natural.

Probably the most common contractions are here's, there's, where's, what's, let's, haven't, hasn't, hadn't, won't, wouldn't, can't, couldn't, mustn't, don't, doesn't, didn't, aren't, isn't, and weren't. Then there are the pronoun contractions: I'll, I'm, I'd, I've, he's, he'll, he'd, and so forth.

Using these and other contractions when they facilitate the flow of words will do much to give your writing a quality of spontaneity and warmth. But they must be used with discretion. Using too many contractions can sometimes make your writing too informal. And used in the "wrong" place, they might not "sound right." Indeed, the key is whether the contraction sounds right when the sentence is read.

Take this section from the Gettysburg Address: "But in a larger sense, we cannot dedicate, we cannot consecrate, we cannot hallow this ground." The passage would undoubtedly have lost its historic tone if *can't* had been used instead of cannot.

On the other hand, the advertising slogan "We'd rather fight than switch" would lose some of its punch if phrased, "We would rather fight than switch." Appropriate usage depends on how the contraction makes the sentence sound when it's read aloud.

Next time you get ready to write a letter, ask yourself, "How would I say this if I were talking to the person?" Then go ahead and write in that vein.

The Royal Bank of Canada

Letters That Sell

Everyone writes letters that sell, and every letter has as its purpose the selling of something: goods, services, ideas or thoughts.

Someone may say that a family letter has no such purpose, but consider this: a letter telling about the children seeks to promote a favourable impression of their welfare and happiness; a letter telling about illness is designed to gain sympathy; the letter that says nothing but "I hope you are well" is selling the idea "I am thinking of you."

Family letters are usually rambling letters. They would be improved both in their readability and their informativeness if they adopted some of the principles that are used to sell goods and services. Business building letters, on the other hand, could with advantage incorporate some of the friendly informality of family letters.

Salesmanship of any kind is basically a person moving goods by persuading another person that he needs them, or winning that person's support or approval of an idea or a plan.

Some noncommercial type sales letters are those that champion good causes, such as community welfare or health standards or national unity. They seek to influence the thinking of individuals or groups.

It is not a simple task to compose a letter designed to sell. Like any other product of value, it calls for craftsmanship. There are techniques to be learned, techniques of conveying ideas, propositions, conclusions or advice appealingly and purposefully.

IN THE BEGINNING

In creating a letter to sell something we need to begin by thinking about the person to whom we are writing. A lawyer studies his opponent's case just as sharply

LETTERS THAT SELL. © The Royal Bank of Canada (*Monthly Letter* May 1974). Reprinted with the permission of The Royal Bank of Canada.

as his client's; the manager of a baseball or hockey team analyses the qualities, good and bad, of members of the opposing team.

The writer must anticipate and answer in his letter questions that will occur to the reader: What is this about? How does it concern me? How can you prove it? What do you want me to do? Should I do it?

People buy goods or services because these will give them a new benefit or extend or protect a benefit they already have, so the writer needs to translate what he offers into owner benefits.

The proffered benefits must be accessible and adapted to the reader's position, environment and needs. No letter is likely to sell sun-bonnets to people who live beyond the Arctic Circle or baby carriages to bachelors. We may classify a potential customer as a man, woman, company or institution that will have use for a product or service, has sufficient money to pay for it, and in whom a desire for possession may be created.

The reader's interest: that is the guiding star in sales letter writing. See his interests, his angle, and accommodate your stance to them. A simple precaution against sending a letter to the wrong person is to ask yourself what use you would have for the commodity if you were in the reader's place.

It is a good rule to spend more time thinking about the reader than about what you have to say. Otherwise you may become wrapped up in the virtues of your product so that you forget that the decision to buy rests with your prospect.

The self-interest of the person to whom you write is a major factor to consider in successful sales communication. When you remember it, you give the impression that you have singled out this reader as being an important individual, and that is an excellent introduction.

It is not to be expected that the writer of letters that sell will know every person to whom he writes, but he must know certain facts: approximate income and age, occupational level, his business, and things like that. Then he is able to slant his sales points accurately toward the reader's needs, interests and purchasing power.

KNOW YOUR PRODUCT

The reader's attention should be attracted to the product or service, not to the grand style or picturesque phraseology of your letter. When you catch a person's attention you are focusing his consciousness on something. Concentrate on your commodity. The best magnet to draw and hold attention is what you say about the product, showing it to be useful and the means of fulfilling a desire.

It is no small accomplishment to analyse and marshal into order the facts about a product so as to win the thoughtful consideration of a person who has plenty of other things on his mind.

In purchasing almost any sort of commodity, the buyer has a choice between what you are offering and what others are selling. Your sales job is to show the superiority of your product. Tell why what you offer is necessary or desirable, what it will accomplish in your reader's business, and how it can be fitted into his present layout and his plans. Do not content yourself with telling about the article as it sits on display: picture it in use in the reader's home or factory.

Your letter needs to convey the assurance that you are telling the truth about your goods. It is not a sensational offer that makes a letter convincing, but the feeling that the reader can depend upon what is said. He should feel assured that he will be buying what he thinks he is buying. Customer dissatisfaction caused by misleading sales talk can cause shock waves that affect the whole selling organization.

LET YOUR PERSONALITY SHOW

Make your letter sound friendly and human: put your personality on paper. Your letter is you speaking. Some of the features in your personality that you can display are: friendliness, knowledge, keen-mindedness, trustworthiness and interest in the prospect's welfare.

What you have in your mind about the good quality, appearance and usefulness of your product has to be communicated to your reader so as to arouse his interest, create a desire to possess, and induce him to buy.

Communication is not the easiest thing in the world to attain in writing, in art, or in music. Dr. Rollo May wrote in *Man's Search for Himself:* "We find in modern art and modern music a language which does not communicate. If most people, even intelligent ones, look at modern art without knowing the esoteric key, they can understand practically nothing."

It is not enough to write something so that it can be read. The degree to which communication occurs depends upon the degree to which the words represent the same thing for the reader as they do for the writer.

The recipient of a letter that is not clear is likely to blame its opacity on the lack of intelligence of the writer.

The art of composing sales letters is not one to be mastered by minds in which there is only a meager store of knowledge and memories.

The art consists in having many mental references and associating them with new thoughts. Consider a poem. Its theme will likely have arisen from a single event, but the images used in its construction will have been drawn from the total life experience of the poet.

Put some flavour into your letters so that they taste good. Your letter will not be like anyone else's. That is a virtue, just as being an individual is a virtue in conversation. Who wishes to be a carbon copy of a textbook letter or to parrot phrases that other people use?

Practise talking on paper as if you were on the telephone. First write down the imagined questions asked by the person on the other end of the line and then your answers, given in simple, direct and pleasing words. To humanize your letters in this way with the natural idiom of conversation does not mean that you use cheap slang or clever verbal stunting.

SHOW SOME STYLE

The style in which you write is not a casual feature of your letter. It is vital to your reader's understanding of what you are saying to him. It is not your job to please the reader's sense of the aesthetic, but to tell and explain plainly what is necessary to introduce your goods or your idea to his favourable attention. This may be done in a way that has grace and comeliness.

Never "talk down" to a reader. Make him feel that he knows a great deal, but here is something he may have missed. There is a big difference, when trying to build business, between making a suggestion and preaching a sermon.

It is highly important in writing a letter to sell something that it should be appropriate. Whatever your writing style may be, it will fit the occasion if it gives this particular correspondent information that will be useful to him, conveys to him a feeling of your interest in him and his business, and assures him of your goodwill.

Besides being grammatically correct, language should be suitable. At one extreme of unsuitability is the language that is too pompous for its load, and at the other is the language of the street which belittles the receiver's intellectual level.

Your words should be the most expressive for their purpose that the language affords, unobstructed by specialty jargon, and your sentences should be shaken free of adjectives—the most tempting of forbidden fruit to a person describing something.

Properly chosen words will convey your appreciation of the addressee as a person, and such friendliness is contagious. Some people are afraid to be friendly in their letters. They fear they will be thought of as "phonies" who have disguised themselves as Santa Claus for the occasion. Being friendly and showing it should not raise this scarecrow. It would be a grave mistake, indeed, for any of us to indulge in flowery language foreign to our natural talk; but it is no mistake at all to incorporate in our letters the warm, personal language that comes naturally to us in person-to-person social contacts.

Letter writing invites us to use the same etiquette as we use in courteous conversation. We look at the person with whom we are talking, converse on his level of understanding, speak gently, and discuss matters he considers important or interesting.

What the reader of your letter will notice is not its normal courtesy, but the extra touch that demonstrates care and understanding, a genuine interest in the reader's wants, a wish to do what is best for him, and the knowledge you show of how it can be done.

Everyone who writes a letter has a moral as well as a business reason to be intelligible. He is placing his reader under an obligation to spend time reading the letter, and to waste that time is to intrude upon his life plan.

There is an eloquence of the written as well as of the spoken word. It consists in adapting a statement to the receptive system of the reader so that he will have maximum help against confusion, against mistaking what is incidental from what is fundamental. A familiar device to use in this effort is to relate the new commodity you are offering to something that is familiar.

USE SUITABLE FORMULAS

There are formulas you may wish to make use of. Your letter must conform in some respects to what letters are expected to be. This does not mean pouring all letters into the same mould. Within the accepted pattern you are free to develop your talent for expression.

Skill is needed in the use of formulas. A form letter reveals itself to the reader and gets short shrift. It is possible to make use of the form as a guide to what points to cover, and then speak your piece on paper in a natural way.

Here are three formulas for letters. The first may be called the sales formula, the second the logical formula, and the third the rhetorical formula.

1. Get attention, provoke interest, rouse desire, obtain decision. Attention is curiosity fixed on something; interest is understanding of the nature and extent of what is new and its relationship to what is old; desire is the wish to take advantage of the proffered benefits; decision is based on confidence in what the writer says about his goods.
2. This is summarized: general, specific, conclusion. You start with a statement so broad and authoritative that it will not be disputed; you show that the general idea includes a specific idea; the conclusion is that what has been said about the general idea is also true of the specific.
3. This is very simple: picture, promise, prove, push. You write an attractive description of what you are selling; you promise that it will serve the reader well in such-and-such a way; you give examples of the commodity in use, proving that it has utility and worth; you urge the reader to take advantage of the promised values.

SELLING NEEDS IDEAS

Selling is done with ideas, so never throw away an idea even if it is of no use at the moment. Put it into your idea file where it will rub against other ideas and perhaps produce something new. The file is like an incubator. Thoughts and fancies you put into it will hatch out projects and plans.

Imagination helps in this operation. A correspondent of ordinary ability may never write anything that is not absolutely accurate and yet fail to interest his readers. This is a real weakness: to be perfect as to form but lacking in imagination and ideas.

Imagination should be given priority over judgment in preparing your first draft of a letter designed to sell. Then put reason to work: delete what is unnecessary; marshal your sentences into logical form so that your ideas advance in an orderly way; revise your words so that your thought is conveyed exactly as you wish it to be.

When you tell the advantages of your product or service or idea, and show how it will fill a need in the reader's life or job, in clear, truthful words placed in easily understood sentences brightened by ideas and imagination, you have done a good job of writing a sales letter.

Desire of the reader to do what you want done is created, just as in conversation, by both rational and emotional means, by proof and by persuasion, by giving reasons. Some goods and some buyers need nothing more than facts. An office manager buying pencils or pens for his staff will respond to an informative, factual, statistical sales letter. He is already sold on the idea of using pencils and pens, so you do not have to coach him about their usefulness; in fact, you may lose a sale if you give the impression of "teaching grandmother to suck eggs." What is needed is to catch his attention, give pertinent information about your product, and show him why buying from you will be profitable to him.

Try to make the information you give really enlightening. Comparing something unknown with something already known makes it possible to talk about the unknown. The analogy (like that between the heart and a pump) can be used as an aid in reasoning and in explaining or demonstrating.

THE SOFT SELL

The tone of a letter designed to sell something should be persuasive rather than insistent. It should seek to create a feeling of wanting, or at least an urge to "let's see."

People do not want to be told how to run their affairs, but anyone who shows them how to do things more economically or faster or better will find keen listeners. Soft sell gives the prospect credit for knowing a good thing when it is shown him, and acknowledges his right to make up his own mind.

The soft sell is recognition of the Missouri mule in human nature. Try to push a mule and he lashes out with his heels. Try to pull him by the halter rope and he braces his legs and defies you to budge him. "In the old cavalry," says A. C. Kemble in *Building Horsepower into Sales Letters,* "they said all it took to get a mule working for you was to recognize that he was an individualist who hated nagging and needed a chance to make up his own mind about things."

One hears a lot in advertising circles about "appeal." It is, according to the dictionary, "the power to attract, interest, amuse, or stimulate the mind or emotions."

Obviously, when you wish to influence someone you must take into account the kind of person you are addressing and what you want him to do. Your appeal must touch his feelings, needs and emotions. It strengthens your position if you can relate your own experience to that of the person you are addressing and write your message around the overlap of that experience.

The sort of mistake to be avoided with great care is slanting your appeal in a way that runs counter to the feelings of those whom you wish to influence.

It has been found in recent years that the advertising messages addressed to older people *as* older people did not win the desired response. In travel, for example, only a very small minority who want or need a sheltered situation are attracted by the semicustodial "trips for the elderly."

The Swiss Society for Market Research decided that "to sell anything to the over-65 age group it is important to keep one concept in mind: most senior citizens are vigorous and independent. Don't try to reach them with a head-on approach to the senior market. It probably won't work."

Writing a letter that pleases the recipient is not enough: it must be designed to lead to action. Do not fear to be explicit about what you want. Coyness in a letter is not attractive, and it exasperates the reader. Answer the reader's questions:

"What has this to do with me?" and "Why should I do what this person is asking me to do?"

You may answer these questions and encourage a purchase by appealing to emotional motives like pride, innovation, emulation, or social prestige; or to rational motives like money gain, economy, security, time-saving or safety.

READ YOUR LETTER CRITICALLY

Imagine your letter to be your garden upon your return from vacation. You have to get into it and prune, clean up, tie up, and trim the edges.

Read the letter as if you were the recipient. How does it strike you? What can be added to attract attention? Is there anything irrelevant in it? Read the letter aloud to capture the conversational rhythm.

If you are not satisfied, do not crumple up the paper on which your draft is written. Try rearranging the paragraphs, the sentences, the words. Give the letter a new twist. Change the shape of your appeal. Delete anything that is distracting.

Be careful when trying to shorten a letter that seems to be too long. While a letter should be as short as possible, consistent with clearness and completeness, it is not the length that counts, but the depth. Since clearness and brevity sometimes get in the way of each other, remember that the right of way belongs to clearness. It will make a good impression if you find occasion to write: "I can be quite brief because this letter deals with a topic already well known to you."

The end of your letter, like the end of your pencil, should have a point. It should answer the reader's natural question: "So what?"

FOLLOW THROUGH

Do not let your customer forget you. When you produce a piece of copy that hits the bull's-eye, that is not the time to sit back and take things easy. It is a time to imagine what you would do if you were in your competitor's chair . . . and then do it first.

Competition is a fact of life. Wherever there are two wild animals trying to live on the same piece of land or two persons depending upon the same source of sustenance, there is competition. The customer who was a prospect to you before he bought your goods is now a prospect to your competitor. With the proper follow-through attention, he will turn to you when he needs up-dating of equipment or new goods.

Writing a letter that sells goods or ideas, and following through so as to retain the customer, requires just as much specialized talent and mental ability as any other kind of advertising, if not more.

When you run into difficulties, composition of the sales or follow-up letter may give you a feeling of confusion. You may feel like throwing up your hands in despair of finding the exactly right slant or the perfect array of words. That is not unnatural. Nietzsche, the German philosopher, said in *Thus Spake Zarathustra:* "I tell you, one must still have chaos in one, to give birth to a dancing star."

The effort is worthwhile. When you set yourself to snap out of the depressing pedestrian type of letter that is so commonplace, you are raising yourself and your firm to a place where people will sit up and pay attention. As a student of sales letter writing you will generate ideas, as a philosopher you will assess the letters as to their purpose and usefulness, and as a writer you will energize them.

To summarize: the backbone of the principles of writing letters that sell is made up of these vertebrae—know why you are writing and what about; believe in what you are writing; be tactful and friendly and truthful; base your appeal on the prospect's interests. . . . and check your letter and revise it.

Allan A. Glatthorn

"I Have Some Bad News for You"

A consultant in management communications and the author of more than thirty books on writing, Allan A. Glatthorn recently retired from the position of Senior Faculty Member in the Teaching of Writing Program at the University of Pennsylvania.

One of the most difficult letters or memos to write is the one containing bad news. A subordinate requests a salary increase which you cannot grant. A community organization asks for a contribution which you do not wish to make. An unqualified applicant asks for an appointment to discuss a position with your company, and you do not wish to take the time. You decide to have to dismiss a good worker because declining business mandates cutbacks. In each case you have some unpleasant news to deliver—but you wish to deliver it in a way that does not offend or alienate. How do you accomplish this difficult task? Let me offer some general guidelines and then explain the specific techniques.

The first guideline is to remind you that the successful manager is people-sensitive, able to empathize with others. When you receive a request, you stand in the shoes of the petitioner. You realize, to begin with, that from the standpoint of the asker there are no foolish requests. People ask for things that they need, that they believe they are entitled to, that they hope to receive. And you also realize that there is really no good way to break bad news. No matter how empathic and tactful you may be, your bad news will inevitably cause disappointment and will often cause distress.

The second guideline derives from the first: remember that bad news is best delivered face to face. People getting bad news have some strong needs. They want an opportunity to express their negative feelings. They want the chance to press the request or appeal the decision. They want to explore the reasons more fully to be sure that there is no hidden message. These needs can best be met in a face-to-face meeting. So if you have bad news to give to an employee or a valued client, the best method is to confer with the individual, deliver the bad news in person, and then, if desirable, follow up with a written statement for the record. Don't avoid the unpleasant confrontation by writing a memo, hoping that the petitioner will go away quietly.

Finally, remember that everyone values honesty and forthrightness, especially when being disappointed. Typically, the receiver of bad news is inclined to be cynical, often resenting your attempts to be tactful and politic. You therefore should be sure that your desire to soften the blow does not beguile you into distorting the truth. Explain as tactfully as possible the real reasons for the bad news, instead of offering lame excuses.

With these general guidelines in mind, you should be able to send two kinds of "bad news" messages: the indirect and the direct.

THE INDIRECT BAD NEWS MESSAGE

The indirect message of bad news uses the soft and gentle approach. It tries to cushion the blow by burying the bad news in the middle of the letter or memo, surrounded by positive expressions of appreciation. You send the indirect message under one or more of these circumstances:

- You want further contact with the petitioner.
- You want to project the image of a caring individual.
- You believe that the petitioner won't be able to handle a more direct statement.

The formula is a simple one: THANKS . . . BECAUSE . . . SORRY . . . THANKS.

Thanks

You begin with a positive statement. You express appreciation for the idea offered, the interest expressed, the petition received. Here's how an indirect bad news letter from an insurance company begins:

> Thank you very much for your recent inquiry about our automobile insurance coverage. We are pleased that you have considered transferring your account to our company.

Because

You then continue with the reasons for the bad news. The strategy here is that stating the reasons first cushions the shock, preparing the reader for the bad news to come. Remember to use reasons that will make sense to the petitioner and that will at least project an image of sincerity. If at all possible, state reasons that will depersonalize the rejection, as the next paragraph of that letter does:

In order to keep our premiums as low as possible, we find it necessary to accept only a small number of new accounts. And in fairness to our present customers, we accept applications only from those whose claims records are comparable to those of drivers we presently insure.

Sorry

You next present the bad news itself, but you state it in a positive fashion. If you can suggest an alternative, do so. If you can find a way to leave the door open, make that clear—but do not give false hopes. Notice how the insurance letter continues:

Even though we would like to include you among our insured drivers, we will not be able to do so at the present time. It seems to us that your best choice at this time is to remain with your present company. If your claims record improves during the coming year, we would be happy to have you reapply for coverage with us.

Thanks

The indirect message of bad news closes with another expression of appreciation—and ends on a positive note, as this example illustrates:

We do appreciate your considering our company. And we hope you will be able to reapply under more positive circumstances.

Figure 1 shows another example of this formula at work.

```
To: Bill Harkins
From: Joanne Clemens
Subject: Flex-Time Proposal
Date: February 19,1985    File: A-342

    Bill, thanks very much for forwarding your proposal for a
flex-time schedule for your department. I appreciate the cre-
ative energy—I can always count on getting good ideas from you.
    Our data suggest that flex time would result in an uneven
distribution of worker hours at a time when uniformity seems
desirable. As you are aware, the recent increase in customer
demand is presenting us with the right kind of problem: our
people and equipment are working at maximum capacity. My expe-
rience with other companies using flex time indicates that our
productivity would suffer if we instituted it now.
    And my concern for maintaining productivity makes me reluc-
tant to implement your excellent suggestion now. However, I
have asked our personnel department to review the data. Because
of my respect for your leadership, I want to give all of your
ideas the most careful consideration.
    Thanks again for taking the time to share your ideas with
me. I appreciate your developing a very sound proposal at a
time when I know you are quite busy.
```

FIGURE I   The Indirect Bad News Message

THE DIRECT MESSAGE OF BAD NEWS

The direct message is a tough no-nonsense statement. While courteous, it gets right to the point and does not try to bury the bad news. It would be used under one or more of these circumstances:

- You want to slam the door shut, discouraging any other request from that petitioner.
- You want to project an image of toughness and directness.
- You are addressing an individual who prefers forthrightness and equates indirectness with softness or dishonesty.

The formula for the direct message is THANKS . . . SORRY . . . BECAUSE . . . THANKS. It uses the same ingredients as the indirect message, but it changes the order. And the language is less subtle and more direct.

Thanks

You begin with a courteous expression of appreciation, since courtesy is expected even in the most direct communication. So a firm and direct letter rejecting an applicant might begin like this:

> Thank you for your letter and resume. We appreciate your interest in joining Mutual Life.

Sorry

The direct message moves quickly to the bad news, to be sure that the message is heard as intended. The bad news is stated forthrightly—but without offending. So the rejection letter continues:

> Unfortunately, we do not have a position available that would match your qualifications, and it seems unlikely that such a position will develop in the future.

Because

Now the reasons come, after the bad news has been delivered. The reasons are stated directly but not offensively, as in this example:

> We have found that our most successful middle-level managers are those who have had the benefit of working with Mutual in a variety of nonmanagerial positions. We therefore tend to promote from within and do not encourage applications for managerial positions from those who have not previously worked with us.

Thanks

The message ends courteously—but the door is firmly closed. There must be no mistake about it: the news is bad. So the rejection letter ends like this:

> We hope you will be able to find employment with a company that can make use of your excellent qualifications, and we do appreciate your interest in applying with us.

The successful manager projects both people-sensitivity and toughness. The indirect message emphasizes sensitivity; the direct message affirms the toughness.

Harold K. Mintz

How to Write Better Memos

Harold K. Mintz was Senior Technical Editor for the RCA Corporation when he wrote this article.

Memos—interoffice, intershop, interdepartmental—are the most important medium of in-house communication. This article suggests ways to help you sharpen your memos so that they will more effectively inform, instruct, and sometimes persuade your coworkers.

Memos are informal, versatile, free-wheeling. In-house they go up, down or sideways.* They can even go to customers, suppliers, and other interested out-siders. They can run to ten pages or more, but are mostly one to three pages. (Short memos are preferable. Typed single-space and with double-space between paragraphs, Lincoln's Gettysburg Address easily fits on one page, and the Declaration of Independence on two pages.) They can be issued on a one-shot basis or in a series, on a schedule or anytime at all. They can cover major or minor subjects.

Primary functions of memos encompass, but are not limited to:

- Informing people of a problem or situation.
- Nailing down responsibility for action, and a deadline for it.
- Establishing a file record of decisions, agreements and policies.

Secondary functions include:

- Serving as a basis for formal reports.
- Helping to bring new personnel up-to-date.

---

*We will return to this sentence later.

HOW TO WRITE BETTER MEMOS. By Harold K. Mintz. Excerpted by special permission from *Chemical Engineering* (January 26, 1970). Copyright © 1970, by McGraw-Hill, Inc., New York, N.Y. 10020.

- Replacing personal contact with people you cannot get along with. For example, the Shubert brothers, tyrannical titans of the American theatre for 40 years, often refused to talk to each other. They communicated by memo.
- Handling people who ignore your oral directions. Concerning the State Department, historian Arthur Schlesinger quoted JFK as follows: "I have discovered finally that the best way to deal with State is to send over memos. They can forget phone conversations, but a memorandum is something which, by their system, has to be answered."

Memos can be used to squelch unjustified time-consuming requests. When someone makes what you consider to be an unwarranted demand or request, tell him to put it in a memo—just for the record. This tactic can save you much time.

ORGANIZATION OF THE MEMO

Memos and letters are almost identical twins. They differ in the following ways: Memos normally remain in-house, memos don't usually need to "hook" the reader's interest, and memos covering a current situation can skip a background treatment.

Overall organization of a memo should ensure that it answer three basic questions concerning its subject:

1. What are the facts?
2. What do they mean?
3. What do we do now?

To supply the answers, a memo needs some or all of the following elements: summary, conclusions and recommendations, introduction, statement of problem, proposed solution, and discussion. Incidentally, these elements make excellent headings to break up the text and guide the readers.

In my opinion, every memo longer than a page should open with a summary, preferably a short paragraph. Thus, recipients can decide in seconds whether they want to read the entire memo.

Two reasons dictate placing the summary at the very beginning. There, of all places, you have the reader's undivided attention. Second, readers want to know, quickly, the meaning or significance of the memo.

Obviously, a summary cannot provide all the facts (see Question 1, above), but it should capsule their meaning, and highlight a course of action.

When conclusions and recommendations are not applicable, forget them. When they are, however, you can insert them either right after the abstract or at the end of the memo. Here's one way to decide: If you expect readers to be neutral or favorable toward your conclusions and recommendations, put them up front. If you expect a negative reaction, put them at the end. Then, conceivably, your statement of the problem and your discussion of it may swing readers around to your side by the time they reach the end.

The introduction should give just enough information for the readers to be able to understand the statement of the problem and its discussion.

LITERARY QUALITIES

A good memo need not be a Pulitzer Prize winner, but it does need to be clear, brief, relevant. LBJ got along poorly with his science adviser, Donald Hornig, because Hornig's memos, according to a White House staffer, "were terribly long and complicated. The President couldn't read through a page or two and understand what Don wanted him to do, so he'd send it out to us and ask us what it was all about. Then we'd put a short cover-memo on top of it and send it back in. The President got mad as hell at long memos that didn't make any sense."

Clarity is paramount. Returning to the asterisked sentence in the second paragraph of the introduction, I could have said: "Memoranda are endowed with the capability of internal perpendicular and lateral deployment." Sheer unadulterated claptrap.

To sum up, be understandable and brief, but not brusque, and get to the point.

Another vitally important trait is a personal, human approach. Remember that your memos reach members of your own organization; that's a common bond worth exploiting. Your memos should provide them with the pertinent information they need (no more and no less) and in the language they understand. Feel free to use people's names, and personal pronouns and adjectives: you-your, we-our, I-mine. Get people into the act; it's they who do the work.

Lastly, a well-written memo should reflect diplomacy or political savvy. More than once, Hornig's memos lighted the fuse of LBJ's temper. One memo, regarded as criticizing James E. Webb (then the head of NASA), LBJ's friend, infuriated the President.

Another example of a politically naive memo made headlines in England three years ago. A hospital superintendent wrote a memo to his staff, recommending that aged and chronically ill patients should not be resuscitated after heart failure. Public reaction exploded so overwhelmingly against the superintendent that shock waves even shook Prime Minister Wilson's cabinet. Result? The Health Ministry torpedoed the recommendation.

Two other courses of action would have been more tactful for the superintendent: make the recommendation orally to his staff or, if he insisted on a memo, stamp it "private" and distribute it accordingly.

Literary style is a nebulous subject, difficult to pin down. Yet if you develop a clear, taut way of writing, you may end up in the same happy predicament as Lawrence of Arabia. He wrote "a violent memorandum" on a British-Arab problem, a memo whose "acidity and force" so impressed the commanding general that he wired it to London. Lawrence noted in his *Seven Pillars of Wisdom* that, "My popularity with the military staff in Egypt, due to the sudden help I had lent . . . was novel and rather amusing. They began to be polite to me, and to say that I was observant, with a pungent style . . . ."

FORMAT OF THE MEMO

Except for minor variations, the format to be used is standard. The memo dispenses with the addresses, salutations, and complimentary closes used in letters. Although format is a minor matter, it does rate some remarks.

To and From Lines—Names and departments are enough.

Subject—Capture its essence in ten words or less. Any subject that drones on for three or four lines may confuse or irritate readers.

Distribution—Send the memo only to people involved or interested in the subject matter. If they number less than say, ten, list them alphabetically on page 1; if more than ten, put them at the end.

Text—Use applicable headings listed after the three questions under "Organization."

Paragraphs—If numbering or lettering them helps in any way, do it.

Line Spacing—Single space within paragraphs, and double space between.

Underlines and Capitals—Used sparingly, they emphasize important points.

Number of Pages—Some companies impose a one-page limit, but it's an impractical restriction because some subjects just won't fit on one page. As a result, the half-baked memo requires a second or third memo to beef it up.

Figures and Tables—Use them; they'll enhance the impact of your memos.

CONCLUSIONS

Two cautions are appropriate. First, avoid writing memos that baffle people, like the one that Henry Luce once sent to an editor of *Time*. "There are only 30,000,000 sheep in the U.S.A.—same as 100 years ago. What does this prove? Answer???"

Second, avoid "memo-itis," the tendency to dash off memos at the drop of a pen, especially to the boss. In his book, *With Kennedy,* Pierre Salinger observed that "a constant stream of memoranda" from Professor Arthur Schlesinger caused JFK to be "impatient with their length and frequency."

Marvin H. Swift

Clear Writing Means Clear Thinking Means . . .

When he wrote this article, Marvin H. Swift was Associate Professor of Communication at the General Motors Institute.

If you are a manager, you constantly face the problem of putting words on paper. If you are like most managers, this is not the sort of problem you enjoy. It is hard to do, and time consuming; and the task is doubly difficult when, as is usually the case, your words must be designed to change the behavior of others in the organization.

But the chore is there and must be done. How? Lets take a specific case.

Let's suppose that everyone at X Corporation, from the janitor on up to the chairman of the board, is using the office copiers for personal matters; income tax forms, church programs, children's term papers, and God knows what else are being duplicated by the gross. This minor piracy costs the company a pretty penny, both directly and in employee time, and the general manager—let's call him Sam Edwards—decides the time has come to lower the boom.

Sam lets fly by dictating the following memo to his secretary:

```
To: All Employees
From: Samuel Edwards, General Manager
Subject: Abuse of Copiers

    It has recently been brought to my attention that many of the peo-
ple who are employed by this company have taken advantage of their
positions by availing themselves of the copiers. More specifically,
these machines are being used for other than company business.
```

Obviously, such practice is contrary to company policy and must cease and desist immediately. I wish therefore to inform all concerned—those who have abused policy or will be abusing it—that their behavior cannot and will not be tolerated. Accordingly, anyone in the future who is unable to control himself will have his employment terminated.

If there are any questions about company policy, please feel free to contact this office.

Now the memo is on his desk for his signature. He looks it over, and the more he looks, the worse it reads. In fact it's lousy. So he revises it three times, until it finally is in the form that follows:

To: All Employees
From: Samuel Edwards, General Manager
Subject: Use of Copiers

We are revamping our policy on the use of copiers for personal matters. In the past we have not encouraged personnel to use them for such purposes because of the costs involved. But we also recognize, perhaps belatedly, that we can solve the problem if each of us pays for what he takes.

We are therefore putting these copiers on a pay-as-you-go basis. The details are simple enough. . . .

Samuel Edwards

This time Sam thinks the memo looks good, and it is good. Not only is the writing much improved, but the problem should now be solved. He therefore signs the memo, turns it over to his secretary for distribution, and goes back to other things.

FROM VERBIAGE TO INTENT

I can only speculate on what occurs in a writer's mind as he moves from a poor draft to a good revision, but it is clear that Sam went through several specific steps, mentally as well as physically, before he had created his end product.

- He eliminated wordiness.
- He modulated the tone of the memo.
- He revised the policy it stated.

Let's retrace his thinking through each of these processes.

Eliminating Wordiness

Sam's basic message is that employees are not to use the copiers for their own affairs at company expense. As he looks over his first draft however, it seems so long that this simple message has become diffused. With the idea of trimming the memo down, he takes another look at his first paragraph:

> It has recently been brought to my attention that many of the peo-
> ple who are employed by this company have taken advantage of their
> positions by availing themselves of the copiers. More specifically,
> these machines are being used for other than company business.

He edits it like this:

ITEM: "recently"
COMMENT TO HIMSELF: Of course; else why write about the problem? So delete the word.

ITEM: "It has been brought to my attention"
COMMENT: Naturally. Delete it

ITEM: "the people who are employed by this company"
COMMENT: Assumed. Why not just "employees"?

ITEM: "by availing themselves" and "for other than company business"
COMMENT: Since the second sentence repeats the first, why not coalesce?

And he comes up with this:

> Employees have been using the copiers for personal matters.

He proceeds to the second paragraph. More confident of himself, he moves in broader swoops, so that the deletion process looks like this:

> Obviously, such practice is contrary to company policy and ~~must cease and desist immediately. I wish therefore to inform all concerned—those who have abused policy or will be abusing it—that their behavior cannot and will not be tolerated. Accordingly anyone in the future who is unable to control himself will have his employment terminated.~~ will result in dismissal.

The final paragraph, apart from "company policy" and "feel free," looks all right, so the total memo now reads as follows:

> To: All Employees
> From: Samuel Edwards, General Manager
> Subject: Abuse of Copiers
>
> Employees have been using the copiers for personal matters.
> Obviously, such practice is contrary to company policy and will
> result in dismissal.
>
> If there are any questions, please contact this office.

Sam now examines his efforts by putting these questions to himself:

QUESTION: Is the memo free of deadwood?
ANSWER: Very much so. In fact, it's good, tight prose.
QUESTION: Is the policy stated?
ANSWER: Yes—sharp and clear.
QUESTION: Will the memo achieve its intended purpose?
ANSWER: Yes. But it sounds foolish.
QUESTION: Why?
ANSWER: The wording is too harsh; I'm not going to fire anybody over this.
QUESTION: How should I tone the thing down?

To answer this last question, Sam takes another look at the memo.

Correcting the Tone

What strikes his eye as he looks it over? Perhaps these three words:

- Abuse . . .
- Obviously . . .
- . . . dismissal . . .

The first one is easy enough to correct: he substitutes "use" for "abuse" But "obviously" poses a problem and calls for reflection. If the policy is obvious, why are the copiers being used? Is it that people are outrightly dishonest? Probably not. But that implies the policy isn't obvious; and whose fault is this? Who neglected to clarify policy? And why "dismissal" for something never publicized?

These questions impel him to revise the memo once again:

```
To: All Employees
From: Samuel Edwards, General Manager
Subject: Use of Copiers

Copiers are not to be used for personal matters. If there are
any questions, please contact this office.
```

Revising the Policy Itself

The memo now seems courteous enough—at least it is not discourteous—but it is just a blank, perhaps overly simple, statement of policy. Has he really thought through the policy itself?

Reflecting on this, Sam realizes that some people will continue to use the copiers for personal business anyhow. If he seriously intends to enforce the basic policy (first sentence), he will have to police the equipment, and that raises the question of costs all over again.

Also, the memo states that he will maintain an open-door policy (second sentence)—and surely there will be some, probably a good many, who will stroll in and offer to pay for what they use. His secretary has enough to do without keeping track of affairs of that kind.

Finally, the first and second sentences are at odds with each other. The first says that personal copying is out, and the second implies that it can be arranged.

The facts of organizational life thus force Sam to clarify in his own mind exactly what his position on the use of copiers is going to be. As he sees the problem now, what he really wants to do is put the copiers on a pay-as-you-go basis. After making that decision, he begins anew:

```
To: All Employees
From: Samuel Edwards, General Manager
Subject: Use of Copiers

We are revamping our policy on the use of copiers . . . .
```

This is the draft that goes into distribution and now allows him to turn his attention to other problems.

THE CHICKEN OR THE EGG?

What are we to make of all this? It seems a rather lengthy and tedious report of what, after all, is a routine writing task created by a problem of minor importance. In making this kind of analysis, have I simply labored the obvious?

To answer this question, let's drop back to the original draft. If you read it over, you will see that Sam began with this kind of thinking:

- "The employees are taking advantage of the company."

- "I'm a nice guy, but now I'm going to play Dutch uncle."

- "I'll write them a memo that tells them to shape up or ship out."

In his final version, however, his thinking is quite different:

- "Actually, the employees are pretty mature, responsible people. They're capable of understanding a problem."

- "Company policy itself has never been crystallized. In fact, this is the first memo on the subject."

- "I don't want to overdo this thing—any employee can make an error in judgment."

- "I'll set a reasonable policy and write a memo that explains how it ought to operate."

Sam obviously gained a lot of ground between the first draft and the final version, and this implies two things. First, if a manager is to write effectively, he needs to isolate and define, as fully as possible, all the critical variables in the writing process and scrutinize what he writes for its clarity, simplicity, tone, and the rest. Second, after he has clarified his thoughts on paper, he may find that what he has written is not what has to be said. In this sense, writing is feedback and a way for the manager to discover himself. What are his real attitudes toward that amorphous, undifferentiated gray mass of employees "out there"? Writing is a way of finding out. By objectifying his thoughts in the medium of language, he gets a chance to see what is going on in his mind.

In other words, *if the manager writes well, he will think well.* Equally, the more clearly he has thought out his message before he starts to dictate, the more likely he is to get it right on paper the first time round. In other words, *if he thinks well, he will write well.*

Hence, we have a chicken-and-the-egg situation: writing and thinking go hand in hand; and when one is good, the other is likely to be good.

REVISION SHARPENS THINKING

More particularly, rewriting is the key to improved thinking. It demands a real openmindedness and objectivity. It demands a willingness to cull verbiage so that ideas stand out clearly. And it demands a willingness to meet logical contradictions head on and trace them to the premises that have created them. In short, it forces a writer to get up his courage and expose his thinking process to his own intelligence.

Obviously, revising is hard work. It demands that you put yourself through the wringer, intellectually and emotionally, to squeeze out the best you can offer. Is it worth the effort? Yes, it is—if you believe you have a responsibility to think and communicate effectively.

Lee Clark Johns

The File Cabinet Has a Sex Life: Insights of a Professional Writing Consultant

Since 1978, Lee Clark Johns has headed Professional Writing Consultants in Tulsa, Oklahoma, teaching writing seminars for the oil and computer industries, banks, government agencies, and other businesses.

Ask anyone "When you were given your first writing assignment on the job, how did you know what to do?"

"I looked in the file cabinet." "I asked how they had done it before."

The exact answers differ slightly, but the sense is the same. When facing new writing tasks—in new jobs, new companies, or new professions—most employees "go to the file." They are looking for writing models that have succeeded in the past, models that they can efficiently copy, models that their supervisor will approve. Because of this practice, however, the organizational patterns and style of documents in the file cabinet resist change. Old formats and stylistic preferences continue to thrive long after they have outlived their usefulness. The evidence is overwhelming: The file cabinet has a sex life; it reproduces itself.

This article examines the nature of that reproductive cycle. What are the models that are being reproduced? What is their parentage? Are these useful models or have they outlived their usefulness? How are the models evolving and what are the sources of change? Who is responsible for "cleaning up the bloodline"? I do not claim to have all the answers to these questions. In fact,

THE FILE CABINET HAS A SEX LIFE: INSIGHTS OF A PROFESSIONAL WRITING CONSULTANT. By Lee Clark Johns. From Carolyn B. Matalene, ed., *Worlds of Writing: Teaching and Learning in the Discourse Communities of Work*. Random House: New York, 1989. Reprinted by the kind permission of the author and the editor.

the greatest danger in an article of this sort is to overgeneralize. But I do want to identify patterns and cite typical examples that illustrate the nature of the writing done in the workplace.

How do I know about these issues? Since 1978, I have been teaching writing seminars to employees in a wide variety of organizations—in both the private and the public sectors. I have worked with oil companies and banks, with engineering firms and accounting firms, with the police department and university administrators, with the water and sewer department, a federal agency, a computer company and a hotel/motel management company. The individuals within these organizations make up almost a what's what of professions: executives, secretaries, attorneys, accountants, computer specialists, manufacturing supervisors and managers, marketing specialists, pipeliners, engineers, technical writers, police officers, juvenile justice social workers, city government employees, bankers, architects, and so forth. The article draws on this broad experience to share my understanding—and very informal research—of how people write in the workplace. In my opinion, they all face the *same* writing problems. And they seem to solve their writing problems in amazingly similar ways.

Instead of looking specifically at the nature of writing in a particular discourse community, I am interested in the writing patterns and behaviors that cut across discourse community boundaries. I see far more similarities than differences in the writing that different professionals produce. Of course, people do write within the narrow constraints of their professional fields and produce documents that satisfy those communities. But more frequently, they write for people outside the discourse community. Or more accurately, they operate within several "discourse communities" at the same time. For example, oil company accountants write to other accountants, work with computer programmers to plan and understand on-line accounting systems, address legal and taxation issues—all within the oil industry. And they write to gain management approval, to inform royalty owners, and to negotiate with other oil companies. Just which is their "discourse community"?

That's where the file cabinet comes in. Instead of operating within the narrowly defined boundaries of one discourse community, employees operate within several. So the writing they do relies on a variety of models. Some models literally come out of the departmental file cabinet: company formats, organizational patterns imposed by the particular department or by a manager, stylistic preferences dictated by a reviewing supervisor. Other models are drawn from the writer's own experience: from formats learned in school or from thinking patterns developed in professional studies. Whatever the source, these writers turn to what's been done before to produce an adequate document efficiently, get it approved, and move on to the next task.

Unfortunately, many of these old models have outlived their usefulness. They simply do not meet the needs of the readers who must use them or the organization that must move information productively and efficiently. They are dinosaurs. To stretch the analogy, many business documents—from short memos and letters to very long research reports—have very tiny heads, huge bodies, and weak tails.

But they are prolific. Instead of becoming extinct, they multiply. No change in climate or reduction in food supply threatens to destroy them. In fact, these dinosaur models are nurtured in their academic infancy, groomed in professional adolescence, and rewarded in working adulthood. And they will continue to reproduce unless significant changes take place in that nurturing environment. These are fairly strong claims. But the thousands of documents that I read each year testify that when people go to the file cabinet for models, they find these outmoded, sometimes ineffective dinosaurs. Let us examine the old models themselves, their sources, and the problems they cause in modern business and industry.

PARENT: THE ACADEMY

The proto-parent is the academic essay and its longer version, the research paper. Most people learn to write in school. Thus, the models taught there become extremely powerful in shaping their perception of "what writing ought to be." In English classes, these models are primarily personal essays, arguments on public issues, literary analyses, and term papers. Most secondary school teachers are getting their students "ready for college," and college teachers are preparing students for other college writing. Thus, writing academic papers becomes an end in itself.

Leslie Moore and Linda Peterson, in describing their writing-across-the-curriculum program at Yale, reveal this goal: that students should be prepared to write within their academic disciplines. As is typical, students in both freshman and advanced writing classes "are expected to observe a series of conventions agreed upon by members of a scholarly community" (Moore and Peterson, 474). While the move toward writing across the curriculum broadens the previously narrow concerns of English department classes, it is only a first step because it is still inner-directed.

However, very few students remain in academic life. When they graduate, most take jobs in organizations that are nonacademic in nature. Unfortunately, many feel they have not been prepared for the writing they do on the job. Recent surveys and leaders in our field increasingly question the adequacy of college writing courses in preparing students for on-the-job writing tasks (Harwood, 1982). My consulting experience supports this criticism. In every seminar, I hear, "Why wasn't I taught this in school?" "This" includes alternatives of organization and style that differ from traditional academic models.

Part of the answer is that the writing requirements in an academic environment usually do not match the requirements of the workplace. J. C. Mathes and Dwight Stevenson, in their 1976 article in *Engineering Education,* outline these differences and suggest an alternative approach for engineering professors (Mathes and Stevenson, 154–156). But progressive teachers in all departments need to recognize the differences. In school, students learn to write to one audience (the teacher), to work alone, to reveal what they know about a subject, and to achieve a high grade as the only result. In the workplace, the same students must write to many different audiences, often work in groups or must receive a supervisor's approval for the document, should focus on what the readers *need* to know about the subject, and hope to achieve very specific results from the document. Thus, the total context—audience, method, content, and purpose—differs significantly between academic and workplace writing. And even though many business and technical writing classes recognize and address these differences, the fact is that most students do not take upper-division writing classes.

The Models

Of course, traditional academic writing is an important first step in learning a model. It teaches students to shape incomplete and unruly thoughts. Some students also learn to value clarity and conciseness. Finally, the academic organizational model—introduction, body, conclusion—is perfectly adequate for short memos and letters. Writers who have stored that model produce clear, short documents that satisfy their readers.

The problem arises when the same model is imposed on a long report. The term paper approach simply does not meet the needs of workplace readers. A typical banking document illustrates the problem. In the banking industry, loan decisions are based on documents called credit analyses. In these normally lengthy documents, the analyst presents this information about the company requesting the loan: background, operating results (earnings, profits, losses), financial position (debts, assets, liquidity), and industry comparisons. Many of these analyses run over eight pages. Generally they are written by inexperienced employees, credit analysts who are essentially management trainees in their first banking position. The credit decisions are made by the senior loan committee, often the most experienced executives.

The combination of writers who are insecure about their technical understanding (but confident in their college-learned writing ability) and readers who must make important decisions in a short period of time produces the problem. The writers, having learned their lessons well in undergraduate school and often in their M.B.A. studies, produce "data dumps." Unsure about what information is significant, they include everything. With no alternative format to follow, they rely on the academic model: short introduction, background, a story about the company's operations and financial position, etc., until they reach the conclusion about credit risks involved in the loan. Such documents do not carry specific recommendations; they are essentially resource documents that help the loan committee make a decision.

But senior executives do not have time to sort through all this data. One executive of a major regional bank holding company explained what happens:

> The Senior Loan Committee meets every Friday morning. I usually receive the analyses we will discuss about 4:00 P.M. on Thursday and take them home for review that night. We may be looking at six to ten potential loans, and there is no way I can read these long, rambling reports. So I do the best I can. I review the ones that I already know something about and hope for the best on the others.

Every Thursday evening he is frustrated, and every Friday morning decisions of major importance to the organization are being made on the basis of poorly organized reports. Other banks have developed a credit analysis format that has an executive summary at the beginning (a model that appears later in this article). But my point here is that the writers—using the only model they know—are doing the best they can.

The Thinking Processes

In the absence of an academic model, writers fall back on their thinking processes, often the same process that the old file models reveal. For example, minutes and trip reports are almost invariably narratives: "What did we do?" Within an organization, these reports can be very important because problem solving and interdepartmental negotiations often occur in meetings or on trips to the regions. But the narrative approach buries the significant decisions and becomes an exercise in frustration for the writer and an ineffective reference for future readers. A classic three-page meeting report shows the problem; in outline form, it reads:

On March 3, 1982, a meeting was held at the Research Center . . . The purpose of the meeting was . . .

Personnel in attendance were: . . .

DISCUSSION

Morning Session
John Smith reviews his presentation of March 2 . . .
At this point, Jones says . . .
Thompson and White leave to attend interviewee seminar. Johnson addresses questions four and eight . . .

page 2
The physical properties questions of four and eight were taken up . . .
Thompson and White return. Philosophical discussion of _____ the problem resumes . . .
Frank takes up question six . . .

Afternoon Session
Frank continues . . .

page 3
The following outline in the form of major tasks along with dates is presented . . .
Charge to Research . . .
The discussion turns to how to proceed with . . . There is no plan put forth . . .

This report, written by a senior research scientist, is all too typical. I rarely see either meeting minutes or trip reports that use substantive headings or any sort of executive summary. Most are narratives. In the absence of alternative models, writers fall back on "the way it's been done before" or on their most primitive communication skill—telling a story.

The *academic discourse community* rewards students for documents that reveal their thinking processes. But in the workplace, the decision makers are not interested in "how the watch was made." They want to know "what to do with it."[1]

Both senior employees and novices often *think* they are writing to others within the discourse community, and they follow the models of that community. For example, several years ago I began to notice the many inductive paragraphs within long research reports from one oil industry laboratory. Most followed this pattern in the discussion section: "Figure 1 presents the findings of the x test. (several lines of factual data) Therefore, it is concluded that. . . . "

The pattern was too prolific to ignore because it consistently violated the topic-sentence-first model for writing effective business paragraphs. I finally realized that these scientists were modeling the scientific method on paper—paragraph after paragraph. When I pointed that out to them, most answered, "But that is what our readers [other scientists] expect."[2]

However, the assumption that documents are read only by those within the narrow discourse community is simply wrong.[3] Other scientists are the primary audience for the discussion section of a report. But so are operational engineers and other nonscientists. *All* could read a deductive paragraph more efficiently.

The Results

Employees who are blindly committed to the academic discourse model face frustration and sometimes career disappointment. Recently, one scientist told me, wistfully:

> Before I came to Company X, everyone always praised my writing. My professors said I did an excellent job on my papers. Then I went to work for Company Y and they loved the way I write. One boss even said, "Finally they give me someone who knows how to write." But then I came to X and have been criticized ever since. I didn't know what they wanted until I took your writing seminar. At Y, Ph.D.'s become top management. They understand the scientific paper. But Ph.D.'s don't become the Director at X.

His assessment of promotion opportunities may be wrong, but he has identified the source of his writing frustrations. His current management wants results-oriented research reports, not the methods-oriented scientific paper. Obviously, the "discourse community" in each of the companies—both in the same industry—is different.

Gregory Colomb and Joseph Williams sum up these differences in addressing the needs of different readers:

> Reader's who are pressed for time or who for other reasons are unwilling to give their time to the writer tend to prefer Point-first structures, since these structures make the reading process as quick and efficient as possible . . . . On the other hand, readers who

are willing or who must give their time to the writer and who expect in return the kinds of pleasures we associate with fine, belletristic writing generally feel more rewarded by Point-last structures, since only in such structures are they accorded the pleasures of the chase. . . . Very few readers in professional settings are willing to allow writers the kind of claim on their time and energy that is inherent in Point-last structures. [Colomb and Williams, 111]

Thus, when scientists *assume* their expert readers will enjoy "discovering" the conclusion with them, they miss the needs of management and even of other experts who could read more effectively with point-first structures. The result of relying on these academically based thinking and writing models is frustration— for both reader and writer.

PARENT: THE PROFESSION

The first-generation offspring of the academic proto-parent are the models students learn for their professions. These, too, are academically based. Some, such as legal formats, are learned in the professional school itself (where students also learn a method of reasoning that may later appear in documents). Others, such as accounting models, are learned in preparation for certifying examinations such as the CPA exam. Finally, some are set by the professional system itself. Court report formats in the juvenile justice system are established by statewide guidelines and reinforced by judges, essentially the chief executive officers of the system. Whether required or simply traditional, these models powerfully shape the specific types of written discourse within the field. Many that I have seen look amazingly like the proto-parent academic report. These three examples from different professional discourse communities illustrate the resemblance.

The Patent Application

Figure 1 shows a patent application that is traditionally arranged. The most important part of the application is the claims section. The applicant is not required to

Abstract	Brief statement of the invention.
Background	Sometimes called Prior Art; what previous related inventions exist.
Summary	The uses of the invention.
Description	Detailed explanation of the invention's components.
Figures	Drawings (may precede the description).
Claims	Summary of major advantages of the invention; each claim must be a single sentence; the sentences are *extremely* long (several paragraphs).

FIGURE I

tell *how* the invention works, just that it does. In a sense, the claims are the "bottom line" of the document.

The Internal Audit Report

The format for internal audit reports is also set by tradition. Irvin N. Gleim's *CIA Examination Review* (for Certified Internal Auditors) acknowledges that "internal audit reports do not have a prescribed format" (Gleim 318–319). However, it presents samples from Sawyer's 1981 textbook entitled *The Practice of Modern Internal Auditing* as the model to follow in preparing for the certification examination. Young auditors, without experience in the companies whose internal operations they will audit, learn the textbook model of Figure 2 as they prepare for their first professional hurdle. This model is closer to an executive summary, but it still places information of primary interest to the *accountant* before information of primary interest to *management*. The accountant is concerned with the status, goals, and scope of the audit. Management wants the results. In practice, this audit format *is* useful to management, but not as useful as it could be. (A more useful format for internal audit reports appears later in the article.)

The Certification Report

The final example of a prescribed professional format comes from the Oklahoma juvenile justice system. A certification report, written by a probation officer, is used by the judge in deciding whether or not a juvenile offender should stand trail (be certified) as an adult. The format is established in the state guidelines

Summary	Brief overview of findings.
Introduction (Foreword)	Status of the audit. (Has it been completed?)
Statement of Purpose	Goals of the audit.
Statement of Scope	What was examined.
Statement of Opinion	Major conclusions.
Audit Findings	The body of the audit report discusses each opinion in detail.
	These are presented in this order: Summary Criteria Facts Cause Effect Recommendation Corrective action taken

FIGURE 2

for court-related social workers and reinforced by most judges in the system. In Figure 3 the prescribed format is on the left and the heading outline of a typical report appears on the right. Both follow a modified chronological pattern, with recommendations coming at the end of the document: The 4 1/2 page report reads like a story—just as the prescribed format intends.[4] Oklahoma social workers testify that the judges want to accumulate detail and form their own opinions as they read.

These three examples of professionally prescribed formats confirm several points. First, they exist and they are extremely powerful. Most people who write such documents insist that they *must* follow the format—even if it is not as useful for readers as it could be. Indeed, the expert readers in each discourse community expect information in the prescribed order and read accordingly. Second, most of these professional models are direct descendants of the academic term paper—background first, conclusions last. Only the internal audit report comes close to a point-first structure, and even it presents the auditor's information (scope and objectives) first. Finally, these models spawn offspring in the other documents the same professionals write. Because these writers define their discourse community narrowly, they assume that clients, management, or other departments will understand and tolerate their professional model.

Prescribed Format	*Actual Report*
A. Any involvement in the juvenile justice system:	*Nature of Offense*

1. Adjudications
  2. Deferred prosecutions
  3. Informal probation
	"Prosecutive merit found indicating that _____ did commit an act, which if committed by an adult, would constitute the offense or murder . . ."

B. Psychological evaluation	*Previous Record*
C. Assessment of family and community environment:	*Psychological Evaluation*
1. Family relations 2. Peer associations	*Family Background* (Parents and siblings)

3. Academic experience
  4. Employment history
  5. Outside interests
	Academic History
	Work History

D. Physical or mental impairments	*Psycho/Social Summary* (two full pages)
E. Diagnostic impression the worker may have of the child	
F. Worker's opinion of the reasonable prospects of rehabilitation and the protection of the public	*Summary Analysis and Diagnostic Impressions* *Treatment Potential* (almost a full page)

	cc: Judge
	      D.A.
	      Attorney (for youth)
	      Family
	      File—agency office

FIGURE 3

The Result

The frustration this narrow definition of audience produces is illustrated by a fascinating, heated exchange that occurred one day in a seminar. The spark was the use of jargon, but I hear that the format of computer documentation is a comparable problem. The two employees served on a task force that brought together accountants and computer specialists to design computer programs for the accounting department of their oil company. When we began to discuss some computer jargon, the accountant exploded, "You throw all those terms at us in the meetings and don't explain what they mean. Or we become invisible. It doesn't seem to make any difference that we're there." The systems analyst shot back, "You don't need to understand the technicalities. Besides, explaining everything would take too much time. Just tell us what you want and we'll do it." The moral: Discourse-community boundaries blur in organizations. Nonexperts *want* to understand the information that affects them, and they want the information to be easily accessible.

PARENT: THE ORGANIZATION

Because of the blurring of these professional boundaries, the immediate progenitor of the models in the file cabinet is the organization itself. In some cases, the models are set by company fiat. The famous Proctor and Gamble one-page memo is an example. Corporate style sheets are another. However, I find that sort of top-down control to be rare. Most of the models are set at the department level or by supervisors who impose their personal preferences on the writing of their subordinates.

Companies do have unique—often unrecognized—conceptions about what "writing" should be. Richard Freed and Glenn Broadhead argue that different companies have different norms: "Each organization is a different culture and each has different rules. And though each will use the English language and write the English language, the writing (and the attitudes about and behaviors during the writing) may very well be different" (Freed and Broadhead, 157). The earlier story about the scientist who moved from Company Y to Company X illustrates the difficulties these differences produce. In her 1987 CCCC presentation, Carol Lipsom told a similar story. In her "Company X," the rule was, "Do you want to get fired? Don't use 'I.'" The source of the rule was a 1955 memo. When a new young manager took over a lab unit, he tried to impose a deductive organizational format and an informal, lively style. The staff refused to change and began to seek transfers. They were concerned that outside readers would object. Furthermore, they had made the company a success, and the young upstart was going too far when he attacked their writing. They knew how to write *proper* scientific reports (Lipsom, 1987).

I hear the same story, although generally it is from subordinates who would like to use a deductive, informal approach that their older *supervisors* will not permit. But I have *never* seen a *companywide* prescribed format or style. In fact, the diversity of formats within a company can be the source of communication problems and frustration.

The Research Report

A sequence of memos establishing the research-report format for technical laboratory divisions illustrates the historical resistance to standardization and the autonomy of group managers in establishing their personal preferences. In 1960, the head of the Production Research Division issued a memo establishing "a revised format" for their research reports (documents that may run up to 100 pages). His goal was to make the reports more useful to their primary audience: "Many operating engineers only need and use certain portions of our reports as now prepared. This new format is intended to assist the operating engineer in using our research developments." Figure 4 summarizes his prescribed format.

In 1982, a different manager of the same research division issued a memo, again prescribing a format for the section's reports and memorandums. He attached a copy of the older memo and other reprints of articles about writing research reports. Figure 5 shows this format.

The first laboratory employees who attended my seminars *insisted* that this was the prescribed company format: introduction, summary, discussion. They were convinced that they could not use more descriptive headings or vary this format in the slightest. Later someone gave me a copy of the format from a

Headings	*Paraphrase*
Purpose and Introduction	This will orient the reader as to why the study . . . was conducted. Brief, pertinent background material can be included. (6 to 8 lines)
Summary (or) Summary and Conclusions	This would briefly recapitulate the highlights of the Discussion part of the report. (A few lines to perhaps a page)
Conclusions	Certain specific conclusions should be set forth and enumerated. General conclusions can be incorporated under "Summary and Conclusions" in a short report.
Recommendations	If recommendations are made, they should be specifically set forth under this heading.
Pertinent Assumptions	The basic or controlling assumptions which limit the applicability of the conclusions should be listed.
Discussion	This is intended to elaborate on the "Summary and Conclusions" presented with the idea of emphasizing applications and limitations.
Appendix	The Appendix will include theory or mathematical development, procedure, and results.

FIGURE 4

different division (see Figure 6). Written in 1983, it defines the continuing problems with research reports and closely models the 1960 memo. Because of its concern for both organization and style, it is the format I suggest in my seminars for this company.

One would think that this model was widely known and used in the organization. Unfortunately, that is not the case. Many scientists discover it for the first time in my seminar (and a few still insist that the model would not be accepted in their groups). Just as the 1960 format did not permanently standardize the company or even the division format, so too this model is prescribed only for the particular manager's group. Thus, despite the long existence of a superior model, the "format of choice" often remains the point-last structure of the academic model.

Headings	Paraphrase
Introduction	The "Introduction" should contain (1) the problem definitions in terms of the "big picture" to the company, (2) background information necessary to support the content of the report, and (3) a clear, concise purpose of the report. References to work or written documents closely related to the subject should be included.
Summary	The "Summary" section includes summary and concluding statements. It conveys to the reader in nontechnical terms . . . the highlights or most significant results of the work being reported. It shows how these results are of benefit to the company and therefore *sells* the report to the reader.
Discussion	One purpose of the "Discussion" is to provide less relevant and/or more detailed background information required by the reader but not included in the "Introduction" . . . The most important purpose of the "Discussion" is to provide sufficient information through verbiage, references, tables, figures, and appendices to technically support the statement within the "Summary" section.
Signature	The author signs the report . . .
References	Typed in the proper format used by SPE for technical publications.
Figures	

FIGURE 5

Subject: Organization of Research Reports

During the last meeting we had some forthright feedback on our recent research reports. Perhaps the comments can best be paraphrased as "Some reports read like a mystery novel: one is left guessing until the very end."

I believe that this is a legitimate complaint and we should realize that many of our reports, while being given a wide distribution to promote accessibility, are often designed for a fairly limited audience. It is therefore not unreasonable to expect that a reader can, by scanning the abstract and early sections of a report, obtain a capsulated understanding of the entire volume. I request that reports be organized along the following lines:

Abstract	Two or three main points of significance that result from the work being reported. . . . Abstracts which merely outline the nature of the work without providing conclusions are inappropriate.
Brief Introduction	This should cover, in summary form, the reason why the work was undertaken and a clear statement of who, we think, will be interested in the results.
Summary	This should be as brief as possible but include the key conclusions, recommendations and results. It should also include one to several sentences on procedures unless these would be obvious or of little or no significance.
Conclusions	This should be an itemized list of conclusions and should not contain any lengthy discussion material. It should be written in language that anyone, including managers, can understand.
Recommendations	This should be an itemized list of recommendations for action by Operations and also for continued or new research. Again, it should be brief, to the point, and in plain English.
Discussion	This is the body of the report and may be subdivided as appropriate. . . . This is the place you can write for other research scientists, specialized users, etc., as appropriate. Clarity is still a virtue. By all means keep in mind the objective of shortening the "Discussion" by using appendices when appropriate.

FIGURE 6

PARENT: THE SUPERVISORY REVIEW

The most powerful control of organizational writing comes from the widespread practice of supervisory review. In many companies, the immediate supervisor must approve a document before it is sent. Thus, the personal preferences of supervisors (and/or their superiors) determine both the organization and the style of many employees' documents. The most local file cabinet literally rules.

There are several good reasons for having supervisors review subordinates' documents. Often the supervisor's name is at the bottom because protocol demands that people of equal rank correspond with each other. Thus, employees who ghostwrite for their supervisors try to emulate the writing practices of the "author." Also, supervisors want to ensure at least minimal uniformity in the documents that issue from their groups. Finally, they bring important company experience to the review; they understand corporate issues or politics that subordinates—especially new employees—may not know. One manager, in a memo on "Effective Writing," reveals the department's need for "better planning of reports, reviewing the content, and critiquing the document before it is issued":

> Our written product is our sales tool to explain results, persuade others, defend our work, and to ask for money. Your success, and mine, is based firmly on the quality of your effort and the clarity with which that effort is communicated to management. Unreported or poorly documented work is lost forever; hence, my insistence that we report our work completely. To paraphrase General Electric's corporate motto: "In Research, paper is our most important product."

Many employees report that they have learned their successful writing techniques from strong supervisors who literally taught them how to write in the workplace.

However, the critiques on supervisory review are usually less glowing. Far too many employees are told, "This is the way we've always done it. You will, too." So outdated formats and style proliferate, protected by the supervisory review process itself. Several problems result. Because each supervisor has a different idea of what constitutes good writing, employees must change to meet the expectations of each new supervisor. One young man told me that in the eighteen months be had been with the company, he had had six different supervisors—each with a different style. "I don't know who I am," he said. Some supervisors insist on rewriting everything. Thus, the employee loses ownership of the document, doesn't learn revision techniques, and often develops an "I-don't-care" attitude. The time wasted in recycling perfectly adequate documents because of personal preferences of supervisors is a major cost to the organization. Ultimately, employees learn simply to accept the changes, get the needed approval, and send the document. The frustrations for both writer and reviewer are obvious.

This detailed explanation of the problems of supervisory review reveals why the file cabinet reproduces unrestrained. First, the *system* of writing in the workplace usually prevents change. It is safer and more efficient for writers to follow the old models rather than to suggest new ones. Therefore good models survive, but so do poor ones. Second, although employees draw on their own academic and professional experience in defining the nature of good writing, they are also *instructed* by their superiors. Thus, the real training in workplace writing is being conducted by "teachers" untrained in either rhetoric or pedagogy. It is a journeyman system in which the "master" controls powerful incentives for the "apprentice."

THE EVOLUTION OF NEW MODELS

Since the system of writing in the workplace both creates and protects the models in the file cabinet, changing those models requires employee initiative and the rare willingness to take risks. Permanent change of writing throughout the organization also requires power. Thus, the solutions to writing problems evolve from some of the same sources as the old models: from individual choice, from strong managers, occasionally from a central corporate group. Change also can come from outside writing consultants. However, change rarely comes from academic or professional sources. Nor do I see it imposed from the top down in large companies. Therefore, most new formats remain localized. They are used within a small group which accepts them either because all agree that the new way is better or because a manager or supervisor has decreed that change will occur.

Most changes are solutions to audience problems; they are reader-based. Usually they are born because an individual is dissatisfied with "the way it's been done before." Sometimes the birth occurs because outside readers are complaining that they cannot understand documents. The recent, hesitant changes in computer documentation attest that eventually someone does listen to readers.

The creative models that follow reveal this reader orientation. All are examples of creative problem solving by individuals or managers. Except for one example from a central policies and procedures group, they are not widely used formats within the company (even though they may be widely read). Thus, their life spans within the file cabinet family may be limited by the longevity of their creators. These models do, however, illustrate the potential for improvement that exists in the workplace. They also reveal what we can learn from the problem-solving abilities of these writers.

The Paragraph

One academic model employees carry reflects what a paragraph should look like. They "see" information presented in blocks of black type. Yet when asked how they prefer to receive information, everyone says, "with white space." Figure 7 shows what happened when a manufacturing supervisor was not satisfied with a simple memo he was writing. He went to his manager for help, who suggested listing the alternatives. Within thirty minutes, the employee produced the revised version, which is not only easier to read but also more complete. (Obviously it *does* still need some sentence-level revision.) The employee not only had not thought of using white space but also needed "permission" to change what a paragraph looked like.

PRODUCTS, Inc.

Date:

To:

From:

SUBJECT: Kit Alternatives

The original Log Plan for the _____ controller used _____ 6
Regional offices for the _____ Kit distribution centers due to the
high cost of the Third Program stated below. The 6 Regional Kits
would cost $ _____ dollars. But, one exposure in regards to that
plan is establishing only 6 sites for _____ Kits [which] would be
insufficient saturation of spare parts in the field.

Attached are three alternative Kit distribution cost analyses. The
first program would consist of 12 centers, located at key Air Freight
Hub Distribution sites at a cost of $ _____ . Second program
consists of 14 centers, located at our own 6 Regional sites plus 8
addition Zones. Cost to establish this program would be
$ _____ . The third program is the most costly at $ _____ .
They would be distributed to all 129 Service Centers forecasted in
the _____ Maintenance Plan.

PRODUCTS, Inc.

Date:

To:

From:

SUBJECT: Kit Alternatives

The original Log Plan for the _____ controller used 6 Regional
offices for the Kit distribution centers due to the high cost of the
Third Program stated below in "Alternatives." But, one exposure in
regards to that plan is establishing only 6 sites for Kits [which]
would be insufficient saturation of spare parts in the field.

Original Plan:

    6 Regional Kits = $ _____

Alternatives:

    First Program = $ _____
    Consists of 12 centers, located at Key Air Freight Hub
    Distribution sites.

    Second Program = $ _____
    Consists of 14 centers, located at 6 Regional offices plus 8
    addition Zones.

    Third Program = $ _____
    Consists of all 129 Service Centers forecasted in the
    Maintenance Plan.

You will find attached cost analyses of the three Alternative
programs for Kit distribution.

FIGURE 7

The Problem Report

The next examples show improved computer problem-report formats. Both formats were created by individuals dissatisfied with the old models. Figure 8 shows one approach. Written for an expert audience, its jargon is dense, but the format is extremely useful to the readers who need to know in a glance what has happened.

Figure 9 reveals the evolution of a problem/solution format. The author, who is not at all reticent about changing inadequate models, had already modified the file-cabinet model to produce the document on the left. After the discussion of audience, purpose, and format in my writing seminar, he reworked his own improvements. He said, "When I looked at my report, I realized there wasn't any reason to have all those introductory lines. They weren't there for any purpose." He also made some stylistic changes, tightening and clarifying his instructions by replacing passive verbs. The result is a much cleaner and clearer report.

Current Status of Reported DFHSM Problems

Below is a short description of the current DFHSM-reported problems, along with their status. I have also provided an impact statement which indicates the severity of the problem. If you require more information, please contact me.

PROBLEM: Abend S878 during automatic backup of VSAM volumes.

STATUS: This problem has been reported to IBM and they have accepted an APAR (OYO5894). The documentation has been forwarded to IBM. The problem has been recorded as #00169 in the TDC problem database.

IMPACT: Approximately every 72 hours, DFHSM will terminate with an Abend S878, typically during automatic backup of our VSAM volumes. At that time Operations will restart DFHSM, with little impact to the Operating System or client applications.

PROBLEM: Message ARC1139I and ARC1239I (GDG Rollover problem and HMIGRATE failure respectively)

STATUS: This problem has been reported to SSS/SKK. Providing the required documentation has proven to be difficult due to the amount of data required. Escalation of this problem has just begun, with the actual problem assignment to SSS. For more detailed information, refer to problem #01165 in the TDC problem database.

IMPACT: These two error messages began after installation of the new release of ACF2 and have been nothing more than a nuisance. The user with the ARC1139I (GDG Rollover) would not terminate but just receive the message, and the ARC1239I message (HMIGRATE) was found during testing, and widespread user command awareness has yet to be accomplished.

FIGURE 8

FILE: TLS   DOC0409   A VM/SP CONVERSATIONAL MONITOR SYSTEM

UPDATE DESCRIPTION:- -
        DATE: 04/01/87 – CHECKED OUT
        DATE: 04/09/87 – TO PRODUCTION TEST
        DATE: 04/09/87 – THIS DOCUMENTATION FILE.
        FIX-ID #: - - - - - -
        CSS PROB #:   08347
        CSS PROJ #: _____
        TYPE: SOURCE FIX (FOR TLS).
        CHANGES BY:
        CHANGES TO:
        TLS.TSK- (UPDATED).
        OTHER MODULES AFFECTED: (NONE).
        PROBLEM:
            TLS sometimes gets into 'LOGGING' mode because of bad data in its copy of tape
            request information.
            This requires manual intervention to patch the TLS 'LOG' file in order to proceed out of
            LOGGING mode.
        SOLUTION:
            TLS.TSK has been modified to validate the tape MCB information that it receives. (This data
            is checked only to verify that it is in the valid ASCII character range.)
            If invalid data is detected, the request is handled as follows:
                (1) If the request is for a 'CLOSE', a good return code is given back to TMS, but the
                     information is NOT transmitted to the VM system.
                (2) For any other request, a bad return code is given back to TMS indicating that the
                     request is INVALID.
            In either case (1) or (2) error messages are sent to the operator's console indicating that
            this situation has occurred.
            The following console messages are produced:
                .TLS: !!! –INVALID MCB DETECTED xxxxxxxxxxxxxx   – !!!
                .TLS: !!! –FD=ffffffffff TSK=tttttttt   – !!!
                .TLS: !!! –REPORT THIS TO SSS MINI/MICRO,TULSA   – !!!
            where . . .
                xxxxxxxxxxxxxx =   BEFORE VM MSG
                                or AFTER VMREPLY
                                or AFTER FILEDEF
                ffffffffff   =   the file descriptor of the request
                tttttttt   =   the task id initiating the request
        IMPACT: USERS – (none)
        IMPACT: PGMRS – (NONE)
        IMPACT: OPERATIONS –
            If the above messages appear, operators are to record the information displayed in the
            first 2 messages.
            A copy of the console log should be retained.
            this information is to be reported to Software Support Services, Mini/Micro Systems,
            Tulsa.
            (No operator intervention is required.)
        SPECIAL INSTRUCTIONS FOR INSTALLATION:
            (A) It is merely necessary to replace the old TLS.TSK by this new version.

FIGURE 9a   Original Version

FILE: TLS   DOCO824   A   VM/SP CONVERSATIONAL MONITOR SYSTEM

UPDATE DESCRIPTION: - 8/20/87
   CHANGES TO: TLS.TSK - (UPDATED).
   CHANGES BY:
   OTHER MODULES AFFECTED: (NONE).
   DATES:
      04/01/87 - Checked out for changes.
      08/19/87 - Sent to production test.
   PROBLEM:
      TLS sometimes gets into 'LOGGING' mode because of bad data in its copy of tape request
      information.
      This requires manual intervention to patch the TLS 'LOG' file in order to proceed out of
      LOGGING mode.
   SOLUTION:
      TLS.TSK has been modified to validate the tape MCB information that it receives. (This data
      is checked only to verify that it is in the valid ASCII character range.)
      If TLS detects invalid data, it handles the request as follows:
         (1) If the request is for a 'CLOSE', a good return code is given back to TMS, but the
              information is NOT transmitted to the VM system.
         (2) For any other request, a bad return code is given back to TMS indicating that the
              request is INVALID.
      In either case (1) or (2), TLS writes error messages to the operator console, as follows:
         .TLS: !!! -INVALID MCB DETECTED xxxxxxxxxxxxx   - !!!
         .TLS: !!! -FD=ffffffffff TSK=tttttttt   - !!!
         .TLS: !!! -REPORT THIS TO SSS MINI/MICRO,TULSA   - !!!
      where . . .
         xxxxxxxxxxxxx   =   BEFORE VM MSG
                          or AFTER VMREPLY
                          or AFTER FILEDEF
         ffffffffff   =   the file descriptor of the request
         tttttttt   =   the task id initiating the request
IMPACT: OPERATIONS -
          If the above messages appear, operators should record the information displayed in the
          first 2 messages.
          A copy of the console log should be retained.
          Report this information to Software Support Services, Mini/Micro Systems, Tulsa.
          (TLS processing will proceed normally; no operator intervention is required.)
IMPACT: USERS - (none)
IMPACT: PGMRS - (NONE)

SPECIAL INSTRUCTIONS FOR INSTALLATION:
   It is merely necessary to replace the old TLS.TSK by this new version.

FIGURE 9b   Revised Version

The Management-Oriented Report

The next two samples function in similar company contexts and offer similar reader-based solutions. The sources of poor models for both the credit analysis and the internal audit report are discussed earlier in this article. In both improved models, the managers established the format for reports emanating from their departments. In both cases, the documents are read by higher managers and executives who base important decisions on the information contained in the report. Therefore, both formats begin with an executive summary, just as the successful research report format did.

The Credit Analysis

This credit-analysis format entered the bank's file cabinet when a new manager of the Credit Department arrived. He had used it in the bank he came from.[5] The four-page description of the format closely details the contents of each section. Figure 10 is an abbreviated outline, with my comments in parentheses.

Analysis Format

I. **PURPOSE** (of the loan)
   A. Name of Borrower
   B. Purpose of Loan
   C. Amount of Loan
   D. Participation Details
   E. Description of the Commitment
   F. Price of Loan
   G. Date of Maturity
   H. Secured or Unsecured
   I. Source of Repayment
   J. Guarantee
   K. Anticipated Level of Usage (If Credit Line)

II. **BASIS OF ANALYSIS**
   A. State financial statements reviewed
   B. Did these statements provide the proper basis of analysis for this request? . . .

III. **SUMMARY AND SIGNIFICANT CREDIT ISSUES** (Executive Summary that covers key points from the discussion)
   A. Operating Performance
   B. Strengths and Weaknesses of Financial Condition assessed in terms of cash flow
   C. Debt Serviceability
   D. Significant Factors which may affect the company's future performance and/or ability to service its debt obligations . . .

IV. **CONCLUSION**
   Determination of risk based on the company's financial strength and repayment ability as reviewed in your analysis and in consideration of the type of credit request.

FIGURE 10

V. **BACKGROUND**

(Information about the company's history and products.)

VI. **OPERATIONS**

Begin with a conclusive topic sentence pertaining to the performance of sales and of income. The first part of operations should give main reasons for changes in the income statement.

A. Sales/Revenues
B. Cost of Goods Sold
C. Changes in General, Selling and Administrative Expenses
D. Interest Expense
E. Operating Profit/Net Profit After Tax
F. Recap of Major Factors Leading the Net Income Changes
G. Appraisal of Operations

VII. **FINANCIAL POSITION**

The cash flow should be used to explain the liquidity and leverage sections. Each of these sections should begin with a conclusive topic sentence. If each of the sections is long, then a beginning summary of financial position could be a lead into this part of the analysis.

A. Liquidity (several detailed sections)
B. Leverage (several detailed sections)

VIII. **DEBT SERVICE**

Interpret results of the debt service analysis in view of the credit request . . .

IX. **COLLATERAL QUALITY**

X. **STRENGTH OF GUARANTEE**

XI. **INDUSTRY OUTLOOK**

XII. **APPENDICES**

(financial worksheets and calculations)

FIGURE 10   *(Continued)*

This format not only permits the senior loan committee to understand the significant credit issues at a glance but also helps the young analysts who write the reports. As participants in a management training program, they are inundated with new information about the banking industry and about credit issues. Deciding what information is important from the mass of data is difficult enough. They would be lost without a report format. In their training, their supervisor and I use a case-study approach to teach them how to analyze the data concurrently with how to write the reports. We work through the *process* of writing a credit analysis. Thus, when they write their first real reports, they have already practiced with a solid model rather than having to grope for their own.

The Internal Audit Report

The process of writing is also important in the following internal audit report format. The format itself improves upon the traditional pattern outlined in the *CIA Examination Review* cited earlier. But the manager of this audit department has also set up a system in which the auditors meet with key "clients" in the department being audited to go over the rough draft of the report. His goal is to eliminate the image of auditors as policemen who raid, looking for crime. Discussing the report at a draft stage reveals possible misunderstandings and permits revisions. Since the "clients" are one of the primary audiences of the report and have a major stake in what it says, the review step is the ultimate in reader sensitivity. Interestingly, this is the second environment where I have heard the statement, "Paper is our most important product." The other was the Research Center manager.

Figure 11 shows the improved format of these audit reports. With permission from the manager, I am reproducing several pages of an actual report. Notice the one-page, action-oriented summary, the second-page introduction and conclusion, the recommendation-first organization of each finding, and the delay of scope and objectives until the last page.

This report format meets the needs of all its readers: the executives who need to know the big picture, the departmental managers who need to improve methods of operation within their responsibility, individuals within the audited department who are responsible for only one of the findings, and the team of auditors who need a consistent report format that their supervisor and manager will approve. Even with the good model, these reports go through a thorough review process—including the preliminary review with the "client" department—before being issued.

CONFIDENTIAL
INTERNAL AUDIT REPORT

COMPANY	Petroleum, Inc.	AUDIT SUBJECT AND LOCATION	Petroleum, Inc. — Supply and Distribution and Wholesale Marketing
AUDIT NO.			
AUDIT DATE	August 26, 1986		City, State

SUMMARY AUDIT RESULTS:

Our audit of the Petroleum, Inc., Supply and Distribution (S&D) and Wholesale Marketing Departments indicated excellent compliance by both departments to internal and operational controls in the areas of exchange and spot contract administration and documentation, supply forecast and inventory control information systems and wholesale product pricing.
Our review, however, indicated the need for Management attention to:

Utilization of data base software to operate current department information systems more effectively,

Implementation of a refined products quality control program,

Expansion of retail sales forecast procedures, and

Establishment of freight verification procedures.

Mr. _____ and Mr. _____ agreed to the audit comments and recommendations. Mr. _____ indicated that the implementation of improvements relative to quality control, sales forecasting and freight verification procedures would depend on involvement by the Retail Division and the Accounting Department. Presently, the S&D Department has been assigned neither the specific responsibility nor the manpower to direct such activities.

FIGURE 11

Please read and reply as indicated **using the response format set out on the reverse side.** Forward your reply and other comments to: (Please provide copies of the response to each report recipient.)		**GENERAL AUDITOR**	
COPIES OR EXCERPTS TO:	**LOCATION**	**PARAGRAPHS**	**REPLY REQUIRED**
Name	City, State	All	No
Name	City, State	All	No
Name	City, State	All	No
Name	City, State	All	Yes
Name	City, State	All	No
Deloitte Haskins & Sells	**Tulsa, Oklahoma**	**All**	**No**
—Central Files	**Tulsa, Oklahoma**	**All**	**No**

▶ **YOUR COPY**	**APPROVED BY**	Audit Manager	**DATE**	August 26, 1986

TO: Audit Manager   DATE: August 26, 1986

FROM: Auditors

SUBJECT: Petroleum, Inc.   Reference: PE–00–00
Supply and Distribution
and Wholesale Marketing of
Finished Product

I.   *Audit Summary*

Introduction

An internal audit of the Supply and Distribution and Wholesale Marketing Departments of Petroleum, Inc., was performed during the period of June 9 through July 25, 1986. The audit was completed by _____ and _____ and included three weeks of fieldwork at the _____ Refinery.

The audit scope and objectives are summarized in Section II of this report.

Audit Issues: Control Practices and Procedures

The results of our audit indicated excellent compliance by the two departments, Supply and Distribution and Wholesale Marketing, in carrying out their functions in accordance with Management's directives.

During the first quarter of 1986, the efforts of both departments were plagued significantly by ESI software problems affecting the truck rack reporting system, unfavorable variances in exchange, wholesale, and retail demand, volatile market prices, and the performance of a refinery turnaround. Continuing attention, however, appears to be directed toward each of these areas as well as toward developing integrated strategies of inventory control and marketing.

FIGURE 11   *(Continued)*

Audit issues requiring Management's consideration and action noted during the audit were as follows:

Subject:	Supply, Distribution &	To: Audit Manager
	Wholesale Marketing	Prepared by: Auditors
	of Finished Product	Date: August 26, 1986
	Audit	
Ref:	PE–oo–00	
Page 2		

A. *The Supply and Distribution Department should evaluate the feasibility of networking department microcomputers or using Data Base software to operate current department information systems more effectively.*

During our audit, we reviewed the Supply and Distribution (S&D) Department's various information systems relating to forecasting, inventory control, and product scheduling. We noted several instances in which the same information is used repeatedly and input to different microcomputers.

Presently, these S&D systems are not integrated and an information data base does not exist. Consequently, all communication of information between staff members must be done manually. The data used at each microcomputer workstation must be input separately into each spreadsheet or report used by that system.

In the absence of an integrated system, the potential exists for the following:

- Input error
- Inconsistent information
- Inefficient time usage
- Inadequate response time needed to analyze changes in production or the marketplace

Consolidation of data across the functional areas of forecasting, inventory control, and scheduling should improve the quality of information in addition to streamlining information gathering procedures.

Discussions with Management

Since the audit, Mr. _____ has initiated discussions with a consultant currently contracted to establish data base systems for the _____ Transportation Department. Mr. _____ has also emphasized input quality control procedures to improve information generated by the current systems.

Subject:	Supply, Distribution &	To: Audit Manager
	Wholesale Marketing	Prepared by: Auditors
	of Finished Product	Date: August 26, 1986
	Audit	
Ref:	PE–00–00	
Page 4		

II. Audit Scope and Objectives

Our audit scope encompassed _____ supply, distribution and marketing activities during the twelve months preceding the audit, with particular emphasis on January through May 1986. Also included was a limited review of _____ International spot market transactions.

The primary objectives of the audit were to assess the reasonableness of:

- Supply forecast and inventory control information systems,
- Exchange and spot contract administration and documentation,

FIGURE 11 *(Continued)*

　　　　– Wholesale product pricing and marketing procedures, and
　　　　– Exchange accounting procedures.
III.　Audit Conclusion

A draft copy of the report was reviewed by Messrs. _____ and _____. General concurrence to the audit recommendations was expressed. To complete the audit process on a timely basis, a written response is requested from Mr. _____ by September 26, 1986.

We wish to express our appreciation to the staff members in both the Supply and Distribution and the Wholesale Marketing Departments as well as in _____ International for the excellent cooperation and assistance we received during our audit.

| _____ | _____ |
| Internal Audit Manager | Senior Auditor |

_____
Associate Auditor

FIGURE 11　*(Continued)*

The Procedure

The final example of a successful format is the offspring of a central policy and procedures department. It is based on Playscript, a technique that I encounter only occasionally. Most procedures I see follow the old block-paragraph format, with no introduction and the action buried in passive verbs in midsentence. However, this company is imposing a different format because it is in a unique position to do so. The company is growing rapidly. It is no longer an everybody-knows-everybody size. As a result of this growth—both in geographical distance and in corporate size—extensive policies and procedures are being written for the first time. (However, the same standardization does not extend to other company documents.) Although I have serious problems with the clarity of the headings, Figure 12 reveals a very useful format.

These examples show that change can occur in the file cabinet. The documents are more effective because they meet readers' needs to acquire information efficiently. Not one of them looks like an academic paper, nor do they follow the prescribed professional formats. But they are good because they work within the organization.

The Role of the Writing Consultant

One source of new models deserves comment: Some companies are beginning to turn to writing consultants for improved models. With the increasing recognition of the importance of moving information efficiently in an information society (almost a cliché these days), the demand for writing consultants is increasing.

Prepared by:

Date:

Subject:

Reviewed by:

Statement:	This procedure outlines the steps to be followed for a Policy Replacement unit.
Applicability:	This procedure applies to all sales office locations, Central Order Processing, Customer Administration, and Traffic.
Definition:	A Policy Replacement is the replacement of units, features, or parts at _____ 's option at no cost to the customer, when the replacement is not required by warranty or other contractual obligation.
Related Documents:	Procedures Manual Form SO33 — Field Equipment Deinstallation and Move Authorization

Procedure:	Responsibility	Step/Action
	Customer	1. Notifies Sales Office to request a change of equipment.
	Sales Office	2. Obtains necessary approvals from Central.
		3. Cuts a new Sales Order for requested equipment per OEP Procedures.
		4. Makes a note in the Special Comments section of the sales order, "Policy Replacement" and records the serial numbers of the units being replaced.
		5. Forwards sales order package to Central Order Processing. NOTE: The Sales Office DOES NOT release the sales order to Manufacturing.
	Central Order Processing	6. Processes sales order according to OEP Procedures.

FIGURE 12

But the demand usually takes the form of seminars for employees, not a communications audit or redesign of departmental formats. Janice Redish, herself a proto-parent of document design, echoes the resistance to change that I, too, see in companies.

Our initial projects with new clients are often conducted as exceptions to company standards. This requires a strong project leader on the client's side who will justify the exception in the first place and will then work with us to convince reviewers throughout the project. . . . Although we have written manuals that are exceptions to company or agency standards in several fields, the new reader-oriented style has caught on only in the computer field. Marketplace pressures operate here, but they do not operate in the field of military technical manuals, government policy manuals or employee benefit handbook. (Redish, 149)

Even when a consultant is hired to improve the communication system, the resistance to change is enormous.

Thus, the influence of consultants is limited, often because we work at the same local level that the problem-solving employees do. We generally work with individuals in seminars rather than with the corporate system. Top management may pay lip service to improvements in company writing, but they rarely take the time either to attend the training itself or to set up management briefings. Even in reviewing sample documents, we see only pieces of the whole communication system. We are hired to improve "skills" rather than to attack productivity issues. Finally, we have little influence within the organization because good writing is not quantifiable; it is not seen as a bottom-line issue. Thus, our influence in solving company writing problems remains personal and localized.

The File-Cabinet Style

So far my primary concern has been the sources of document formats that the file cabinet reproduces. However, the same forces establish the style of workplace documents. In fact, the resistance to change in style is even more powerful than to change in format. From one industry to another, in private and public sectors, I see conformity to the pedantic style: formal vocabulary, excessive nominalization, passive verbs, long sentences. The plain style is the refreshing exception rather than the rule. Thus, I was amazed to read, in *The Journal of Business Communication,* that "There is no doubt that the plain style of short words in short sentences is the 'house character' of the contemporary business community" (Mendelson, 15). That simply is not the case.

The resistance to the plain style does *not* come from the top. I know of no corporate executive who says, "Impress me with big words and long sentences." I know of many who complain about the complexity of the documents they must read. Yet employees often think that writing that goes up the authority ladder must be impressive—that is, formal and complex. No company style book (or composition textbook, for that matter) promotes a wordy, difficult style. Quotes from the Air Force Academy's Executive Writing Course are typical:

> Obscure, pretentious, wordy, indirect language obscures thought and fact. Use plain ordinary English. Be economical with words. Use active voice.
> —General William C. Moore, Jr.

> Be selective. Be concise. Don't tell someone what you know; tell them what they need to know, what it means, and why it matters.
> —General David C. Jones

> Informal writing is now the Air Force writing style.
> —AFP 1302, p. 34.

Most company directives echo these statements. They assert, as the research manager did, "Write in language that anyone, even managers, can understand."

But the pedantic style prevails. A few examples will suffice.

From the oil industry:

Gentlemen:
The necessary attention has been rendered the division order afforded this office affecting the captioned, and the same is returned herewith for your further handling. The extra copy of the form has been retained for our file completion.

From an attorney:

Therefore please find enclosed herewith a copy of our authorization to examine those accounts. It is requested that we be allowed to examine, from a period beginning January 1, 1981, through and including July 30, 1981, any and all accounting statements of account, record of transactions, etc., of the above referenced and numbered accounts. Further, it is requested that we be provided with a list of all accounts which have or are maintained at your institution, together with the account numbers thereof, for either _____ and _____ Corporation.

From a lobby sign in a New York bank:

All checks presented for encashment must be endorsed in the teller's presence.

Fortunately none of these examples comes from my clients. Seminar participants have passed them along for my amusement. But they are not much worse than some samples I see every day.

Stylistic Parents

Why, then, do people write this way? They *think* they are supposed to. Lee Odell says that the writer may be influenced by the culture of the organization, by internalized values, attitudes, knowledge, and ways of acting (Odell, 250). Indeed, companies often have a perceived "semantic environment" that maintains the formal, wordy style. For example, one middle manager's defense of his almost unintelligible prose was typical: "I know that the vocabulary muddies the meaning, but that's what my superiors want. I have to write that way." Although I doubt the accuracy of his assessment, his *perception* remains. I hear the same sort of comments from individuals in every field.

What are the sources of that perception?

1. *Individuals want to sound professional.* They fear that they will not be taken seriously if they use short words in short sentences. The younger or more insecure they are, the more likely they are to cling to the safety of the impressive pedantic style.
2. *The academic world rewards complexity.* Students learn early on that big words and long sentences produce longer papers and higher grades. J. Scott Armstrong, a Wharton School professor, tested his theory that written complexity increases academic acceptance. He gave easy or difficult versions of four passages to management professors to rate "the competence of the research that is being

reported." The professors were not told the name of the journal or the author. Consistently, the professors rated the easy versions lower than the more difficult ones (Horn, 12).

3. *The professional discourse community requires the style.* For example, the patent application claims cited earlier must be one sentence each, even if each sentence runs several paragraphs long. However, professional guidelines sometimes encourage the plain style, as does the *CIA Examination Review* for internal auditors. "The writing should be kept simple, letting nothing get in the way of the transfer of ideas from writer to reader. . . . Writing should be kept lively, using action words to command attention" (Gleim, 322). But the prevailing use of professional boilerplate convinces most accountants that professional standards require a ponderous style.

4. *The supervisory review process reinforces the older, more formal style.* Even if employees want to use a more direct, concise style, their supervisors change it. Sometimes, supervisors fear clarity; they cannot be held accountable if no one understands the document. More often, they insist on the formal style because it is "professional." After all, they have gotten to their current positions using it. Why change now?

Thus, in spite of widespread recognition that the plain style is more efficient and, therefore, increases productivity, the file cabinet continues to reproduce an inflated, verbose style. However, stylistic change is also possible—but not without management support. According to one technical writer, for at least eight years users had requested computer documentation that was easier to read. The style began to change only after the manager attended my writing seminar and accepted the idea that documentation could be clear *and* professional. Now the programmers who write the manuals use "you," active verbs, etc. The resulting "new" style is much more reader-sensitive:

> This being your first time to log on, you will have no files in your library. Type "list" and hit enter to see if you have any files (this is just for your information.) . . .
>
> Type in your log-on password.
>
> The screen will go blank for a few minutes. Then you will see a screen full of PF key selections . . .

In the past the programmers followed an established "computer style." A manager had to give them permission to change. Changes in style, as well as in format, often require a champion.

The Future: Selective Reproduction

Who is responsible for cleaning out the file cabinet?
Everyone.

Because teachers, the academic world, individuals in the workplace, supervisors and managers, and the corporate culture all contribute the models that thrive in the file cabinet, they must all cooperate to replace archaic models with superior descendants. Consultants, too, must accept the challenge.

The Challenge for Teachers

Teachers of composition should examine their academic heritage, course content, and teaching methods. Most of us chose to be English teachers because we loved literature and valued the liberal arts tradition. However, according to Freed and Broadhead, our own "culture" imposes blinders: "We taught and perhaps even believed [the sacred rhetoric text] because it was given, there before us, and there before our being there, though we rarely recognized its sacredness because it was a code without saying" (Freed and Broadhead, 163). Many of us teach the way we were taught. But we do not have to abandon these values to recognize the broader needs of our students.

Increasingly, voices in our profession are promoting change (Flower, 1981; Holcombe and Stein, 1981; Odell, 1985; Redish, 1985; Halpern, 1985; Hairston, 1986). These writers suggest that the goal of writing instruction should be broader than simply preparing students for academic writing. We should teach strategies for solving both academic *and* professional writing problems. Students should become the creative problem solvers who improve the content of the file cabinet.

How can we achieve that broader goal?

1. *Students should learn a variety of models instead of the limited models of the academic audition.* They should learn to use executive summaries, white spaces, visual cues such as headings and figures, the icons that are coming with the electronic revolution, etc.
2. *The classroom should duplicate the context of the workplace.* I do *not* mean that students should be writing letters and memos. But they should write to real audiences, for real purposes. They should learn techniques for collaborative writing. They should practice constructive critiquing techniques to prepare for the writing review process. They should pay real penalties for poor or sloppy writing. As David Lauerman and his colleagues report, such a change in classroom practices "made us alter some preconceptions: writing became practice, rather than a test; prose models (i.e., essays?) were replaced by samples; rules gave way to strategies. The classroom became a busier and noisier place" (Lauerman et al., 450).
3. *Writing assignments should test students' problem-solving skills.* Students are pragmatic; they value realistic assignments. Lauerman et al., Couture et al., and Anderson report considerable success with writing assignments that duplicate workplace problems. I concur. After I began consulting, the methods I used in teaching freshman composition changed significantly for the better. So did the results.

4. *Teachers need to understand the writing requirements of the workplace.* Experience is the best teacher, but many teachers have only academic experience. Some colleges are instituting writing internships for their writing majors. Perhaps future teachers of English—both secondary and college level—should have practice in the workplace as well as practice in teaching.

These are not new or startling recommendations. Most have been proposed elsewhere. But I must add my voice: We *must* change the traditional methods of teaching writing. They are failing our students.

The Challenge for the Academy

The academic world should examine its own models. The academic essay, term paper, and thesis enjoy a distinguished history. They have worked well in shaping students' ideas. But the old models assume a narrow result: the scholar graduate. How can we broaden (not replace) that limited goal?

1. *Writing requirements should change throughout the academic community.* Writing-across-the-curriculum programs have been an important beginning, but they have not gone far enough. Armed with new understanding of the writing students will do when they graduate, we should help other departments design writing assignments and contexts that duplicate the workplace environment. For example, M.B.A. theses and scientific lab reports that focus only on method rather than results should be placed on the endangered species list.

2. *The academic community has much to learn from the creative writers in the workplace about more effective ways to communicate ideas.* Perhaps the traditional scholarly paper needs to be examined. Recent changes in the MLA Style Sheet are a first step, but what about different uses of visual clues, the appearance of paragraphs, summary pages, and readable style? Just as parents can learn modern ideas from their mature children, so, too, the proto-parent can learn from its precocious offspring.

The Challenge for the Workplace

Companies and agencies should understand that clear writing is a bottom-line issue. Many companies worry about correct expression as a reflection of the corporate image. But their corporate culture does not promote effective formats or efficient style. They fail to recognize that productive information transfer is as important as a productive assembly line.

1. *Companies should examine the models in their file cabinets.* These models should face the same standards of productivity as do the manufactured or service product of the company: standards of efficient production, high quality, and cost-effectiveness. The resulting better models will lower communication barriers instead of creating them. How can we become a modern information society if we are operating with archaic equipment?

2. *Companies should also evaluate the supervisory review process, not to abolish it but to improve it.* If the real "teaching" of writing in the workplace occurs at the supervisory level, that teaching should focus on shared corporate goals and objectives, the standards mentioned above. Also, supervisors who review subordinates' writing should learn techniques for productive reviews. Reducing the length of the rewriting cycle and the frustration of all participants will increase employee efficiency.

The Challenge for Consultants

Writing consultants should change their self-image from trainers to true consultants. Since many of us come from an English department background, we carry a service-course mentality. We must recognize the value of our knowledge and, thus, "see" ourselves differently. A consultant is a change agent.

1. *Our role will change if we understand that writing is central, not peripheral, to a company's success.* We will still present writing seminars for employees. But we will also dare to approach top management about improving the communication system. This step requires the courage to push for significant change; it is the same courage required of the problem-solving employees who change the models in the file cabinet. Certainly, we better serve our clients if we apply all our expertise, not just our teaching skills.
2. *Consultants should become the pipeline of improved models.* We have the opportunity to draw on the exciting research being conducted in rhetorical theory and practice. We also have the rare advantage of access to the creative but localized models such as those shown in this article. But we need a mechanism for sharing these new models without violating our ethical responsibility of confidentiality to our clients. Last year, at a leadership conference of the Association of Professional Writing Consultants, we discussed publishing an annual digest of "good models" submitted by consultants around the country. Such a digest, when it becomes a reality, will allow access to new models for writing consultants and for their clients.
3. *A word of caution: Consultants should not become too comfortable within one discourse community.* In her CCCC presentation, Carol Lipsom cautioned, "Consultant beware." But her advice was that we need to listen to and attend to the local culture, to gauge the internal corporate politics in which employees refuse to change. Although I share her concern for self-preservation, I reject the notion that my job is simply to fit into the local system. In a sense, I'm in trouble in my consulting role if I am too comfortable with the company jargon, formats, and style. As change agents, we should be *improving* the old models, not simply nurturing the reproductive cycle.

The file cabinet will continue to reproduce. Employees—both novice and experienced—will turn to it for models of what has been done before. Our concern as teachers of writing, as members of the academic community, or as writing consultants in the workplace should be the quality of the models being reproduced. Will the old dinosaurs continue to thrive, or will creative evolution occur?

NOTES

1. Paradis et al. explain the problems created for a new employee who tried to repro-duce on paper the *process* of his problem solving; his manager and supervisor crit-icized his eight-page report because the recommendations were buried in the final section. But the employee wanted his reasoning to be reviewed; "his logic was as important to him as the recommendation itself." Because students learn in college writing assignments that the quality of the writing *effort* counts, they are left with no alternative model oriented to the organization's needs (300–302).
2. Lester Faigley, in defining "discourse communities," seems to agree with the sci-entists. He strongly implies that documents are used only within restricted pro-fessional or company contexts. "Texts are almost always written for persons in restricted groups" (238).
3. Robert Bataille's questionnaire results from Iowa State engineering and indus-trial administration alumni underlines the importance of audiences outside the writer's field of expertise. These alumni said they wrote 33 percent of their doc-uments to coworkers or superiors somewhat or more outside their field. After adding the documents they wrote to the public or to customers, they estimated that 54 percent of all their writing was directed to a lay audience (278).
4. Russell Rutter reports a similar format for presentence reports in Illinois. He states that some judges do not *want* evaluations and recommendations (291).
5. I have not seen a similar format in my other banking clients' files. In fact, credit analysts at a rival bank resisted adopting it because "that's not the format we use here." As far as I could tell, the rival *had* no established format. In contrast, the regional bank executive I quoted earlier liked the approach because it would solve his Thursday evening reading frustration.

WORKS CITED

Anderson, Paul V. "What Survey Research Tells Us About Writing at Work." *Writing in Nonaca-demic Settings.* Ed. Lee Odell and Dixie Goswami. New York: Guilford, 1985, pp. 3–83.

Bataille, Robert R. "Writing in the World of Work: What Our Graduates Report." *College Composition and Communication 33* (1982):276–280.

Colomb, Gregory G., and Joseph M. Williams. "Perceiving Structure in Professional Prose: A Multiply Determined Experience." *Writing in Nonacademic Settings.* Ed. Lee Odell and Dixie Goswami. New York: Guilford, 1985, pp. 87–128.

Couture, Barbara, Jone Rymer Goldstein, Elizabeth Quiroz. "Building a Professional Writing Program Through a University-Industry Collaborative." *Writing in Nonacademic Settings.* Ed. Lee Odell and Dixie Goswami. New York: Guilford, 1985, pp. 391–426.

Faigley, Lester. "Nonacademic Writing." *Writing in Nonacademic Settings.* Ed. Lee Odell and Dixie Goswami. New York: Guilford, 1985, pp. 231–248.

Flower, Linda. *Problem-Solving Strategies for Writing.* New York: Harcourt, Brace, Jovanovich, 1981.

Freed, Richard C., and Glenn J. Broadhead. "Discourse Communities, Sacred Texts, and Insti-tutional Norms." *College Composition and Communication 38* (1987): 154–165.

Gleim, Irvin N. *CIA Examination Review,* 2nd ed. Gainesville, Fla.: Accounting Publications, 1984.

Hairston, Maxine. "Different Products, Different Processes: A Theory About Writing." *College Composition and Communication 37* (1986): 442–452.

Halpern, Jeanne W. "An Electronic Odyssey." *Writing in Nonacademic Settings.* Ed. Lee Odell and Dixie Goswami. New York: Guilford, 1985, pp. 157–201.

Harwood, John T. "Freshman English Ten Years After: Writing in the World." *College Composition and Communication 33* (1982): 281–283.

Holcombe, Marya W., and Judith K. Stein. *Writing for Decision Makers: Memos and Reports with a Competitive Edge.* Belmont, Calif.: Lifetime Learning, 1981.

Horn, Jack C. "Bafflegab Pays." *Psychology Today* May 1980: 12.

Lauerman, David A., Melvin W. Schroeder, Kenneth Sroka, and E. Roger Stephenson. "Workplace and Classroom: Principles for Designing Writing Courses." *Writing in Nonacademic Settings.* Ed. Lee Odell and Dixie Goswami. New York: Guilford, 1985, pp. 427–450.

Lipsom, Carol. Conference on College Composition and Communication, Atlanta, Ga. March 1987.

Mathes, J. C., and Dwight W. Stevenson. "Completing the Bridge: Report Writing in 'Real Life' Engineering Courses." *Engineering Education* Nov. 1976: 154–158.

Mendelson, Michael. "Business Prose and The Nature of the Plain Style." *The Journal of Business Communication 24* (1987): 3–18.

Moore, Leslie E., and Linda H. Peterson. "Convention as Connection: Linking the Composition Course to the English and College Curriculum." *College Composition and Communication 37* (1986): 466–477.

Odell, Lee. "Beyond the Text: Relations Between Writing and Social Context." *Writing in Nonacademic Settings.* Ed. Lee Odell and Dixie Goswami. New York: Guilford, 1985, pp. 249–280.

Paradis, James, David Dobrin, and Richard Miller. "Writing at Exxon ITD: Notes on the Writing Environment of a Research and Development Organization." *Writing in Nonacademic Settings.* Ed. Lee Odell and Dixie Goswami. New York: Guilford, 1985. pp. 281–307.

Redish, Janice C., Robbin M. Battison, and Edward S. Gold. "Making Information Accessible to Readers." *Writing in Nonacademic Settings.* Ed. Lee Odell and Dixie Goswami. New York: Guilford, 1985, pp. 129–153.

Rutter, Russell. "Teaching Writing to Probation Officers: Problems, Methods, and Resources. *College Composition and Communication 33* (1982): 288–295.

U.S. Air Force Academy. *Executive Writing Course.* USAF Academy, Co., 79/0726.

John S. Fielden and Ronald E. Dulek

How to Use Bottom-Line Writing in Corporate Communications

When they wrote this article, John S. Fielden and Ronald Dulek were professors of management communications at the University of Alabama. Jointly, they also authored a series of books on effective business writing.

Every top executive complains about "wordy" memos and reports. From Eisenhower to Reagan, stories have circulated about their refusal to read any memo longer than one page. The CEO of one of the largest companies in the United States actually demands reports so short they can be typed on a three-by-five card. And J.P. Morgan is reputed to have refused audience to anyone who could not state his purpose on the back of a calling card.

"Don't be wordy!" "Be brief, brief, brief!" "Be succinct!" One writing expert after another exhorts business people with these slogans. And who will disagree? We do.

As a result of an in-depth study of the writing done at the division headquarters of a very large and successful company, we are absolutely convinced that advice such as "Be brief!" is not only useless, it does not even address itself to the real writing problem.

What causes trouble in corporate writing is not the length of communications (for most business letters, memos, and reports are short), but a lack of efficiency in the organizational pattern used in these communications. And, as you will see, it is for the most part a lack of organizational efficiency on the part of writers that is often deliberate, or, if not consciously deliberate, so deeply ingrained in their behavioral programming that it causes an irresistible impulse to beat around the bush.

Put simply, people organize messages backwards, putting their real purpose last. But people read frontwards and need to know the writer's purpose immediately. That purpose is what we eventually came to call the message's bottom line.

At the beginning of the study we wondered, why do people write backwards? One possibility is that they are writing histories of their mental processes as they think their way through a problem. Since their conclusion could only be arrived at after analysis, the report therefore would state its conclusions last. But if that were the reason, it would only be analytical memos and reports that would be organized backwards.

Such was not the case in the study we did. Almost *all* memos and reports put their purpose last. Why? We determined to study the problem to see if we could design a cure for such blatantly inefficient writing.

COMPREHENSION IS THE KEY

In the division headquarters we worked with, 9,000,000 (internal and external) messages of all types are distributed annually. Our study began with an intense analysis of a sample of 2,000 letters, memos, and reports randomly drawn from company files. The typical communication was one page. Only in rare cases did any memo or report exceed three pages. These various documents were well-written in the sense of being above average in terms of mechanical correctness and aptness of word choice.

Yet only one in twenty of these communications was organized efficiently. Below is one of the actual memos we analyzed (disguised, of course). Look at your watch before you read it. Keep a record of how many times you have to read it before you really *comprehend* its message, and how long it takes to understand the report's purpose.

Memo A

The Facilities people have been working on consolidating HQ marketing Functions into the new building at Pebble Brook. As presently envisioned, Marketing Research will remain in its current location but be provided with additional space for expansion. The following functions will be moved into the new facility—Business Analysis, Special Applications, and Market Planning. It is expected that Public Sector will be relocated in a satellite location. The above moves will consolidate all of Marketing into the Pebble Brook location with the exception noted above.

Attached is a preliminary outline of the new building by floor and whom it will house. I am interested in knowing if this approach is in agreement with your thoughts.

This memo seems brief on the surface. It contains only 115 words. But let's measure brevity not by words but by the length of time it takes a reader to comprehend a message.

We feel that if you were really the addressee, you would have had to read this memo twice. Why? Because you didn't know *why* you were reading it until the last paragraph. Once you discover the purpose—that you are being asked to approve a plan—you want to reread the memo to see if you do, in fact, agree with the moves. The memo suddenly (and, unfortunately, at its end) informed you that you were on the hook. Obviously, there is a big difference between the way you will read a memo containing information of general (and casual) interest and one which requires you to make a decision involving the physical moving and reshuffling of hundreds of powerful and sensitive people. Yet organizationally this memo as presented does not show even a foggy awareness of this difference.

Now read the revision. Notice how your comprehension time would have dropped significantly had you received this revision instead of the original.

Memo B

Attached is a preliminary outline—by floor—of the new building at Pebble Brook and a statement of whom it will house. I am interested in knowing if this approach is in agreement with your thinking.

Our suggestion is that we make the following changes:

1. Business Analysis, Special Applications, and Market Planning will. . . .
2. Public Sector will be. . . .
3. Marketing Research will remain. . . .

In terms of comprehension, the revision lets you know right away that you are expected to make a decision. Whether or not you would mull over that decision, we cannot tell. But we do know this: in terms of actual time expended in comprehending what is being asked of the reader, Memo B can be comprehended in one-third the time required by Memo A. And, if we measure brevity in terms of comprehension time, rather than number of words, Memo B produces a 66 percent savings in comprehension time.

We learned immediately in our study that comprehension time drops dramatically when a memo states its purpose—why it is being written for the reader—at the very beginning.

Confirm this point by reading Memo C, another disguised memo drawn from the division headquarters. Time your comprehension as before.

Memo C

The first of a series of meetings of the Strategic Marketing planning group will be held on Thursday, September 7, from 1 to 4 P.M. in Conference Room C. These important meetings are for the purpose of monitoring and suggesting changes in overall market strategies and product support. Attached is a list of those managers who should attend on a regular basis. These managers should specifically be prepared to review alternative strategies for the new product line. The purpose of this reminder is to ask your help in encouraging attendance and direct participation by your representatives. Please have them contact Frank Persons for any further information and to confirm their attendance.

Now read Memo D and compare comprehension time once again.

Memo D

Please encourage those of your managers whose names are listed on the attachment to attend regularly and directly participate in the meetings of the Strategic Marketing Planning group.

The next meeting is to be held on Thursday, September 7, from 1 to 4 P.M. in Conference Room C.

Please have your representative(s):

1. Contact Frank Persons for any further information and to confirm attendance.
2. Be prepared specifically to review alternative strategies for the new F-62 line.
3. Be ready to discuss changes in overall market strategies and product support.

Memo D is obviously more efficient. Why? Not only because it begins by stating its purpose but also because it itemizes the actions requested of the reader, organizing them in an easy-to-digest checklist. The original gives extensive background about the meetings but buries the requested action in a fat paragraph. Most readers would have to read the original at least twice just to be able to ferret out exactly what is being asked of them.

The time being saved, of course, seems insignificant on communications as short as Memos A and C. But consider how significant the savings would be corporate-wide if every manager's comprehension time in reading all messages could be reduced by even a small percentage.

HIGH COST OF COMPREHENSION

A recent study done by International Data Corporation states that managers in information industries spend an astonishing 60 percent of their time reading and writing; professionals spend 50 percent.[1] If cost accounted, how much would, say, a 20 percent to 30 percent savings in reading and writing time amount to for companies in this industry alone? The possibilities are arresting.

We made some cost estimates for the communications undertaken by the division headquarters we studied. The 9,000,000 messages distributed annually by the division headquarters included, of course, all sorts of mailings and multiple copies of such things as new product announcements, price changes, and the like, often running into the thousands of copies. Therefore, it was not fair to assume all 9,000,000 messages mailed were individually composed. Instead, we determined through conservative estimates that 12 percent of the 9,000,000 mailings were individual communications. And for each of these we will assume, for the purpose of this article, the ridiculously low figure of $10 to be the cost of creating, typing, and distributing. Based on these estimates, the minimum total composition cost for this one divisional headquarters would be $10,800,000 a year (see Table 1).

---

[1] *Automated Business Communications: The Management Workstation.* (Framingham, Mass.: International Data Corporation, 1981): 21.

But writing time is, of course, only part of the story. What about the cost of comprehending all 9,000,000 of these messages? For while we estimate that only 12 percent of these messages were individually composed (that is, not copies), all 9,000,000 messages were presumably intended to be read. Again, in an attempt to dramatize through understatement, we will use a low salary figure: $20,000 per year. You can, of course, substitute the actual salary and other figures for your own company and determine for yourself at least roughly the magnitude of your company's reading costs.

As Table 2 shows, we approximated the division's minimum reading costs to be over $4,500,000. Of course, this figure is bound to be far below actual costs. Not only are our salary estimates unrealistically low, but we haven't taken into account the fact that many of the documents were read by multiple readers. In fact, many memos urged recipients to pass information on to colleagues and subordinates.

USING DIRECT PATTERNS

While these dollar figures were somewhat astonishing and the possibilities of dollar savings enticing, the company under study evidenced the greater concern about the waste of productivity involved in such inefficient communications having become the norm. The specter of hard-working employees' time being wasted by an inundation of inefficiently organized memos and letters was distressing. The company asked us to teach people how to report in a "bottom-line" fashion. Therefore, we taught people to tell readers immediately what was their

TABLE 1   Estimated Division Headquarters Writing Costs

Number of messages sent annually	9,000,000
Percent individually composed	12%
Total individually composed	1,800,000
Composition cost (per message)	$10
Minimum total composition cost	$10,800,000

TABLE 2   Estimated Division Headquarters Reading Costs

Low-median salary	$20,000
	$10/hour
	17¢/minute
Average* reading time	3 minutes
Reading cost per document	$0.51
Reading cost for 9,000,000 messages	$4,590,000

*Assumed that some messages read in depth; some not given more than a glance; some barely looked at.

purpose in writing and what they expected of the reader, if anything. If people had no purpose in writing, they probably shouldn't write in the first place. If they didn't expect anything of the reader and were just offering possibly useful information, we told them to say so right away.

In short, we were teaching people to use a direct organizational pattern. We were urging them to eschew the circuitous pattern in which writers, because of some sensitively (real or imagined), withhold their purpose and do not let their readers know why they are being written to and what is being asked of them until their minds have been conditioned to accept the points the writers are trying to get across.

Obviously, there is nothing wrong with a circuitous organizational pattern in certain circumstances. But in this company, and we suspect in many other companies across the country, the circuitous pattern has become the norm for all types of communication in all situations.

Just ask yourself: how sensible is it to always write backwards, in a way that is just the opposite of how people comprehend information? Look at one more illustrative letter from our study:

Memo E
This is in reference to the letter sent you by Joe Smith of ABC Materials, Inc.

Mr. Smith requested information available from Product Analysis Reports (PAR's). As soon as information was made available to me from this source, I orally relayed the response to Mr. Smith.

Making use of the Planning Application Model, I was able to respond to Mr. Smith's request for further information about potential new products of possible interest to ABC. It was not until I received the copy of Mr. Smith's letter that I was aware that the data provided for him was not sufficient.

I have used all the resources that I am aware of to resolve Mr. Smith's concerns. Mr. Smith has informed me he is more than satisfied with the work done and considers the project completed. He has also announced an intention of doing further business with us.

Attached are copies of all the requests that I've been asked to submit during the six months that I have been assigned to this account. Also attached is a copy of an ABC analysis, submitted by my predecessor, which related to one of the items referenced in his letter. Upon request, I will forward copies of all of the relevant analysis that are in my files.

Where's the bottom line? What's this writer trying to get across? Isn't it the following?

Memo F
I have reviewed and acted upon the letter sent to you by Joe Smith of ABC Materials. Mr. Smith has informed me he is more than satisfied with the work done and considers the project completed. He has also announced an intention of doing further business with us.

Here in some detail are the steps I have taken for Mr. Smith. . . .

Since what this writer is reporting is good news, the communication is not sensitive. There's no need to report this information as circuitously as we might well be tempted to do if we had bungled the situation with Mr. Smith and had lost his business.

What percentage of all communications would you estimate to fall into the sensitive category that may call for a circuitous organizational pattern? Ten percent is the outer limit of possibility, unless one has a specialized job such as handling complaints, writing sales letters, dealing with shareholders, or the like. Why, then, upon analyzing these 2,000 sample letters, memos, and reports from this corporate division, did we find that almost all documents were organized circuitously? Why, in this extremely well-managed and successful company, was it the exceedingly rare letter or memo that did not bury its purpose somewhere in the third or fourth paragraph, and most frequently in the last sentence?

At the time, we had no idea of the etiology of the disease, but we felt we had a simple cure. We would tell people how to organize their thoughts so that the bottom line of their message would be immediately highlighted and promptly presented in everything they wrote. To facilitate this goal, we invented a series of bottom-line reporting principles which, if followed, would enable writers to communicate in a direct, straightforward, no-nonsense fashion in all situations that were not fraught with sensitivity. This would, we thought, save writing time and expense (see Table 3).

And readers, too, would benefit—instead of having to search through a memo to find out what purpose the writer had in writing to them and what the writer wanted or expected them to do, they could look at the first paragraph and see the answer to these vital questions. Moreover, if the purpose and topic seemed irrelevant to their interests or needs, they could reject reading it, or merely give it a glance. Significant reading time and dollar savings should certainly result.

The program was instituted throughout the division. Did it work? Yes, in terms of getting the principles across and in terms of getting intellectual acceptance of these principles.

TABLE 3   Principles of Bottom-Line Reporting

Principle 1:	State your purpose first unless there are overriding reasons for not doing so.
Principle 2:	State your purpose first, even if you believe your readers need a briefing before they can fully understand the purpose of your communication.
Principle 3:	Present information in order of its importance to the reader.
Principle 4:	Put information of dubious utility or questionable importance to the reader into an appendix or attachment.
Principle 5:	In persuasive situations, where you do not know how your reader will react to what you ask for, state your request at the start in all cases except:
	a. Those where you don't (or barely) know the reader, and to ask something immediately of a relative (or absolute) stranger would probably be perceived as being "pushy."
	b. Those where the relationship between you and your reader is not close or warm.
Principle 6:	Think twice before being direct in negative messages upward.

But getting emotional commitment to these principles was quite another story. We sensed in discussions that writers' commitment to being circuitous was not merely a bad habit. It was something else, something that was so ingrained that forcing personnel to be direct actually caused disquiet in many people. What could have been the reason?

PROGRAMMING FOR INEFFICIENCY

Obviously, people who work in large organizations were not born there. They have come to those organizations programmed by their social upbringing and by their educational experiences. And both of these earlier programmings strongly contribute to resistance to bottom-line reporting.

Social Upbringing

People seldom are conscious of how their social upbringing programs them to be indirect. Yet almost every sensitive social situation reinforces the wisdom of being circuitous, of not being direct. Aren't most brief answers to sensitive questions regarded as brusqueness or curtness, as being short with someone?

It begins early. The children are asked, "Do you want to go to Aunt Alice's house?" The children answer, honestly and directly, "No!" Unacceptable! The children are scolded and soon learn to beat around the bush the next time, all the while searching for some plausible excuse to forestall the visit. It is not surprising that as adults, the same children, when asked by the boss, "What do you think of my new plan? Think it'll work?", think twice before responding, "No!"

Educational Programming

Having been thoroughly programmed by their families that being direct is being impolite, the children now go to school. Here they soon learn that a twenty-page term report gets a high grade; a two-page report gets a low grade. A five-page answer to a test question is good; a one-paragraph answer is bad. Regardless of what teachers may profess, they invariably give extra credit for "effort." And effort is most easily measured by numbers of words or by pounds of pages. A premium is placed on long-windedness, and long-windedness is achieved by being circuitous rather than direct.

Indoctrination into Anxiety

On their first jobs in a large organization, young people are naturally nervous. They are very concerned that whatever they write or say not make people upset. They are also very concerned about "getting good grades." Therefore, they fall back upon the same behavior that was rewarded in school. They are going to do everything possible not to look lazy. They are going to be thorough in everything they write. Every chance to write a report to a superior is a chance to write that blockbuster of a term paper that could not fail to impress the boss. They are going to get that "A."

The fact that young people enter organizations at the bottom provides a final step in their programming for being circuitous and indirect. Everybody knows that writing *up* in an organization is far different from writing *down*. When young people enter an organization, the only direction they *can* write is up. Therefore, all the early experiences received in corporations consist of writing situations where they have to write information to people who are in fact, or may someday be, their superiors. Naturally, they become very uneasy.

Now let's suppose the company institutes a program to encourage personnel to be more direct. Imagine yourself as that newly hired young person in the organization. Are you going to believe any program suggesting that you be blunt and direct in your upward or lateral communications, when your entire lifetime programming has proved to you over and over again that bluntness is all too frequently suicidal? No chance!

Young people may give lip service to such a program, but in any real-life situation in which they feel threatened (in actuality, almost all situations) they will avoid coming to the point with an almost religious passion. And in negative or sensitive situations, the last thing they are going to do is state their purposes and requests directly.

By contrast, their higher level superiors, having enjoyed years of power positions in the hierarchy (from whence they could write down to anyone in any fashion they pleased), take a far different view of writing. The superiors now pride themselves on directness and bewail long-windedness on the part of their subordinates. But the subordinates' desires for self-preservation (reinforced by all their preorganizational programming) force them to give lip service at best to corporate attempts to "get to the point."

And, let's face it, in the corporate pyramid almost everybody is somebody's subordinate and, perhaps because of files, one never knows who is going to read what has been written. Therefore, circuitous writing is partly the habit of a lifetime and partly CYA.

WHAT YOUR ORGANIZATION CAN DO

Is a cure then impossible? Is inefficient, circuitous writing simply to be endured and its costs in lost productivity and wasted dollars merely written off?

No; a cure for inefficient writing is possible. But a thorough organization-wide cure requires that:

- People recognize and reject their social and educational programming for being circuitous in all non-sensitive writing situations. This deprogramming is the responsibility of the individual.

- People learn to write efficiently; that is, learn to organize their messages in such a way as to make it easy (and fast) for readers to comprehend the message. Teaching the bottom-line principles will impart this skill. But implementing the bottom-line principles requires strong high-level management support.
- People must develop the self-confidence necessary to send bottom-line messages upward in nonsensitive (or slightly sensitive) messages. A long-range cure depends to a great extent on attitude, on reducing the tensions inherent in superior-subordinate communications. Most writing insecurities stem from real, not imagined, failures on the part of superiors to communicate clearly and unequivocally their willingness to accept bottom-lined messages from subordinates.

Higher level executives have to appreciate how threatening directness can be to subordinates. Superiors, therefore, need to be persuaded not only to have the following credo taped to the wall above their desks but also communicated to all subordinates with whom they relate:

The Superior's Credo

1. I will ask all subordinates to be direct in their messages to me and I will not become angry if subordinates do so politely—even when those thoughts run counter to mine.
2. I will recognize and appreciate subordinates' attempts to conserve my time (and other readers' time) in all memos and reports they write to me, or for my signature.
3. I will work out with subordinates some general understanding of how much detail I require in various circumstances.
4. I will make clear to subordinates that I judge their communications not by length and weight, but by directness and succinctness.
5. And when I myself report up in the organization, I will be as direct as I expect my subordinates to be.

Once senior executives have adopted and put this credo into practice, all subordinates should recognize that there is now no excuse for them not to live by the following credo:

The Subordinate's Credo

1. I will have the courage (in all but the most sensitive or negative situations) to state at the beginning of messages my purpose in writing, exactly what information I am trying to convey, and/or precisely what action(s) I want my reader to take (if possible, itemized in checklist form for easy comprehension).
2. My readers, especially if they are my superiors, are extremely busy. I must not waste their time by making them read unnecessary undigested detail any more than I would waste their time chattering on in a face-to-face interview.
3. I will make a judgment as to how much my readers need to know in order to take the action required by the communication.
4. If I am in doubt as to whether specific information is necessary to my readers, I will either put this information in summary form in attachments, or tell readers that I stand ready to offer more information if so requested.
5. I will avoid the arsenal of the con man. If I want something of a superior, I will ask for it forthrightly.

M. Jimmie Killingsworth

E-Mail: A New Medium for Technical Correspondence

A member of the faculty at Texas A&M, M. Jimmie Killingsworth has published widely on business and technical communication.

To understand fully the opportunities and constraints that the emerging media have introduced, it's a good idea to think about how electronic communication relates to older modes of communication such as personal conversations, telephone calls, and written correspondence. Consider the relative advantages of each mode.

During a personal conversation you can watch your correspondent's eyes and bodily movements, checking for signs of attention, understanding, and emotional responses. You can use gestures and alter your pitch and tone for emphasis. You can assume a mutual understanding of the background of the conversation and can condense your speech accordingly. If the correspondent doesn't understand, he or she can ask a question. You are right there to answer it.

What do you lose in a telephone call? Instead of having a physical being before you, you have only a voice transmitted electronically, so you miss gestures and body language. Still, you can convey information to your partner quickly in a condensed form because he or she can always stop you to ask a question. Again, you can vary your tone of voice and listen for oral cues from your partner in conversation. Moreover, you can increase your efficiency. You can sit at your desk and have conversations with hundreds of people without having to go out and meet each one.

With a letter, not only do you lose the correspondent's physical presence; you also lose the feedback provided by an immediate voice response. Now you must calculate more closely the effects of your words and information. You must analyze the knowledge that your audience is likely to possess, cover your topic's background extensively (or at least remind the reader of previous contacts with the information), and provide full and clearly stated reasons for all of your interpretations.

In other words, shifting to the written mode requires that you provide more *context* for the information you present. You imagine the kinds of questions your readers might ask and the uses to which they will put the information. Then you write in a way that anticipates their questions. The result is that you take longer to explain any point than you ever would in a conversation.

For example, in a face-to-face conversation with a colleague I might say, "I disagree. We need to keep the funding for labs and travel separate. You remember what happened to X Division, don't you?" If my correspondent still looks doubtful, I might add, "And the situations are closer than you may think!" On the telephone, the conversation would run about the same, except that I would have to interpret a nervous silence on the other end of the line rather than a doubtful expression on the face. But in a letter I would probably have to spell everything out much more fully. My explanation might go something like this:

> I'll have to disagree with the notion that we can combine research and travel funds, which you suggested in last Monday's meeting. Even though research often depends upon travel, the two categories should remain separate because the legislature funds them separately. To mix the two types of money would border too closely on misappropriation of funds. You may remember the charges against X Division, which arose under similar circumstances. They, too, wanted to mix funds from legislatively mandated categories (in their case, clerical salaries and equipment funds).

If letters require so much more time and energy, why do we still write them? Because written text continues to offer a number of advantages that are missing in telephone calls:

- *A letter allows you to state your full position on a topic without interruption.* This process can help you clarify your thoughts. It also offers your correspondent a chance to study your position carefully and then frame a response.
- *A letter creates a written record,* a "paper trail" to which you can refer later and that increases the legal power of the exchange. In some circles—in the international diplomatic corps, for example, or in organizations concerned with trademarks and patents—letters may even have the force of a contract. You can tape your conversations, of course, in an effort to achieve the same effects, but tape recording involves a number of legal and ethical uncertainties and usually must be accompanied by some written record (signed permissions or transcripts) before it has the same effect as written correspondence. A letter signifies a seriousness of purpose that is missing even in "on the record" conversations. A letter "puts it in writing."

Electronic mail, or *e-mail*—the transmission of messages between correspondents on a computer network—offers most of the advantages of the written medium while recovering some of the advantages of the phone call and the face-to-face conversation. It provides a written record and allows for reflective reading and writing. It also permits a speed of transmission that puts regular mail to shame—even expensive overnight mail.

Whether you are transmitting internal "memos" within your company using a local area network or sending "letters" halfway around the world via networks such as the Internet,[1] you can transmit in a matter of seconds detailed messages or even whole documents uploaded onto the network. Your correspondent can download and print the document. Then, almost instantaneously, your reader can send you a response, keeping the printed document or the e-mail file for a record of the transaction. This form of communication virtually amounts to a *written conversation*.

On e-mail, my hypothetical conversation with my colleague might begin pretty much as it did in my memo, perhaps with a bit more informality, since most people do not see e-mail as having the formal power that a memo on official letterhead has:

> Your comment last Monday about combining research and travel funds worries me. Research and travel are related, true enough, but the two categories should remain separate because the legislature funds them separately. Could mixing the categories be interpreted as misappropriation of funds? Remember the charges against X Division, which arose under similar circumstances? They, too, wanted to mix funds from legislatively mandated categories (in their case, clerical salaries and equipment funds).

My colleague might then respond:

> I understand your worries about mixing categories, but I don't think we have to be concerned about the plan I'm proposing. Where X Div. went wrong was messing with salaries. That category is sacrosanct to the state auditors. But there are precedents for mixing research and travel. Should I fax you the notes I have from the last audit?

I could come back, including a little of the context because I know that while he can read the response almost instantanesouly, he may not have the opportunity to read and respond right away:

> Yes, I'd like to have a look at those notes on the last audit. They could be useful in the future. But what you've told me about the mixing of categories satisfies me that you're doing the right thing now.

Given the advantages of e-mail—the way it offers the advantages of writing letters and telephoning while minimizing the disadvantages of both—you may wonder why anyone would use the other media at all any more. But e-mail still has some problems, such as the following:

* *Access is not universal.* Your correspondent may not be connected to a network compatible with yours. You will feel seriously frustrated if your screen displays "message

---

[1]On the variety of networks an bulletin boards and their respective histories, functions, clienteles, and advantages and disadvantages, see Howard Rheingold, *The Virtual Community: Homesteading on the Electronic Frontier* (Reading, Mass.: Addison-Wesley 1993).

GUIDELINES AT A GLANCE
Using e-mail for technical correspondence

1. Begin your e-mail correspondence with a fairly formal, detailed delivery—as you would for a letter or a position paper. Your first transmission should supply context for your reader and present a coherent set of points for discussion.
2. The correspondence may then evolve into something more like a conversation, thanks to the rapid transmission of responses and feedback. Your correspondent may ask questions or provide a quick counterpoint, so something like a dialogue begins. Or your correspondent may reflect for a while before replying, and then send a longer, more carefully considered response, which demands the same from you. The interchange will then become more like an exchange of letters. The beauty of e-mail networks is that they provide the flexibility to support both kinds of communication.
3. E-mail may also evolve from a correspondence into a conference. When you share e-mail from one correspondent with other correspondents, be sure to secure permission from the original correspondent and to provide the necessary context for "over-the-shoulder" readers. As in any good technical communication, give shape and meaning to your transmissions, interpreting and reducing the data to clear action-oriented information. Don't just dump files over the network, hoping that someone will get something out of your mass of data. Develop the message so that it accommodates the needs of the people who are operating in the context of use.

returned" or "message undeliverable" after you've spent a long time carefully crafting a message, which you now must download and reformat to print as a letter.

- *The text editors available on many systems are poor and require a great deal of work to produce a clean text.* Because some editing software does not have the power of most simple word processors, users are tempted just to let the errors go, much as they would in fast conversation. But such a course of action is risky in a world where impressions count—in government, business, and industry. Severe miscommunication can also result from carelessness. So it is best to struggle with the bad text editor and send a clean, clear message.
- *Your correspondents may have divergent attitudes and expectations about e-mail.* Because it is a new medium, users are still a bit unsure about how much they can depend upon their correspondents to read e-mail messages and give them the same consideration they give to written correspondence. After all, conventions for writing and reading letters have evolved over several centuries, whereas e-mail has come into wide use only in the last decade or two. Some people treat e-mail like a friendly, informal conversation; others think of it as being more like a letter.
- *Some users clog transmission lines by using e-mail for publishing, creating a genre now widely known as "junk e-mail."* A message sent to more than a few correspondents may read like a form letter. Most readers expect e-mail, like letters, to be crafted for their particular needs. They expect e-mail to be addressed personally to them,

TABLE I Advantages and Disadvantages of Media for Technical Correspondence

Medium	Advantages	Disadvantages
Face-to-face conversation	Partners can use body language, gestures, and variations in vocal delivery; feedback is immediate; they can condense speech and background information.	It leaves no written record, may be treated less seriously than written correspondence, and is subject to inaccuracies. It is also obviously limited by considerations of space and time.
Telephone conversation	The advantages are similar to those of face-to-face conversation except for absence of body language and gestures; but telephone is even faster and easier to enact on many occasions and across great distances.	It leaves no written record, is subject to inaccuracies, and may be taken even less seriously than face-to-face conversations; it is also subject to malfunctions of technology.
Written letters or memos	Written correspondence offers an opportunity for extended reasoning on a position, allows for reflective writing and reading, and leaves a written record of the transaction.	These forms lack the personal warmth and informality of conversations; they require more time and effort for composition and transmission; feedback is slow; and they are subject to malfunctions of technology (word processors, mail delivery, etc.).
E-mail	E-mail allows for full presentation of positions and reflective reading and writing; it provides a written record (or computer file) of the transaction; it permits fast feedback and response.	It can be difficult to edit, difficult to transmit to certain correspondence, and subject to malfunctions of technology. Also, there's some uncertainty about how correspondents will respond; conventions of composition and response are still evolving.

not distributed by means of a long general mail list on which their address appears, perhaps without their knowledge or permission.[2]

Table 1 summarizes some of the advantages and disadvantages of e-mail compared to the other media for correspondence.

Some of the problems of e-mail could be prevented if authors treated the medium as a serious means of correspondence worthy of all the care and attention they give their regear letters. With this goal in mind, you may find it helpful to think of e-mail as an evolving process with a still fluid, but increasingly clear, set of conventions, which are summarized in Guidelines at a Glance.

---

[2]The "virtual journal," an innovative attempt to use e-mail networks as an avenue for serious publication, represents a possible exception to this critique. In this instance, subscribers are actually solicited and not merely placed on a mail list without their permission.

Part 4

Reports and Other Longer Documents

Nearly everyone in business and industry writes reports. The word *report* is really just a generic term for a variety of documents that vary in form and purpose. Some reports are purely informative; others are persuasive or argumentative. Reports can be simple check lists, or they can take the form of interoffice memos, letters to clients, or more full-blown documents that are the results of weeks, if not months, of effort.

Business and technical writing practice sometimes distinguishes between formal and informal reports. Formal reports generally follow a multi-part format and are used primarily to present the results of a detailed project. Such a project often involves a considerable outlay of capital plus time and effort. The format for a formal report may mandate a cover letter or a memo of transmittal attached to a bound document consisting of an abstract, a table of contents, a glossary, an introduction, a detailed discussion of all aspects of the topic, a set of conclusions and recommendations, and pages and pages of attachments.

Informal reports tend to be shorter documents. Their formats are less complex, consisting only of such essential items as an introduction, a discussion, a set of conclusions, and, where appropriate, a list of recommendations.

To ensure that their reports are useful, writers should take the same kind of process approach that they would use when writing any other business or technical document. They should plan their reports carefully from the start so that they can clearly define and stick to their intended purpose whether that purpose be to provide information, to analyze information and draw conclusions, or to make recommendations.

Because they are often action-oriented, reports require writers to analyze their audience—or audiences—carefully. By virtue of their length, some reports can intimidate readers who need to know the information they contain.

A classification of the kinds of audiences that business and technical writing address, which Thomas Pearsall first delineated in his *Audience Analysis for Technical Writing* (1969), is especially relevant to the audiences for reports. Those audiences can include any, or all, of the following:

- the layperson
- the executive
- the expert
- the technician
- the operator.

Each brings a different background and a different set of needs to his or her reading of a report that writers must take into account if they hope to produce an effective document. In general, using abstracts and visual aids to supplement the contents of both formal and informal reports will make those reports more accessible to larger groups of readers of varying degrees of expertise.

Proposals are in some ways simply specialized reports, although they can be written in letter or memo form as well. The primary purpose of a proposal is to persuade readers to do something. Sales letters and requests for adjustments are two fairly simple proposals. Internal proposals, which are aimed at changing policies and procedures with an organization, and external proposals, such as documents seeking grants or funding, are more complex persuasive documents. Because proposals aim at convincing an audience to act in a way the writer wants, a process approach can help lead writers to a successful proposal.

This section of *Strategies* begins with an essay by J. C. Mathes and Dwight W. Stevenson that returns to a familiar theme in this anthology: the importance of audience analysis. Mathes and Stevenson offer a detailed examination of the problem of audience analysis and a solution to that problem designed to meet the needs of the several audiences that reports in particular, and other business and technical documents in general, address.

Underscoring the need for attention to audience, Robert W. Dodge reports on the results of a survey done at Westinghouse in the 1960s to determine the reading habits of busy managers. Those results still have relevance for business and technical writers today. Dodge concludes that, while the body of the report may have required the most time for the writer to produce, busy managers read this section of the report with considerably less frequency than they read the other sections, usually giving more attention to the abstract. Christian K. Arnold further underscores the importance of the abstract in meeting audience needs in his discussion of the key elements of an effective abstract. Vincent Vinci then offers what amounts to a checklist that business and technical report writers can use as a last step in an effective writing process.

Supposedly a picture is worth a thousand words. In creating and using illustrations to supplement written text, business and technical writers need to make sure that those thousand words are the ones they intended. Walter E. Oliu, Charles T. Brusaw, and Gerald J. Alred offer easy-to-follow instructions for the effective use of the most widely used visual aids. It is, however, possible to misuse visual aids, intentionally or unintentionally. In a now classic piece, Darrell Huff suggests ways writers can misuse visual aids and statistics to manipulate the truth.

David W. Ewing then discusses persuasive writing, addressing the wider issue of the different strategies needed to persuade rather than inform readers. Ewing's general comments are then complemented by those of Philip C. Kolin about how to write effective proposals.

J. C. Mathes and Dwight W. Stevenson

Audience Analysis: The Problem and a Solution

Both J. C. Mathes and Dwight W. Stevenson are professors of humanities in the College of Engineering at the University of Michigan.

Every communication situation involves three fundamental components: a writer, a message, and an audience. However, many report writers treat the communication situation as if there were only two components: a writer and his message. Writers often ignore their readers because writers are preoccupied with their own problems and with the subject matter of the communication. The consequence is a poorly designed, ineffective report.

As an example, a student related to the class her first communication experience on a design project during summer employment with an automobile company. After she had been working on her assignment for a few weeks, her supervisor asked her to jot him a memo explaining what she was doing. Not wanting to take much time away from her work and not thinking the report very important, she gave him a handwritten memo and continued her technical activities. Soon after, the department manager inquired on the progress of the project. The supervisor immediately responded that he had just had a progress report, and thereupon forwarded the engineer's brief memo. Needless to say, the engineer felt embarrassed when her undeveloped and inadequately explained memo became an official report to the organization. The engineer thought her memo was written just to her supervisor, who was quite familiar with her assignment. Due to her lack of experience with organizational behavior, she made several false assumptions about her report audience, and therefore about her report's purpose.

The inexperienced report writer often fails to design his report effectively because he makes several false assumptions about the report writing situation. If the writer would stop to analyze the audience component, he would realize that:

1. It is false to assume that the person addressed is the audience.
2. It is false to assume that the audience is a group of specialists in the field.
3. It is false to assume that the report has a finite period of use.
4. It is false to assume that the author and the audience always will be available for reference.
5. It is false to assume that the audience is familiar with the assignment.
6. It is false to assume that the audience has been involved in daily discussions of the material.
7. It is false to assume that the audience awaits the report.
8. It is false to assume that the audience has time to read the report.

Assumptions one and two indicate a writer's lack of awareness of the nature of his report audience. Assumptions three, four, and five indicate his lack of appreciation of the dynamic nature of the system. Assumptions six, seven, and eight indicate a writer's lack of consideration of the demands of day-by-day job activity.

A report has value only to the extent that it is useful to the organization. It is often used primarily by someone other than the person who requested it. Furthermore, the report may be responding to a variety of needs within the organization. These needs suggest that the persons who will use the report are not specialists or perhaps not even technically knowledgeable about the report's subject. The specialist is the engineer. Unless he is engaged in basic research, he usually must communicate with persons representing many different areas of operation in the organization.

In addition, the report is often useful over an extended period of time. Each written communication is filed in several offices. Last year's report can be incomprehensible if the writer did not anticipate and explain his purpose adequately. In these situations, even within the office where a report originated, the author as well as his supervisor will probably not be available to explain the report. Although organizational charts remain unchanged for years, personnel, assignments, and professional roles change constantly. Because of this dynamic process, even the immediate audience of a report sometimes is not familiar with the writer's technical assignment. Thus, the report writer usually must design his report for a dynamic situation.

Finally, the report writer must also be alert to the communication traps in relatively static situations. Not all readers will have heard the coffee break chats that fill in the details necessary to make even a routine recommendation convincing. A report can arrive at a time when the reader's mind is churning with other concerns. Even if it is expected, the report usually meets a reader who needs to act immediately. The reader usually does not have time to read through the whole report; he wants the useful information clearly and succinctly. To the reader, time probably is the most important commodity. Beginning report writers seldom realize they must design their reports to be used efficiently rather than read closely.

The sources of the false assumptions we have been discussing are not difficult to identify. The original source is the artificial communication a student is required to perform in college. In writing only for professors, a student learns to write for audiences of one, audiences who know more than the writer knows, and audiences who have no instrumental interests in what the report contains. The subsequent source, on the job, is the writer's natural attempt to simplify his task. The report writer, relying upon daily contact and familiarity, simply finds it easier to write a report for his own supervisor than to write for a supervisor in a different department. The writer also finds it easier to concentrate upon his own concerns than to consider the needs of his readers. He finds it difficult to address complex audiences and face the design problems they pose.

AUDIENCE COMPONENTS AND PROBLEMS THEY POSE

To write a report you must first understand how your audience poses a problem. Then you must analyze your audience in order to be able to design a report structure that provides an optimum solution. To explain the components of the report audience you must do more than just identify names, titles, and roles. You must determine who your audiences are as related to the purpose and content of your report. "Who" involves the specific operational functions of the persons who will read the report, as well as their educational and business backgrounds. These persons can be widely distributed, as is evident if you consider the operational relationships within a typical organization.

Classifying audiences only according to directions of communication flow along the paths delineated by the conventional organizational chart, we can identify three types of report audiences: *horizontal, vertical,* and *external*. For example, in the organization chart in Figure 1, *Part of organization chart for naval ship engineering center,*[1] horizontal audiences exist on each level. The Ship Concept Design Division and the Command and Surveillance Division form horizontal audiences for each other. Vertical audiences exist between levels. The Ship Concept Design Division and the Surface Ship Design Branch form vertical audiences for each other. External audiences exist when any unit interacts with a separate organization, such as when the Surface Ship Design Branch communicates with the Newport News Shipbuilding Company.

What the report writer first must realize is the separation between him and any of these three types of audiences. Few reports are written for horizontal audiences within the same unit, such as from one person in the Surface Ship Design Branch to another person or project group within the Surface Ship Design Branch itself. Instead, a report at least addresses horizontal audiences within a larger framework, such as from the Surface Ship Design Branch to the Systems

---

[1]A reference in H. B. Benford and J. C. Mathes, *Your Future in Naval Architecture,* Richards Rosen, New York, 1968.

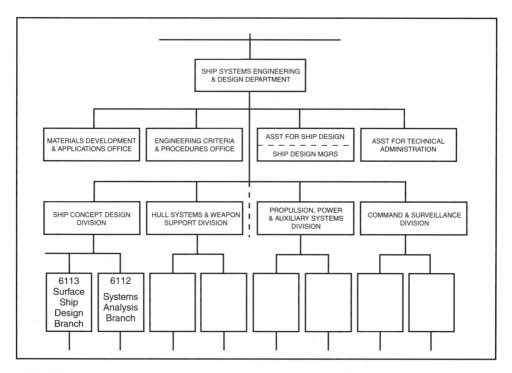

FIGURE I   Part of organization chart for naval ship engineering center

Analysis Branch. Important reports usually have complex audiences, that is, vertical and horizontal, and sometimes external audiences as well.

An analysis of the problems generated by horizontal audiences—often assumed to pose few problems—illustrates the difficulties most writers face in all report writing situations. A systems engineer in the Systems Analysis Branch has little technical education in common with the naval architect in the Surface Ship Design Branch. In most colleges he takes only a few of the same mathematics and engineering science courses. The systems engineer would not know the wave resistance theory familiar to the naval architect, although he could use the results of his analysis. In turn, the naval architect would not know stochastics and probability theory, although he could understand systems models. But the differences between these audiences and writers go well beyond differences in training. In addition to having different educational backgrounds, the audiences will have different concerns, such as budget, production, or contract obligations. The audiences will also be separated from the writer by organizational politics and competition, as well as by personality differences among the people concerned.

When the writer addresses a horizontal audience in another organizational unit, he usually addresses a person in an organizational role. When addressed to the role rather than the person, the report is aimed at a department or a group. This means the report will have audiences in addition to the person addressed. It may be read primarily by staff personnel and subordinates. The addressee ultimately may act on the basis of the information reported, but at times he serves only to transfer the report to persons in his department who will use it. Furthermore, the report may have audiences in addition to those in the department addressed. It may be forwarded to other persons elsewhere, such as lawyers and comptrollers. The report travels routinely throughout organizational paths, and will have unknown or unanticipated audiences as well.

Consequently, even when on the same horizontal organizational level, the writer and his audience have little in common beyond the fact of working for the same organization, of having the same "rank" and perhaps of having the same educational level of attainment. Educational backgrounds can be entirely different; more important, needs, values, and uses are different. The report writer may recommend the choice of one switch over another on the basis of cost-efficiency analysis; his audiences may be concerned for business relationships, distribution patterns, client preferences, and budgets. Therefore, the writer should not assume that his audience has technical competence in the field, familiarity with the technical assignment, knowledge of him or of personnel in his group, similar value perspectives, or even complementary motives. The differences between writer and audience are distinctive, and may even be irreconcilable.

The differences are magnified when the writer addresses vertical audiences. Reports directed at vertical audiences, that is, between levels of an organization chart, invariably have horizontal audience components also. These complex report writing situations pose significant communication problems for the writer. Differences between writer and audience are fundamental. The primary audiences for the reports, especially informal reports, must act or make decisions on the basis of the reports. The reports thus have only instrumental value, that is, value insofar as they can be used effectively. The writer must design his report primarily according to how it will be used.

In addition to horizontal audiences and to vertical audiences, many reports are also directed to external audiences. External audiences, whether they consist of a few or many persons, have the distinctive, dissimilar features of the complex vertical audience. With external audiences these features invariably are exaggerated, especially those involving need and value. An additional complication is that the external audience can judge an entire organization on the basis of the writer's report. And sometimes most important of all, concerns for tact and business relationships override technical concerns.

In actual practice the writer often finds audiences in different divisions of his own company to be "external" audiences. One engineer encountered this problem in his first position after graduation. He was sent to investigate the inconsistent test data being sent to his group from a different division of the company in another city. He found that the test procedures being used in that division were faulty. However, at his supervisor's direction he had to write a report that would not "step on any toes." He had to write the report in such a manner as to have the other division correct its test procedures while not implying that the division was in any way at fault. An engineer who assumes that the purpose of his report is just to explain a technical investigation is poorly prepared for professional practice.

Most of the important communication situations for an engineer during his first five years out of college occur when he reports to his supervisor, department head, and beyond. In these situations, his audiences are action-oriented line management who are uninterested in the technical details and may even be unfamiliar with the assignment. In addition, his audiences become acquainted with him professionally through his reports; therefore, it is more directly the report than the investigation that is important to the writer's career.

Audience components and the significant design problems they pose are well illustrated by the various audiences for a formal report written by an engineer on the development of a process to make a high purity chemical, as listed in Figure 2, *Complex audience components for a formal report by a chemical engineer on a process to make a high purity chemical*. The purpose of the report was to explain the process; others would make a feasibility study of the process and evaluate it in comparison to other processes.

The various audiences for this report, as you can determine just by reading their titles, would have had quite different roles, backgrounds, interests, values,

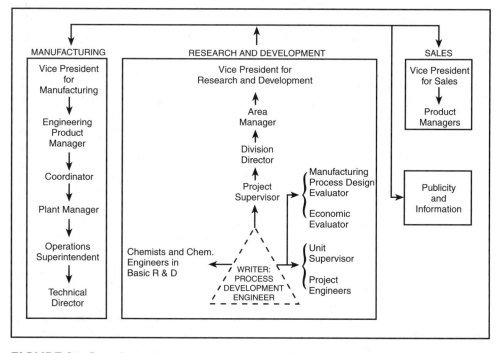

FIGURE 2   Complex audience components for a formal report by a chemical engineer on a process to make a high purity chemical

needs, and uses for the report. The writer's brief analysis of the audiences yielded the following:

> He could not determine the nature of many of his audiences, who they were, or what the specifics of their roles were.

> His audiences had little familiarity with his assignment.

> His report would be used for information, for evaluation of the process, and for evaluation of the company's position in the field.

> Some of his audiences would have from a minute to a half hour to glance at the report, some would take the report home to study it, and some would use it over extended periods of time for process analysis and for economic and manufacturing feasibility studies.

> The useful lifetime of the report could be as long as twenty years.

> The report would be used to evaluate the achievements of the writer's department.

> The report would be used to evaluate the writing and technical proficiencies of the writer himself.

This report writer classified his audiences in terms of the conventional organization chart. Then to make them more than just names, titles, and roles he asked himself what they would know about his report and how they would use it. Even then he had only partially solved his audience problem and had just begun to clarify the design problems he faced. To do so he needed to analyze his audiences systematically.

A METHOD FOR SYSTEMATIC AUDIENCE ANALYSIS

To introduce the audience problem that report writers must face, we have used the conventional concept of the organization chart to classify audiences as *horizontal*, *vertical*, and *external*. However, when the writer comes to the task of performing an instrumentally useful audience analysis for a particular report, this concept of the organization and this classification system for report audiences are not very helpful.

First, the writer does not view from outside the total communication system modeled by the company organization chart. He is within the system himself, so his view is always relative. Second, the conventional outsider's view does not yield sufficiently detailed information about the report audiences. A single bloc on the organization chart looks just like any other bloc, but in fact each bloc represents one or several human beings with distinctive roles, backgrounds, and personal characteristics. Third, and most importantly, the outsider's view does not help much to clarify the specific routes of communication, as determined by audience needs, which an individual report will follow. The organization chart may describe the organization, but it does not describe how the organization functions. Thus many of the routes a report follows—and consequently the needs it addresses—will not be signaled by the company organization chart.

In short, the conventional concept of report audiences derived from organization charts is necessarily abstract and unspecific. For that reason a more effective method for audience analysis is needed. In the remaining portion of this . . . [selection], we will present a three-step procedure. The procedure calls for preparing an egocentric organization chart to identify individual report readers, characterizing these readers, and classifying them to establish priorities. Based upon an egocentric view of the organization and concerned primarily with what report readers need, this system should yield the information the writer must have if he is to design an individual report effectively.

Prepare an Egocentric Organization Chart

An egocentric organization chart differs from the conventional chart in two senses. First, it identifies specific individuals rather than complex organizational units. A bloc on the conventional chart may often represent a number of people, but insofar as possible the egocentric chart identifies particular individuals who are potential readers of reports a writer produces. Second, the egocentric chart categorizes people in terms of their proximity to the report writer rather than in terms of their hierarchical relationship to the report writer. Readers are not identified as organizationally superior, inferior, or equal to the writer but rather as near or distant from the writer. We find it effective to identify four different degrees of distance as is illustrated in Figure 3, *Egocentric organization chart*. In this figure, with the triangle representing the writer, each circle is an individual reader identified by his organizational title and by his primary operational concerns. The four degrees of distance are identified by the four concentric rings. The potential readers in the first ring are those people with whom the writer associates daily. They are typically

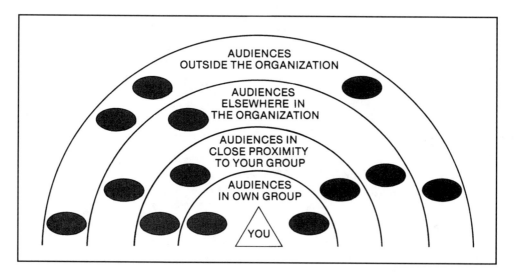

FIGURE 3   Egocentric organization chart

those people in his same office or project group. The readers in the second ring are those people in other offices with whom the writer must normally interact in order to perform his job. Typically, these are persons in adjacent and management groups. The readers in the third ring are persons relatively more distant but still within the same organization. They are distant management, public relations, sales, legal department, production, purchasing, and so on. They are operationally dissimilar persons. The readers in the fourth ring are persons beyond the organization. They may work for the same company but in a division in another city. Or they may work for an entirely different organization.

Having prepared the egocentric organization chart, the report writer is able to see himself and his potential audiences from a useful perspective. Rather than seeing himself as an insignificantly small part of a complex structure—as he is apt to do with the conventional organizational chart—the writer sees himself as a center from which communication radiates throughout an organization. He sees his readers as individuals rather than as faceless blocs. And he sees that what he writes is addressed to people with varying and significant degrees of difference.

A good illustration of the perspective provided by the egocentric organization chart is the chart prepared by a chemical engineer working for a large corporation, Figure 4, *Actual egocentric organization chart of an engineer in a large corporation.* It is important to notice how the operational concerns of the persons even in close proximity vary considerably from those of the development engineer. What these people need from reports written by this engineer, then, has little to do with the processes by which he defined his technical problems.

The chemical engineer himself is concerned with the research and development of production processes and has little interest in, or knowledge of,

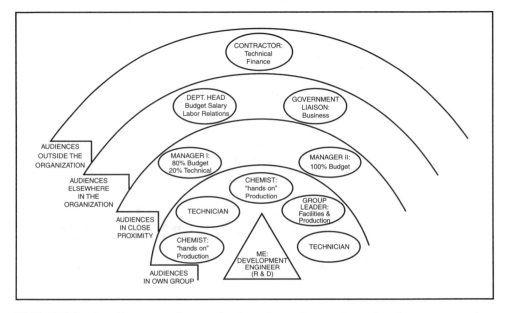

FIGURE 4 Actual egocentric organization chart of an engineer in a large corporation

budgetary matters. Some of the audiences in his group are chemists concerned with production—not with research and development. Because of this they have, as he said, "lost familiarity with the technical background, and instead depend mostly on experience." Other audiences in his group are technicians concerned only with operations. With only two years of college, they have had no more than introductory chemistry courses and have had no engineering courses.

Still another audience in his group is his group leader. Rather than being concerned with development, this reader is concerned with facilities and production operations. Consequently, he too is "losing familiarity with the technical material." Particularly significant for the report writer is that his group leader in his professional capacity does not use his B.S.Ch.E. degree. His role is that of manager, so his needs have become administrative rather than technical.

The concerns of the chemical engineer/report writer's audiences in close proximity to his group change again. Instead of being concerned with development or production operations, these audiences are primarily concerned with the budget. They have little technical contact, and are described as "business oriented." Both Manager I and Manager II are older, and neither has a degree in engineering. One has a Ph.D. degree in chemistry, the other an M.S. degree in technology. Both have had technical experience in the lab, but neither can readily follow technical explanations. As the chemical engineer said, both would find it "difficult to return to the law."

The report writer's department head and other persons through whom the group communicates with audiences elsewhere in the organization, and beyond it, have additional concerns as well as different backgrounds. The department head is concerned with budget, personnel, and labor relations. The person in contact with outside funding units—in this case, a government agency—has business administration degrees and is entirely business oriented. The person in contact with subcontractors has both technical and financial concerns.

Notice that when this writer examined his audiences even in his own group as well as those in close proximity to him, he saw that the natures, the backgrounds, and especially the operational concerns of his audiences vary and differ considerably. As he widened the scope of his egocentric organization chart, he knew less and less about his audiences. However, he could assume they will vary even more than those of the audiences in close proximity.

Thus, in the process of examining the audience situation with an egocentric organization chart, a report writer can uncover not only the fact that audiences have functionally different interests, but also the nature of those functional differences. He can proceed to classify the audiences for each particular report in terms of audience needs.

Preparation of the egocentric organization chart is the first step of your procedure of systematic audience analysis. Notice that this step can be performed once to describe your typical report audience situation but must be particularized for each report to define the audiences for that report. Having prepared the egocentric chart once, the writer revises his chart for subsequent reports by adding or subtracting individual audiences.

Characterize the Individual Report Readers

In the process of preparing the egocentric organization chart, you immediately begin to think of your individual report readers in particular terms. In preparing the egocentric chart discussed above, the report writer mentioned such items as a reader's age, academic degrees, and background in the organization as well as his operational concerns. All of these particulars will come to mind when you think of your audiences as individuals. However, a systematic rather than piece-meal audience analysis will yield more useful information. The second step of audience analysis is, therefore, a systematic characterization of each person identified in the egocentric organization chart. A systematic characterization is made in terms of *operational, objective,* and *personal* characteristics.

The *operational characteristics* of your audiences are particularly important. As you identify the operational characteristics for a person affected by your report, try to identify significant differences between his or her role and yours. What are his professional values? How does he spend his time? That is, will his daily concerns and attitudes enable him to react to your report easily, or will they make it difficult for him to grasp what you are talking about? What does he know about your role, and in particular, what does he know or remember about your technical assignment and the organizational problem that occasioned your report to come to him? You should also consider carefully what he will need from your report. As you think over your entire technical investigation, ask yourself if that person will involve staff personnel in action on your report, of if he will in turn activate other persons elsewhere in the organization, when he receives the report. If he should, you must take their reactions into account when you write your report.

In addition, you should ask yourself, "How will my report affect his role?" A student engineer recently told us of an experience he had during summer employment when he was asked to evaluate the efficiency of the plant's waste treatment process. Armed with his fresh knowledge from advanced chemical engineering courses, to his surprise he found that, by making a simple change in the process, the company could save more than $200,000 a year. He fired off his report with great anticipation of glowing accolades—none came. How had his report affected the roles of some of his audiences? Although the writer had not considered the report's consequences when he wrote it, the supervisor, the manager, and related personnel now were faced with the problem of accounting for their waste of $200,000 a year. It should have been no surprise that they were less than elated over his discovery.

By *objective characteristics* we mean specific, relevant background data about the person. As you try to identify his or her educational background, you may note differences you might have otherwise neglected. Should his education seem to approximate yours, do not assume he knows what you know. Remember that the half-life of engineering education today is about five years. Thus, anyone five to ten years older than you, if you are recently out of college, probably will be only superficially familiar with the material and jargon of your advanced technical courses. If you can further identify his past professional experiences and roles, you might be able to anticipate his first-hand knowledge of your role and technical activities as well as to clarify any residual organizational commitments and value systems he might have. When you judge his knowledge of your technical area, ask yourself, "Could he participate in a professional conference in my field of specialization?"

For *personal characteristics,* when you identify a person by name, ask yourself how often the name changes in this organizational role. When you note his or her approximate age, remind yourself how differences in age can inhibit communication. Also note personal concerns that could influence his reactions to your report.

A convenient way to conduct the audience analysis we have been describing and to store the information it yields is to use an analysis form similar to the one in Figure 5, *Form for characterizing individual report readers* [p. 210]. It may be a little time-consuming to do this the first time around, but you can establish a file of audience characterizations. Then you can add to or subtract from this file as an individual communication situation requires.

One final point: This form is a means to an end rather than an end in itself. What is important for the report writer is that he thinks systematically about the questions this form raises. The novice usually has to force himself to analyze his audiences systematically. The experienced writer does this automatically.

Classify Audiences in Terms of How They Will Use Your Report

For each report you write, trace out the communication routes on your egocentric organization chart and add other routes not on the chart. Do not limit these routes to those specifically identified by the assignment and the addresses of the report. Rather, think through the total impacts of your report on the organization. That is, think in terms of the first, second, and even some third-order consequences of your report, and trace out the significant communication routes involved. All of these consequences define your actual communication.

When you think in terms of consequences, primarily you think in terms of the uses to which your report will be put. No longer are you concerned with your technical investigation itself. In fact, when you consider how readers will use your report, you realize that very few of your potential readers will have any real interest in the details of your technical investigation. Instead, they want to know the answers to such questions as "Why was this investigation made? What is the significance of the problem it addresses? What am I supposed to do with the results of this investigation? What will it cost? What are the implications—for sales, for production, for the unions? What happens next? Who does it? Who is responsible?"

It is precisely this audience concern for nontechnical questions that causes so much trouble for young practicing engineers. Professionally, much of what the engineer spends his time doing is, at most, of only marginal concern to many of his audiences. His audiences ask questions about things which perhaps never entered his thoughts during his own technical activities when he received the assignment, defined the problem, and performed his investigation. These questions, however, must enter into his considerations when he writes his report.

NAME: TITLE:

A. OPERATIONAL CHARACTERISTICS:
    1. His role within the organization and consequent value system:

    2. His daily concerns and attitudes:

    3. His knowledge of your technical responsibilities and assignment:

    4. What he will need from your report:

    5. What staff and other persons will be activated by your report
       through him:

    6. How your report could affect his role:

B. OBJECTIVE CHARACTERISTICS:
    1. His education—levels, fields, and years:

    2. His past professional experiences and roles:

    3. His knowledge of your technical area:

C. PERSONAL CHARACTERISTICS:
    Personal characteristics that could influence his reactions—age,
    attitudes, pet concerns, etc.

FIGURE 5   Form for characterizing individual report readers

Having defined the communication routes for a report you now know what audiences you will have and what questions they will want answered. The final step in our method of audience analysis is to assign priorities to your audiences. Classify them in terms of how they will use your report. In order of their importance to you (not in terms of their proximity to you), classify your audiences by these three categories:

- *Primary audiences*—who make decisions or act on the basis of the information a report contains.
- *Secondary audiences*—who are affected by the decisions and actions.
- *Immediate audiences*—who route the report or transmit the information it contains.

The *primary audience* for a report consists of those persons who will make decisions or act on the basis of the information provided by the report. The report overall should be designed to meet the needs of these users. The primary audience can consist of one person who will act in an official capacity, or it can consist of several persons representing several offices using the report. The important point here is that the primary audience for a report can consist of persons from any ring on the egocentric organization chart. They may be distant or in close proximity to the writer. They may be his organizational superiors, inferiors, or equals. They are simply those readers for whom the report is primarily intended. They are the top priority users.

In theory at least, primary audiences act in terms of their organizational roles rather than as individuals with distinctive idiosyncracies, predilections, and values. Your audience analysis should indicate when these personal concerns are likely to override organizational concerns. A typical primary audience is the decision maker, but his actual decisions are often determined by the evaluations and recommendations of staff personnel. Thus the report whose primary audience is a decision maker with line responsibility actually has an audience of staff personnel. Another type of primary audience is the production superintendent, but again his actions are often contingent upon the reactions of others.

In addition, because the report enters into a system, in time both the line and staff personnel will change; roles rather than individuals provide continuity. For this reason, it is helpful to remember the words of one engineer when he said, "A complete change of personnel could occur over the lifetime of my report." The report remains in the file. The report writer must not assume that his primary audience will be familiar with the technical assignment. He must design the report so that it contains adequate information concerning the reasons for the assignment, details of the procedures used, the results of the investigation, and conclusions and recommendations. This information is needed so that any future component of his primary audience will be able to use the report confidently.

The *secondary audience* for a report consists of those persons other than primary decision makers or users who are affected by the information the report transmits into the system. These are the people whose activities are affected when a primary audience makes a decision, such as when production supervision has to adjust to management decisions. They must respond appropriately when a primary audience acts, such as when personnel and labor relations have to accommodate production line changes. The report writer must not neglect the needs of his secondary audiences. In tracing out his communication routes, he will identify several secondary audiences. Analysis of their needs will reveal what additional information the report should contain. This information is often omitted by writers who do not classify their audiences sufficiently.

The *immediate audience* for a report are those persons who route the report or transmit the information it contains. It is essential for the report writer to identify his immediate audiences and not to confuse them with his primary audiences. The immediate audience might be the report writer's supervisor or another middle management person. Yet usually his role will be to transmit information rather than to use the information directly. An information system has numerous persons who transmit reports but who may not act upon the information or who may not be affected by the information in ways of concern to the report writers. Often, a report is addressed to the writer's supervisor, but except for an incidental memo report, the supervisor serves only to transmit and expedite the information flow throughout the organizational system.

A word of caution: at times the immediate audience is also part of the primary audience; at other times the immediate audience is part of the secondary audience. For each report you write, you must distinguish those among your readers who will function as conduits to the primary audience.

As an example of these distinctions between categories of report audiences, consider how audiences identified on the egocentric organization chart, Figure 4 [p. 206], can be categorized. Assume that the chemical engineer writes a report on a particular process improvement he has designed. The immediate audience might be his Group Leader. Another would be Manager I, transmitting the report to Manager II. The primary audiences might be Manager II and the Department Head; they would ask a barrage of nontechnical questions similar to those we mentioned a moment ago. They will decide whether or not the organization will implement the improvement recommended by the writer. The Department Head also could be part of the secondary audience by asking questions relating to labor relations and union contracts. Other secondary audiences, each asking different questions of the report, could be:

> The person in contact with the funding agency, who will be concerned with budget and contract implications.

> The person in contact with subcontractors, determining how they are affected.

> The Group Leader, whose activities will be changed.

The "hands on" chemist, whose production responsibilities will be affected.

The technicians, whose job descriptions will change.

In addition to the secondary audiences on the egocentric organization chart, the report will have other secondary audiences throughout the organization—technical service and development, for example, or perhaps waste treatment.

At some length we have been discussing a fairly detailed method for systematic audience analysis. The method may have seemed more complicated than it actually is. Reduced to its basic ingredients, the method requires you, first, to identify all the individuals who will read the report, second, to characterize them, and third, to classify them. The *Matrix for audience analysis,* Figure 6, is a convenient device for characterizing and classifying your readers once you have identified them. At a glance, the matrix reveals what information you have and what information you still need to generate. Above all, the matrix forces you to think systematically. If you are able to fill in a good deal of specific information in each cell (particularly in the first six cells), you have gone a long way towards seeing how the needs of your audiences will determine the design of your report.

Characteristics / Types of audiences	Operational	Objective	Personal
Primary	①	④	⑦
Secondary	②	⑤	⑧
Immediate	③	⑥	⑨

FIGURE 6 Matrix for audience analysis

We have not introduced a systematic method for audience analysis with the expectation that it will make your communication task easy. We have introduced you to the problems you must account for when you design your reports—problems you otherwise might ignore. You should, at least, appreciate the complexity of a report audience. Thus, when you come to write a report, you are less likely to make false assumptions about your audience. To develop this attitude is perhaps as important as to acquire the specific information the analysis yields. On the basis of this attitude, you now are ready to determine the specific purpose of your report.

Richard W. Dodge

What to Report

At the time he wrote this article, Richard W. Dodge was the editor of Westinghouse Engineer.

Technical reports *can* be a useful tool for management—not only as a source of general information, but, more importantly, as a valuable aid to decision making. But to be effective for these purposes, a technical report must be geared to the needs of management.

Considerable effort is being spent today to upgrade the effectiveness of the technical report. Much of this is directed toward improving the writing abilities of engineers and scientists; or toward systems of organization, or format, for reports. This effort has had some rewarding effects in producing better written reports.

But one basic factor in achieving better reports seems to have received comparatively little attention. This is the question of audience needs. Or, expressed another way, "*What does management want in reports?*" This is an extremely basic question, and yet it seems to have had less attention than have the mechanics of putting words on paper.

A recent study conducted at Westinghouse sheds considerable light on this subject. While the results are for one company, probably most of them would apply equally to many other companies or organizations.

The study was made by an independent consultant with considerable experience in the field of technical report writing. It consisted of interviews with Westinghouse men at every level of management, carefully selected to present an accurate cross section. The list of questions asked is shown in Table 1. The results were compiled and analyzed, and from the report several conclusions are apparent. In addition, some suggestions for report writers follow as a natural consequence.

WHAT TO REPORT. By Richard W. Dodge. Reprinted from *Westinghouse Engineer,* 22 (July–September, 1962), by permission of the Westinghouse Electric Corporation.

TABLE I   Questions Asked of Managers

1. What types of reports are submitted to you?
2. What do you look for *first* in the reports submitted to you?
3. What do you want from these reports?
4. To what depth do you want to follow any one particular idea?
5. At what level (how technical and how detailed) should the various reports be written?
6. What do you want emphasized in the reports submitted to you? (Facts, interpretations, recommendations, implications, etc.)
7. What types of decisions are you called upon to make or to participate in?
8. What type of information do you need in order to make these decisions?
9. What types of information do you receive that you don't want?
10. What types of information do you want but not receive?
11. How much of a typical or average report you receive is useful?
12. What types of reports do you write?
13. What do you think your boss wants in the reports you send him?
14. What percentage of the reports you receive do you think desirable or useful? (In kind or frequency.)
15. What percentage of the reports you write do you think desirable or useful? (In kind or frequency.)
16. What particular weaknesses have you found in reports?

WHAT MANAGEMENT LOOKS FOR IN ENGINEERING REPORTS

When a manager reads a report, he looks for pertinent facts and competent opinions that will aid him in decision making. He wants to know right away whether he should read the report, route it, or skip it.

To determine this, he wants answers fast to some or all of the following questions:

- What's the report about and who wrote it?
- What does it contribute?
- What are the conclusions and recommendations?
- What are their importance and significance?
- What's the implication to the Company?
- What actions are suggested? Short range? Long range?
- Why? By whom? When? How?

The manager wants this information in brief, concise, and meaningful terms. He wants it at the beginning of the report and all in one piece.

For example, if a summary is to convey information efficiently, it should contain three kinds of facts:

1. What the report is about;

2. The significance and implications of the work; and
3. The action called for.

To give an intelligent idea of what the report is about, first of all the problem must be defined, then the objectives of the project set forth. Next, the reasons for doing the work must be given. Following this should come the conclusions. And finally, the recommendations.

Such summaries are informative and useful, and should be placed at the beginning of the report.

The kind of information a manager wants in a report is determined by his management responsibilities, but how he wants this information presented is determined largely by his reading habits. This study indicates that management report reading habits are surprisingly similar. Every manager interviewed said he read the *summary* or abstract; a bare majority said they read the *introduction* and *background* sections as well as the *conclusions* and *recommendations;* only a few managers read the *body* of the report or the *appendix* material.

The managers who read the *background* section, or the conclusions and recommendations, said they did so ". . . to gain a better perspective of the material being reported and to find an answer to the all-important question: What do we do next?" Those who read the *body* of the report gave one of the following reasons:

1. Especially interested in subject
2. Deeply involved in the project
3. Urgency of problem requires it
4. Skeptical of conclusions drawn.

And those few managers who read the *appendix* material did so to evaluate further the work being reported. To the report writer, this can mean but one thing: If a report is to convey useful information efficiently, the structure must fit the manager's reading habits.

The frequency of reading chart in Figure 1 [p. 218] suggests how a report should be structured if it is to be useful to management readers.

SUBJECT MATTER INTEREST

In addition to what facts a manager looks for in a report and how he reads reports, the study indicated that he is interested in five broad technological areas. These are:

1. Technical problems
2. New projects and products
3. Experiments and tests
4. Materials and processes
5. Field troubles.

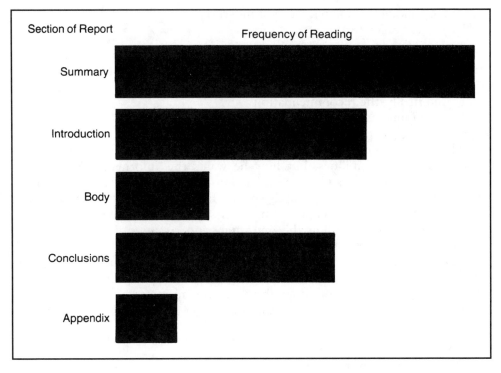

FIGURE I   How managers read reports

Managers want to know a number of things about each of these areas. These are listed in Table 2. Each of the sets of questions can serve as an effective checklist for report writers.

In addition to these subjects, a manager must also consider market factors and organization problems. Although these are not the primary concern of the engineer, he should furnish information to management whenever technical aspects provide special evidence or insight into the problem being considered. For example, here are some of the questions about marketing matters a manager will want answered:

- What are the chances for success?
- What are the possible rewards? Monetary? Technological?
- What are the possible risks? Monetary? Technological?
- Can we be competitive? Price? Delivery?
- Is there a market? Must one be developed?
- When will the product be available?

And, here are some of the questions about organization problems a manager must have answered before he can make a decision:

- Is it the type of work Westinghouse should do?
- What changes will be required? Organization? Manpower? Facilities? Equipment?

TABLE 2   What Managers Want to Know

Problems	Tests and Experiments
What is it?	What tested or investigated?
Why undertaken?	Why? How?
Magnitude and importance?	What did it show?
What is being done? By whom?	Better ways?
Approaches used?	Conclusions? Recommendations?
Thorough and complete?	Implications to Company?
Suggested solutions? Best? Consider others?	**Material and Processes**
What now?	Properties, characteristics, capabilities?
Who does it?	Limitations?
Time factors?	Use requirements and environment?
New Projects and Products	Areas and scope of application?
Potential?	Cost factors?
Risks?	Availability and sources?
Scope of application?	What else will do it?
Commercial implications?	Problems in using?
Competition?	Significance of application to Company?
Importance to Company?	**Field Troubles and Special**
More work to be done? Any problems?	**Design Problems**
Required manpower, facilities and equipment?	Specific equipment involved?
Relative importance to other projects and	What trouble developed? Any trouble history?
products?	How much involved?
Life of project or product line?	Responsibility? Others? Westinghouse?
Effect on Westinghouse technical position?	What is needed?
Priorities required?	Special requirements and environment?
Proposed schedule?	Who does it? Time factors?
Target date?	Most practical solution?
	Recommended action?
	Suggested product design changes?

- Is it an expanding or contracting program?
- What suffers if we concentrate on this?

These are the kinds of questions Westinghouse management wants answered about projects in these five broad technological areas. The report writer should answer them whenever possible.

LEVEL OF PRESENTATION

Trite as it may sound, the technical and detail level at which a report should be written depends upon the reader and his use of the material. Most readers—certainly this is true for management readers—are interested in the significant material and in the general concepts that grow out of detail. Consequently, there is seldom real justification for a highly technical and detailed presentation.

Usually the management reader has an educational and experience background different from that of the writer. *Never* does the management reader have the same knowledge of and familiarity with the specific problem being reported that the writer has.

Therefore, the writer of a report for management should write at a technical level suitable for a reader whose educational and experience background is in a field different from his own. For example, if the report writer is an electrical engineer, he should write his reports for a person educated and trained in a field such as chemical engineering, or mechanical engineering, or metallurgical engineering.

All parts of the report *should preferably* be written on this basis. The highly technical, mathematical, and detailed material—if necessary at all—can and should be placed in the appendix.

MANAGEMENT RESPONSIBILITIES

The information presented thus far is primarily of interest to the report writer. In addition, however, management itself has definite responsibilities in the reporting process. These can be summed up as follows:

1. Define the project and the required reports;
2. Provide proper perspective for the project and the required reporting;
3. See that effective reports are submitted on time; and
4. See that the reports are properly distributed.

An engineering report, like any engineered product, has to be designed to fill a particular need and to achieve a particular purpose within the specific situation. Making sure that the writer knows what his report is to do, how it is to be used, and who is going to use it—all these things are the responsibilities of management. Purpose, use, and reader are the design factors in communications, and unless the writer knows these things, he is in no position to design an effective instrument of communication—be it a report, a memorandum, or what have you.

Four conferences at selected times can help a manager control the writing of those he supervises and will help him get the kind of reports he wants, when he wants them.

Step 1—At the beginning of the project. The purpose of this conference is to define the project, make sure the engineer involved knows what it is he's supposed to do, and specify the required reporting that is going to be expected of him as the project continues. What kind of decisions, for example, hinge upon his report? What is the relation of his work to the decision making process of management? These are the kinds of questions to clear up at this conference.

If the project is an involved one that could easily be misunderstood, the manager may want to check the effectiveness of the conference by asking the engineer to write a memorandum stating in his own words his understanding of the project, how he plans to handle it, and the reporting requirements. This can ensure a mutual understanding of the project from the very outset.

Step 2—At the completion of the investigation. When the engineer has finished the project assignment—but before he has reported on it—the manager should have him come in and talk over the results of his work. What did he find out? What conclusions has he reached? What is the main supporting evidence for these conclusions? What recommendations does he make? Should any future action be suggested? What is the value of the work to the Company?

The broader perspective of the manager, plus his extensive knowledge of the Company and its activities, puts him in the position of being able to give the engineer a much better picture of the value and implications of the project.

The mechanism for getting into the report the kind of information needed for decision making is a relatively simple one. As the manager goes over the material with the engineer, he picks out points that need to be emphasized, and those that can be left out. This is a formative process that aids the engineer in the selection of material and evidence to support his material.

Knowing in advance that he has a review session with his supervisor, the chances are the engineer will do some thinking beforehand about the project and the results. Consequently, he will have formed some opinions about the significance of the work, and will, therefore, make a more coherent and intelligent presentation of the project and the results of his investigation.

This review will do something for the manager, too. The material will give him an insight into the value of the work that will enable him to converse intelligently and convincingly about the project to others. Such a preview may, therefore, expedite decisions influencing the project in one way or another.

Step 3—After the report is outlined. The manager should schedule a third conference after the report is outlined. At this session, the manager and the author should review the report outline step-by-step. If the manager is satisfied with the outline, he should tell the author so and tell him to proceed with the report.

If, however, the manager is not satisfied with the outline and believes it will have to be reorganized before the kind of report wanted can be written, he must make this fact known to the author. One way he can do this is to have the author tell him why the outline is structured the way it is. This usually discloses the organizational weakness to the author and consequently he will be the one to suggest a change. This, of course, is the ideal situation. However, if the indirect approach doesn't work, the more direct approach must be used.

Regardless of the method used to develop a satisfactory outline, the thing the manager must keep in mind is this: It's much easier to win the author's consent to structural changes at the outline stage, i.e., *before* the report is written than afterward. Writing is a personal thing; therefore, when changes are suggested in organization or approach, these are all too frequently considered personal attacks and strained relations result.

Step 4—After the report is written. The fourth interview calls for a review and approval by the manager of the finished report and the preparation of a distribution list. During this review, the manager may find some sections of the report that need changing. While this is to be expected, he should limit the extent of these changes. The true test of any piece of writing is the clarity of the statement. If it's clear and does the job, the manager should leave it alone.

This four-step conference mechanism will save the manager valuable time, and it will save the engineer valuable time. Also, it will ensure meaningful and useful project reports—not an insignificant accomplishment itself. In addition, the process is an educative one. It places in the manager's hands another tool he can use to develop and broaden the viewpoint of the engineer. By eliminating misunderstanding and wasted effort, the review process creates a more helpful and effective working atmosphere. It acknowledges the professional status of the engineer and recognizes his importance as a member of the engineering department.

Christian K. Arnold

The Writing of Abstracts

Christian K. Arnold was an instructor in English and a technical writer and editor before becoming Associate Executive Secretary of the Association of State Universities and Land-Grant Colleges.

The most important section of your technical report or paper is the abstract. Some people will read your report from cover to cover; others will skim many parts, reading carefully only those parts that interest them for one reason or another; some will read only the introduction, results, and conclusions; but everyone who picks it up will read the abstract. In fact, the percentage of those who read beyond the abstract is probably related directly to the skill with which the abstract is written. The first significant impression of your report is formed in the reader's mind by the abstract; and the sympathy with which it is read, if it is read at all, is often determined by this first impression. Further, the people your organization wants most to impress with your report are the very people who will probably read no more than the abstract and certainly no more than the abstract, introduction, conclusions, and recommendations. And the people you should want most to read your paper are the ones for whose free time you have the most competition.

Despite its importance, you are apt to throw your abstract together as fast as possible. Its construction is the last step of an arduous job that you would rather have avoided in the first place. It's a real relief to be rid of the thing, and almost anything will satisfy you. But a little time spent in learning the "rules" that govern the construction of good abstracts and in practicing how to apply them will pay material dividends to both you and your organization.

The abstract—or summary, foreword, or whatever you call the initial thumb-nail sketch of your report or paper—has two purposes: (1) it provides the specialist in the field with enough information about the report to permit him to decide whether he could read it with profit, and (2) it provides the administrator or executive with enough knowledge about what has been done in the study or project and with what results to satisfy most of his administrative needs.

It might seem that the design specifications would depend upon the purpose for which the abstract is written. To satisfy the first purpose, for instance, the abstract needs only to give an accurate indication of the subject matter and scope of the study; but, to satisfy the second, the abstract must summarize the results and conclusions and give enough background information to make the results understandable. The abstract designed for the first purpose can tolerate any technical language or symbolic shortcuts understood at large by the subject-matter group; the abstract designed for the second purpose should contain no terms not generally understood in a semitechnical community. The abstract for the first purpose is called a *descriptive abstract;* that for the second, an *informative abstract.*

The following abstract, prepared by a professional technical abstracter in the Library of Congress, clearly gives the subject-matter specialist all the help he needs to decide whether he should read the article it describes:

> Results are presented of a series of cold-room tests on a Dodge diesel engine to determine the effects on starting time of (1) fuel quantity delivered at cranking speed and (2) type of fuel-injection pump used. The tests were made at a temperature of −10°F with engine and accessories chilled at −10°F at least 8 hours before starting.

Regardless of however useful this abstract might be on a library card or in an index or an annotated bibliography, it does not give an executive enough information. Nor does it encourage everyone to read the article. In fact, this abstract is useless to everyone except the specialist looking for sources of information. The descriptive abstract, in other words, cannot satisfy the requirement of the informative abstract.

But is the reverse also true? Let's have a look at an informative abstract written for the same article:

> A series of tests was made to determine the effect on diesel-engine starting characteristics at low temperatures of (1) the amount of fuel injected and (2) the type of injection pump used. All tests were conducted in a cold room maintained at −10°F on a commercial Dodge engine. The engine and all accessories were "cold-soaked" in the test chamber for at least 8 hours before each test. Best starting was obtained with 116 cu mm of fuel, 85 per cent more than that required for maximum power. Very poor starting was obtained with the lean setting of 34.7 cu mm. Tests with two different pumps indicated that, for best starting characteristics, the pump must deliver fuel evenly to all cylinders even at low cranking speeds so that each cylinder contributes its maximum potential power.

This abstract is not perfect. With just a few more words, for instance, the abstracter could have clarified the data about the amount of fuel delivered: do the figures give flow rates (what is the unit of time?) or total amount of fuel injected (over how long a period?)? He could easily have defined "best" starting. He could have been more specific about at least the more satisfactory type of pump: what is the type that delivers the fuel more evenly? Clarification of these points would not have increased the length of the abstract significantly.

The important point, however, is not the deficiencies of the illustration. In fact, it is almost impossible to find a perfect, or even near perfect, abstract, quite possibly because the abstract is the most difficult part of the report to write. This difficulty stems from the severe limitations imposed on its length, its importance to the over-all acceptance of the report or paper, and, with informative abstracts, the requirement for simplicity and general understandability.

The important point, rather, is that the informative abstract gives everything that is included in the descriptive one. The informative abstract, that is, satisfies not only its own purpose but also that of the descriptive abstract. Since values are obtained from the informative abstract that are not obtained from the descriptive, it is almost always worthwhile to take the extra time and effort necessary to produce a good informative abstract for your report or memo. Viewed from the standpoint of either the total time and effort expended on the writing job as a whole or the extra benefits that accrue to you and your organization, the additional effort is inconsequential.

It is impossible to lay down guidelines that will lead always to the construction of an effective abstract, simply because each reporting job, and consequently each abstract, is unique. However, general "rules" can be established that, if practiced conscientiously and applied intelligently, will eliminate most of the bugs from your abstracts.

1. *Your abstract must include enough specific information about the project or study to satisfy most of the administrative needs of a busy executive.* This means that the more important results, conclusions, and recommendations, together with enough additional information to make them understandable, must be included. This additional information will most certainly include an accurate statement of the problem and the limitations placed on it. It will probably include an interpretation of the results and the principal facts upon which the analysis was made, along with an indication of how they were obtained. Again, *specific* information must be given. One of the most common faults of abstracts in technical reports is that the information given is too general to be useful.

2. *Your abstract must be a self-contained unit, a complete report-in-miniature.* Sooner or later, most abstracts are separated from the parent report, and the abstract that cannot stand on its own feet independently must then either be rewritten or will fail to perform its job. And the rewriting, if it is done, will be done by someone not nearly as sympathetic with your study as you are. Even if it is not separated from the report, the abstract must be written as a complete, independent unit if it is to be of the most help possible to the executive. This rule automatically eliminates the common deadwood phrases like "this report contains . . ." or "this report is a report on . . ." that clutters up many abstracts. It also eliminates all references to sections, figures, tables, or anything else contained in the report paper.

3. *Your abstract must be short.* Length in an abstract defeats every purpose for which it is written. However, no one can tell you just how short it must be. Some authorities have attempted to establish arbitrary lengths, usually in terms of a certain percentage of the report, the figure given normally falling between three and ten per cent. Such artificial guides are unrealistic. The abstract for a 30-page report must necessarily be longer, percentagewise, than the abstract for a 300-page report, since there is certain basic information that must be given regardless of the length of the report. In addition, the information given in some reports can be summarized much more briefly than can that given in other reports of the same over-all dimensions. Definite advantages, psychological as well as material, are obtained if the abstract is short enough to be printed entirely on one page so that the reader doesn't even have to turn a page to get the total picture that the abstract provides. Certainly, it should be longer than the interest span of an only mildly interested and very busy executive. About the best practical advice that can be given in a vacuum is to make your abstract as short as possible without cutting out essential information or doing violence to its accuracy. With practice, you might be surprised to learn how much information you can crowd into a few words. It helps, too, to learn to blue-pencil unessential information. It is perhaps important to document that "a meeting was held at the Bureau of Ordnance on Tuesday, October 3, 1961, at 2:30 P.M." somewhere, but such information is just excess baggage in your abstract: it helps neither the research worker looking for source material nor the administrator looking for a status or information summary. Someone is supposed to have once said, "I would have written a shorter letter if I had had more time." Take the time to make your abstracts shorter; the results are worth it. But be careful not to distort the facts in the condensing.

4. *Your abstract must be written in fluent, easy-to-read prose.* The odds are heavily against your reader's being an expert in the subject covered by your report or paper. In fact, the odds that he is an expert in your field are probably no greater than the odds that he has only a smattering of training in any technical or scientific discipline. And even if he were perfectly capable of following the most obscure, tortured technical jargon, he will appreciate your sparing him the necessity for doing it. T. O. Richards, head of the Laboratory Control Department, and R. A. Richardson, head of the Technical Data Department, both of the General Motors Corporation, have written that their experience shows the abstract cannot be made too elementary. "We never had [an abstract] . . . in which the explanations and terms were too simple." This requirement immediately eliminated the "telegraphic" writing often found in abstracts. Save footage by sound practices of economy and not by cutting out the articles and the transitional devices needed for smoothness and fluency. It also eliminates those obscure terms that you defend on the basis of "that's the way it's always said."

5. *Your abstract must be consistent in tone and emphases with the report paper, but it does not need to follow the arrangement, wording, or proportion of the original.* Data, information, and ideas introduced into the abstract must also appear in

the report or paper. And they must appear with the same emphases. A conclusion or recommendation that is qualified in the report paper must not turn up without the qualification in the abstract. After all, someone might read both the abstract and the report. If this reader spots an inconsistency or is confused, you've lost a reader.

6. *Your abstract should make the widest possible use of abbreviations and numerals, but it must not contain any tables or illustrations.* Because of the space limitations imposed on abstracts, the rules governing the use of abbreviations and numerals are relaxed for it. In fact, all figures except those standing at the beginning of sentences should be written as numerals, and all abbreviations generally accepted by such standard sources as the American Standards Association and "Webster's Dictionary" should be used.

By now you must surely see why the abstract is the toughest part of your report to write. A good abstract is well worth the time and effort necessary to write it and is one of the most important parts of your report. And abstract writing probably contributes more to the acquisition of sound expository skills than does any other prose discipline.

Vincent Vinci

Ten Report Writing Pitfalls: How to Avoid Them

Vincent Vinci was Director of Public Relations for Lockheed Electronics when he wrote this article.

The advancement of science moves on a pavement of communications. Chemists, electrical engineers, botanists, geologists, atomic physicists, and other scientists are not only practitioners but interpreters of science. As such, the justification, the recognition and the rewards within their fields result from their published materials.

Included in the vast field of communications is the report, a frequently used medium for paving the way to understanding and action. The engineering manager whose function is the direction of people and programs receives and writes many reports in his career. And therefore the need for technical reports that communicate effectively has been internationally recognized.

Since scientific writing is complicated by specialized terminology, a need for precision and the field's leaping advancement, the author of an engineering report can be overwhelmed by its contents. The proper handling of contents and communication of a report's purpose can be enhanced if the writer can avoid the following 10 pitfalls.

PITFALL 1: IGNORING YOUR AUDIENCE

In all the forms of communications, ignoring your audience in the preparation of a report is perhaps the greatest transgression. Why? All other forms of communication, such as instruction manuals, speeches, books and brochures, are directed to an indefinable or only partially definable audience. The report, on the other hand, is usually directed to a specific person or group and has a specific purpose. So, it would certainly seem that if one knows both the "who" and the "why," then a report writer should not be trapped by this pitfall.

But it is not enough to know the who and why, you need to know "how." To get to the how, let's assume that the reader is your boss and has asked you to write a trip report. You are to visit several plants and report on capital equipment requirements. Before you write the first word, you will have to find out what your boss already knows about these requirements. It is obvious that he wants a new assessment of the facilities' needs. But, was he unsatisfied with a recent assessment and wants another point of view, or is a new analysis required because the previous report is outdated—or does he feel that now is the time to make the investment in facilities so that production can be increased over the next five years? That's a lot of questions, but they define both the who and why of your trip and, more importantly, your report.

By this time you may get the feeling that I am suggesting you give him exactly what he wants to read. The answer is yes and no. No, I don't mean play up to your boss's likes and dislikes. I do mean, however, that you give him all the information he needs to make a decision—the pros and the cons.

I mean also that the information be presented in a way that he is acclimated to in making judgments. For example, usually a production-oriented manager or executive (even the chief executive) will think in terms of his specialty. The president of a company who climbed the marketing ladder selling solvents will think better in marketing terms. Therefore, perhaps the marketing aspects of additional equipment and facilities should be stressed. You should also be aware that if you happen to be the finance director, your boss will expect to see cost/investment factors too.

A simple method for remembering, rather than ignoring, your audience is to place a sheet of paper in front of you when you start to write your report. On the paper have written in bold letters **WHO, WHY** and **HOW,** with the answers clearly and cogently defined. Keep it in front of you throughout the preparation of your report.

PITFALL 2: WRITING TO IMPRESS

Nothing turns a reader off faster than writing to impress. Very often reports written to leave a lasting scholarly impression on top management actually hinder communication.

Generally, when a word is used to impress, the report writer assumes that the reader either knows its meaning or will take the trouble to look it up. Don't assume that a word familiar to you is easily recognized by your reader. I recall a few years ago, there was a word "serendipity" which became a fashionable word to impress your reader with. And there was "fulsome" and "pejorative," and more. All are good words, but they're often misused or misapplied. They were shoved into reports to impress, completely disregarding the reader. Your objective is that your reader comprehend your thoughts, and there should be a minimum of impediments to understanding—understanding with first reading, and no deciphering.

Unfortunately, writing to impress is not merely restricted to use of obscure words but also includes unnecessary detail and technical trivia. Perhaps the scientist, chemist, chemical engineer and others become so intrigued with technical fine points that the meaningful (to your audience) elements of a report are buried. And quite often the fault is not so much a lack of removing the chaff from the grain but an attempt to technically impress the reader. Of course there exist reports that are full of technical detail because the nature of the communication is to impart a new chemical process, compound or technique. Even when writing this kind of report, you should eliminate any esoteric technical facts that do not contribute to communication, even though you may be tempted to include them to exhibit your degree of knowledge in the field.

PITFALL 3: HAVING MORE THAN ONE AIM

A report is a missile targeted to hit a point or achieve a mission. It is not a barrage of shotgun pellets that scatter across a target indiscriminately.

Have you ever, while reading a report, wondered where or what it was leading to—and even when you are finished you weren't quite sure? The writer probably had more than one aim, thereby preventing you from knowing where the report was heading.

Having more than one aim is usually the sign of a novice writer, but the pitfall can also trip up an experienced engineer if he does not organize the report toward one objective.

It is too easy to say that your report is being written to communicate, to a specific audience, information about your research, tests, visit, meeting, conference, field trip, progress or any other one of a range of activities that may be the subject. If you look at the first part of the sentence, you will see that "specific audience" and "information" are the key words that have to be modified to arrive at the goal of your report. For instance, you must define the specific audience such as the "members of the research council," "the finance committee" or "the chief process engineer and his staff."

Secondly, you need to characterize the information, such as "analysis of a new catalytic process," "new methods of atomic absorption testing" or "progress on waste treatment programs." You should be able to state the specific purpose of your report in one sentence: e.g., "The use of fibrous material improves scrubber efficiency and life—a report to the product improvement committee."

When you have arrived at such a definition of your purpose and audience, you can then focus both the test results and analysis toward that purpose, tempered with your readers in mind.

The usual error made in writing reports is to follow the chronology of the research in the body of the report with a summary of a set of conclusions and recommendations attached. The proper procedure to follow is to write (while focusing on your report goal) the analysis first (supported by test essentials or any other details), then your introduction or summary—sort of reverse chronology. But be sure that your goal and audience are clearly known because they become the basis of organizing your report.

PITFALL 4: BEING INCONSISTENT

If you work for an international chemical firm, you may be well aware of problems in communicating with plant managers and engineers of foreign installations or branches. And I'm not referring to language barriers, because for the most part these hurdles are immediately recognized and taken care of. What is more significant is units of measure. This problem is becoming more apparent as the United States slowly decides whether or not to adopt the metric system. Until it is adopted your best bet is to stick to one measurement system throughout the report. Preferably, the system chosen should be that familiar to your audience. If the audience is mixed, you should use both systems with one (always the same one) in parentheses. Obviously, don't mix units of measure because you will confuse or annoy your readers.

Consistency is not limited to measurements but encompasses terms, equations, derivations, numbers, symbols, abbreviations, acronyms, hyphenation, capitalization and punctuation. In other words, consistency in the mechanics of style will avoid work for your reader and smooth his path toward understanding and appreciating the content of the report.

If your company neither has a style guide nor follows the general trends of good editorial practice, perhaps you could suggest instituting a guide. In addition to the U.S. Government Printing Office Style Manual, many scientific and engineering societies have set up guides which could be used.

PITFALL 5: OVERQUALIFYING

Chemical engineers, astronomers, geologists, electrical engineers, and scientists of any other discipline have been educated and trained to be precise. As a result, they strive for precision, accuracy, and detail. That tends to work against the scientist when it comes to writing. Add to that the limited training received in the arts, and you realize why written expression does not come easily.

Most reports, therefore, have too many modifiers—adjectives, clauses, phrases, adverbs and other qualifiers. Consider some examples: the single-stage, isolated double-cooled refractory process breakdown, or the angle of the single-rotor dc hysteresis motor rotor winding. To avoid such difficult-to-comprehend phrases, you could in the first example write "the breakdown of the process in single-stage, isolated double-cooled refractories," and in the second, "the angle of the rotor winding in single-rotor dc hysteresis motors can cause . . . ," and so on. This eliminates the string of modifiers and makes the phrase easier to understand.

Better still, if your report allows you to say at the beginning that the following descriptions are only related to "single-stage, isolated double-cooled refractories" or "single-rotor dc hysteresis motors," you can remove the cumbersome nomenclature entirely.

In short, to avoid obscuring facts and ideas, eliminate excessive modifiers. Try to state your idea or main point first and follow with your qualifying phrases.

PITFALL 6: NOT DEFINING

Dwell, lake, and barn, all are common words. Right? Right and wrong. Yes, they are common to the nonscientist. To the mechanical engineer, dwell is the period a cam follower stays at maximum lift; to a chemical engineer, lake is a dye compound; and to an atomic physicist, a barn is an atomic cross-sectional area (10^{-24} cm^2).

These three words indicate two points: first, common words are used in science with other than their common meanings; and second, terms need to be defined.

In defining terms you use in a report, you must consider what to define and how to define. Of the two, I consider what to define a more difficult task and suggest that you review carefully just which terms you need defined. If you analyze the purpose, the scope, the direction and your audience (reader/user), you will probably get a good handle on such terms.

"How to define" ranges from the simple substitution of a common term for an uncommon one, to an extended or amplified explanation. But whatever the term, or method of definition, you need to slant it both to the reader and to the report purpose.

PITFALL 7: MISINTRODUCING

Introductions, summaries, abstracts and forewords—whatever you use to lead your reader into your report, it should not read like an exposition of a table of contents. If it does, you might as well let your audience read the table of contents.

The introduction, which should be written after the body of the report, should state the subject, purpose, scope, and the plan of the report. In many cases, an introduction will include a summary of the findings or conclusions. If a report is a progress report, the introduction should relate the current report to previous reports. Introductions, then, not only tell the sequence or plan of the report, but tell the what, how and why of the subject as well.

PITFALL 8: DAZZLING WITH DATA

Someone once said that a good painter not only knows what to put in a painting, but more importantly he knows what to leave out. It's much the same with report writing. If you dazzle your reader with tons of data, he may be moved by the weight of the report but may get no more out of it than that.

The usual error occurs in supportive material that many engineers and scientists feel is unnecessary to give a report scientific importance. The truth is that successful scientific writing (which includes reports) is heavily grounded in reality, simplicity, and understanding—not quantity.

The simplest way to evaluate the relevancy of information is to ask yourself after writing a paragraph, "What can I remove from this paragraph without destroying its meaning and its relationship to what precedes and what will follow?" Then, ask another question, "Does my reader require all that data to comprehend, evaluate or make a decision with?" If you find you can do without excess words, excess description and excessive supportive data, you will end up with a tighter, better and more informative report.

These principles should also be used to evaluate graphs, photographs, diagrams and other illustrations. Remember, illustrations should support or aid comprehension rather than being a crutch on which your report leans. The same should be kept in mind when determining just how much you should append to your report. There is no need to copy all your lab notes to show that detailed experimentation was performed to substantiate the results. A statement that the notes exist and are available will suffice.

PITFALL 9: NOT HIGHLIGHTING

Again, I believe the analogy of the painter applies. A good painter also knows what to highlight and what to subdue in a portrait or scene.

If you don't accent the significant elements, findings, illustrations, data, tests, facts, trends, procedures, precedents, or experiments pertinent to the subject and object of your report, you place the burden of doing so on your reader. As a result, he may consider the report a failure, draw his own conclusions, or hit upon the significant elements by chance. In any event, don't leave it up to your reader to search out the major points of your report.

Highlighting is one step past knowing what goes into your report and what to leave out (see Pitfall 8 above). All the key points of your report should define and focus on the purpose of your report. They must be included in your summary or conclusions, but these sections are not the only places to highlight. Attention should be called to key elements needed for the understanding of your material throughout the body of the report. Several methods may be used: you can underline an important statement or conclusion, you can simply point out that a particular illustration is the proof of the results of an experiment, or, as most professional writers do, you can make the key sentence the first or last sentence of a paragraph.

PITFALL 10: NOT REWRITING

Did you ever hear of an actor who hadn't rehearsed his lines before stepping before an audience? An actor wouldn't chance it—his reputation and his next role depend on his performance. The engineer shouldn't chance it either.

Don't expect the draft of your report to be ready for final typing and reproduction without rewriting.

Once you have judged what your report will contain and how it will be organized, just charge ahead and write the first draft. Don't worry about choosing the precise word, turning that meaningful phrase, or covering all the facts in one paragraph or section. Once you have written your first draft (and the quicker you accomplish this the more time you will have to perfect the text), you are in a better position to analyze, tailor, and refine the report as a whole. Now you are also able to focus all the elements toward your purpose and your audience.

As you begin the rewriting process simply pick up each page of your draft, scan it, and ask yourself what role the material on that page plays in the fulfillment of the report's objective and understanding. You will find that this will enable you to delete, add, change and rearrange your material very quickly.

After you have completed this process, then rewrite paragraph by paragraph, sentence by sentence, and word by word. Your final step is to repeat the procedure of examining each page's contents. When you are satisfied with its flow and cohesion, then you will have a good report, one you know will be well received and acted upon.

Walter E. Oliu, Charles T. Brusaw, and Gerald J. Alred

Creating Tables and Illustrations

Walter E. Oliu is a technical writer with the U.S. Nuclear Regulatory Commission. Charles T. Brusaw is retired from NCR Corporation where he was a senior program instructor. Gerald J. Alred is professor of English at the University of Wisconsin–Milwaukee.

The primary purpose of including tables and illustrations in your writing is to increase your reader's understanding of what you are saying. Tables, graphs, photographs, drawings, charts, and maps—often collectively called *visuals* or *visual aids*—can frequently express ideas or convey information that words alone cannot. Tables allow the easy comparison of large numbers of statistics that would be difficult to understand if they appeared in sentence form. Graphs make trends and mathematical relationships immediately evident. And drawings, photographs, charts, and maps can indicate shapes and relationships in space more concisely and efficiently than can text.

By allowing the reader to interpret data at a glance, these visuals encourage faster decision-making. Tables and illustrations should be functional, not decorative. When using them, consider your purpose and your reader carefully. For example, the drawing of a dental x-ray machine for a high school science class would be different from an illustration provided for the technician who repairs such machines.

Many of the qualities of good writing—simplicity, clarity, conciseness, directness—are just as important in the creation and use of visuals. Presented with clarity and consistency, visuals can help your reader focus on key portions of your document. Be aware, though, that even the best visual only supplements, or supports, the text. Your writing must carry the burden of providing context for the visual and pointing out its significance. The following general guidelines apply to most visual materials.

CREATING ILLUSTRATIONS AND INTEGRATING THEM WITH TEXT

Each type of illustration has unique strengths and weaknesses. . . . The guidelines presented here apply to most visual materials you might use to supplement or clarify the information in your text (including materials produced with computer graphics software). These tips will help you create and present your visual materials to good effect.

1. Make clear in the text why the illustration is included. The amount of description each illustration requires will vary with its importance. An illustration showing an important feature or system may be central to an entire discussion. The complexity of the illustration will also affect the discussion, as will the background your readers bring to the information. Nonexperts require lengthier explanations than experts do, as a rule.
2. Keep the illustration brief and simple by including only information necessary to the discussion in the text.
3. Present only one type of information in each illustration.
4. Keep terminology consistent. Do not refer to something as a "proportion" in the text and as a "percentage" in the illustration.
5. Specify the units of measurement used or include a scale of relative distances, when appropriate.
6. Position the lettering of any explanatory text or labels horizontally for ease of reading, if possible.
7. Give each illustration a concise title that clearly describes its content.
8. Assign a figure or table number, particularly if your document contains five or more illustrations. The figure or table number precedes the title: *Figure 1. Widget Production for Fiscal 19__*. Note that graphics (photographs, drawings, maps, etc.) are generally labeled "figures," while tables are labeled "tables."
9. Refer to illustrations by their figure or table number.
10. If an illustration is central to a discussion, illuminating or strongly reinforcing it, place it as close as possible to the text where it is discussed. However, no illustration should precede its first text mention. Its appearance without an introduction in the text will confuse readers. But, if the illustration is lengthy, detailed, and peripheral to the discussion, place it in an appendix, although even material in an appendix should be referred to in the text.
11. Allow adequate white space on the page around and within the illustration.
12. In documents with more than five illustrations, list them by title, together with figure and page numbers, or table and page numbers, following the table of contents. The figures so listed should be titled "List of Figures." The tables so listed should be titled "List of Tables."

13. If you wish to use an illustration from a copyrighted publication, first obtain a written release to do so from the copyright holder. Acknowledge such borrowings in a source or credit line below the caption for a figure and below any footnotes at the bottom of a table. Illustrated materials in publications of the federal government are not copyrighted. You need not obtain written permission to reproduce them, although you should acknowledge their source in a credit line. (Such a source line appears below Figure 1.)

A discussion of visuals commonly used in on-the-job writing follows. Your topic will ordinarily determine the best visual material to use.

TABLES

A table is useful for showing large numbers of specific, related data in a brief space. Because a table displays its information in rows and columns, the reader can easily compare data in one column with data in another. If such data were presented in the text, the reader would read through groups of numbers and possibly not recognize their significance. Tables typically include the following elements (see Figure 1):

- *Table number.* If you are using several tables, assign each a number; center the number and title above the table. Table numerals are usually Arabic, and they should be assigned sequentially to the tables throughout the text. Tables should

Table number ——→ Table title

Table 1. Recreational Fresh Water Angling by Water-Body Type and Geographical Region*

Boxhead ⟨ Column headings

Geographical Regions	Reservoirs	Manmade Ponds	Natural Lakes & Ponds	Rivers & Streams	Farm Ponds
New England	130	40	570	410	410
Middle Atlantic	710	290	780	1200	630
East North Central	1200	760	3100	1600	1300
West North Central	810	550	1200	970	980
South Atlantic	1100	760	640	1500	1600
East South Central	890	630	190	670	1200
West South Central	1700	610	430	880	1300
Mountain	820	50	280	600	230
Pacific	950	200	820	1400	470
Totals	8300	3900	8000	9200	7800

Stub — Body — Rule

Footnote ——→ *In thousands of anglers. Anglers who fished in more than one water body or region are represented in more than one category.

Source line ——→ SOURCE: U.S. Department of the Interior

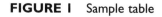

FIGURE I Sample table

be referred to in the text by table number rather than by location ("Table 4" rather than "the above table").

- *Table title.* The title, which is placed just above the table, should describe concisely what the table represents.
- *Boxhead.* It contains the column headings. They should be brief but descriptive. Units of measurement, where necessary, should either be specified as part of the heading or enclosed in parentheses beneath the heading. Standard abbreviations and symbols are acceptable. Avoid vertical lettering whenever possible.
- *Stub.* The lefthand vertical column of a table is called the stub. It lists the items about which information is given in the body of the table.
- *Body.* The body comprises the data below the column headings and to the right of the stub. Within the body, columns should be arranged so that the terms to be compared appear in adjacent rows and columns. Where no information exists for a specific item, substitute a row of dots or a dash to acknowledge the gap.
- *Rules.* These are the lines that separate the table into its various parts. Horizontal rules are placed below the title, below the body of the table, and between the column headings and the body of the table. Tables should not be closed at the sides. The columns within the table may be separated by vertical rules if such lines aid clarity.
- *Source line.* The source line, which identifies where the data were obtained, appears below the table (when a source line is appropriate). Many organizations place the source line below the footnotes.
- *Footnotes.* Footnotes are used for explanations of individual items in the table. Symbols (*, †) or lower case letters (sometimes in parentheses) rather than numbers are ordinarily used to key table footnotes because numbers might be mistaken for numerical data within the table.
- *Continuing tables.* When a table must be divided so that it can be continued on another page, repeat the column headings and give the table number at the head of each new page with a "continued" label ("Table 3, continued").

GRAPHS

Graphs, like tables, present numerical data in visual form. Graphs have several advantages over tables. Trends, movements, distributions, and cycles are more readily apparent in graphs than they are in tables. Further, by providing a means for ready comparisons, a graph often shows a significance in the data not otherwise immediately evident. Be aware, however, that although graphs present statistics in a more interesting and comprehensible form than tables do, they are less accurate. For this reason, they are often accompanied by tables that give exact figures. (Note the difference between the graph and table showing the same data in Figure 2.) If the graph remains uncluttered, the exact data can be added to each column, thereby giving the reader both a quick overview of the data and accurate figures. (See Figures 11 and 12 on pp. 245–246.) There are many different kinds of graphs, most notably line graphs, bar graphs, pie graphs, and picture

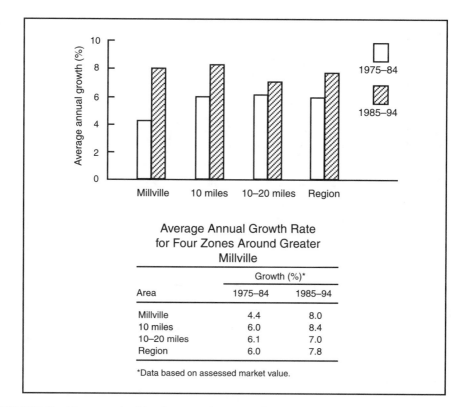

FIGURE 2   Figure and table showing the same data

graphs. All kinds of graphs can now be easily rendered with the aid of computer graphics. (As examples, see Figures 26 and 27 on p. 260.)

Line Graphs

The line graph, which is the most widely used of all graphs, shows the relationship between two sets of figures. The graph is composed of a vertical axis and a horizontal axis that intersect at right angles. Each axis represents one set of figures. The relationship between the two sets is indicated by points plotted along appropriate intersections of the two axes. Once plotted, the points are connected to form a continuous line, and the relationship between the two sets of data becomes readily apparent.

The line graph's vertical axis usually represents amounts (the vertical axis in Figure 3 represents reported cases of Rocky Mountain Spotted Fever), and its horizontal axis usually represents increments of time (the horizontal axis in Figure 3 represents five-year increases).

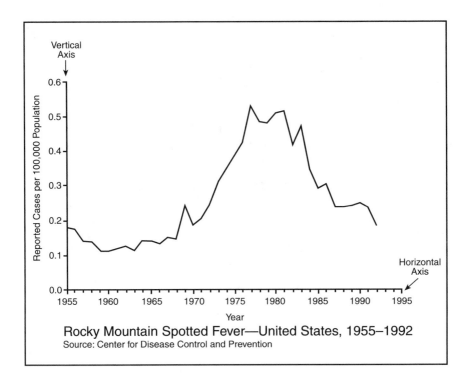

FIGURE 3   Single-line graph

Line graphs with more than one plotted line allow for comparisons between two sets of statistics. In creating such graphs, be certain to identify each plotted line with a label or a legend, as shown in Figure 4. You can emphasize the difference between the two lines by shading the space that separates them. The following guidelines apply to most line graphs:

1. Give the graph a title that describes the data clearly and concisely.
2. If your report includes several visuals, assign a figure number to each one.
3. Indicate the *zero point* of the graph (the point where the two axes meet). If the range of data shown makes it inconvenient to begin at zero, insert a break in the scale, as in Figure 5.
4. Divide the vertical axis into equal portions, from the least amount at the bottom to the greatest amount at the top. The caption for this scale may be placed at the upper left, or, as is more often the case, vertically along the vertical axis, as in Figure 4.
5. Divide the horizontal axis into equal units from left to right, and label them to show what values each represents.
6. The angle at which the curved line rises and falls is determined by the scales of the two axes—that is, by the units into which each axis is divided. Therefore, divide the vertical and horizontal scales so that they give an accurate visual

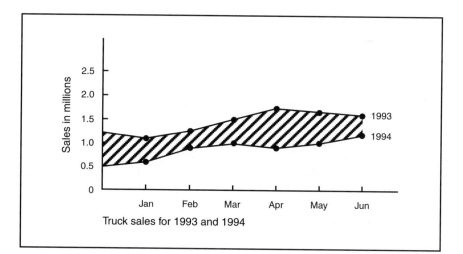

FIGURE 4 Double-line graph with difference between the two years shaded

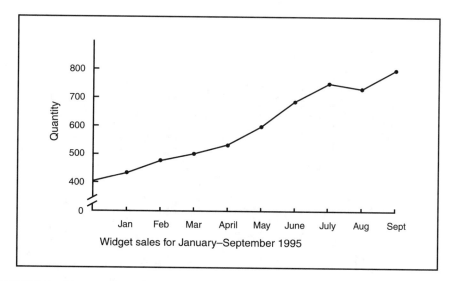

FIGURE 5 Line graph with vertical axis broken

impression of the data. The curve can be kept free of distortion if the ratio between the scales is kept constant. (See Figure 6.)

7. Hold grid lines to a minimum so that the curved lines stand out. Since precise values are usually shown in a table of data accompanying a graph, detailed grid lines are unnecessary. Note the increasing clarity of the three graphs shown in Figures 7, 8, and 9.

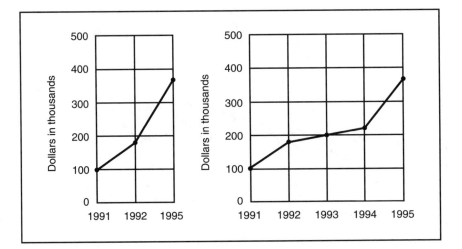

FIGURE 6   Distorted expression of data (left) and distortion-free expression of data

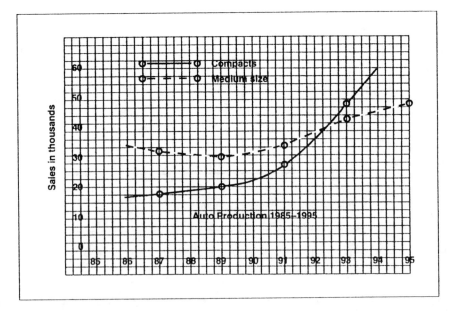

FIGURE 7   A line graph that is difficult to read

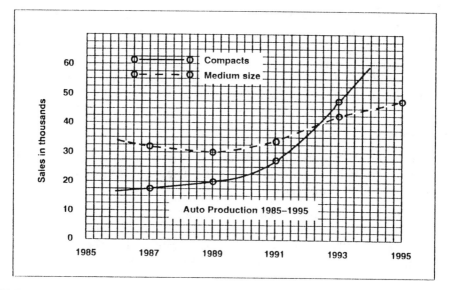

FIGURE 8   A more legible version of Figure 7

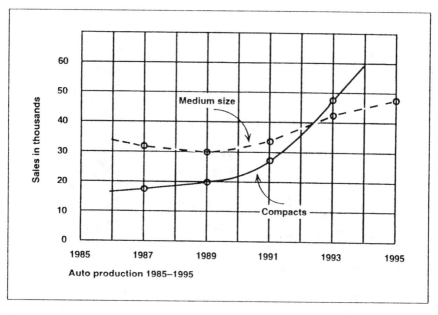

FIGURE 9   A clear version of Figure 7

8. Include the key when necessary, as in Figure 8. Sometimes a label will do just as well, as in Figure 9.
9. If the information comes from another source, include the source line under the graph at the lower left, as in Figure 16 on p. 249.
10. Place explanatory footnotes directly below the figure caption. (See Figure 13.)
11. Make all lettering read horizontally if possible.

Bar Graphs

Bar graphs consist of horizontal or vertical bars of equal width but scaled in length or height to represent some quantity. They are commonly used to show the following proportional relations:

1. Varying quantities of the same item during a fixed period of time (Figure 10)
2. Quantities of the same item at different points in time (Figure 11)
3. Quantities of different items during a fixed period of time (Figure 12)
4. Quantities of the different parts that make up a whole (Figure 13).

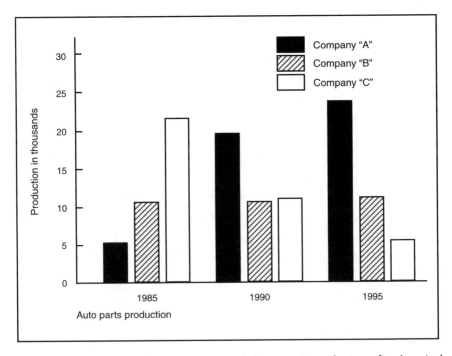

FIGURE 10   Bar graph showing quantities of the same item during a fixed period

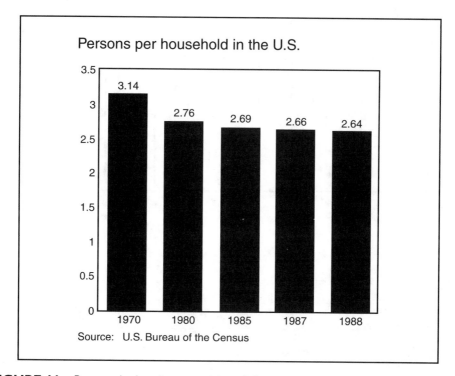

Persons per household in the U.S.

Source: U.S. Bureau of the Census

FIGURE 11 Bar graph showing quantities of the same item at different points in time

Note that in Figure 12, showing refined-petroleum prices, the exact price appears at the top of each bar. This eliminates the need to have an accompanying table giving the price data. If the bars are not labeled, as in Figures 10 and 14, the different portions must be clearly indicated by shading, crosshatching, or other devises. Include a key that represents the various subdivisions.

Bar graphs can also indicate what proportion of a whole the various component parts represent. In such a graph, the bar, which is theoretically equivalent to 100 percent, is divided according to the proportion of the whole that each item sampled represents. (Compare the displays of the same data in Figures 13 and 15.) In some bar graphs, the completed bar does not represent 100 percent, because not all parts of the whole have been included in the sample. (See Figure 14.)

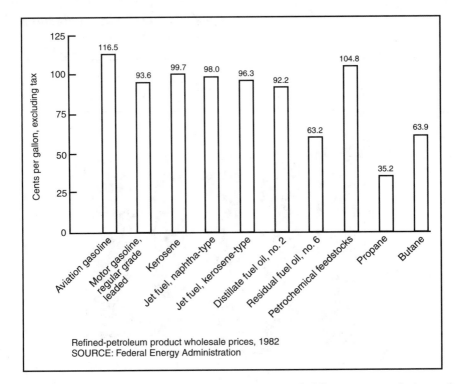

FIGURE 12 Bar graph showing varying quantities of different items during a fixed period (1982)

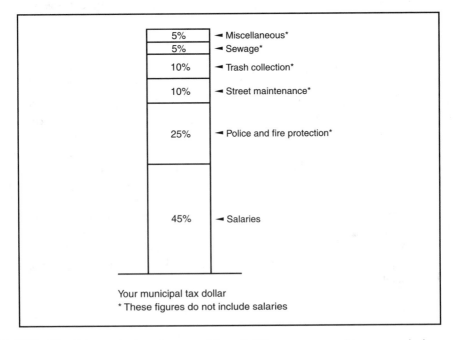

FIGURE 13 Bar graph showing quantities of different parts making up a whole

Pie Graphs

A pie graph presents data as wedge-shaped sections of a circle. The circle equals 100 percent, or the whole, of some quantity (a tax dollar, a bus fare, the hours of a working day), with the wedges representing the various parts into which the whole is divided. In Figure 15, for example, the circle stands for a city tax dollar and is divided into units equivalent to the percentages of the tax dollar spent on various city services.

The relationships among the various statistics presented in a pie graph are easy to grasp, but the information is often rather general. For this reason, a pie graph is often accompanied by a table that presents the actual figures on which the percentages in the graph are based.

When you construct a pie graph, keep the following points in mind:

1. The complete 360° circle is equivalent to 100 percent; therefore, each percentage point is equivalent to 3.6°
2. *When possible,* begin at 12 o'clock position and sequence the wedges clockwise, from largest to smallest. (Adhering to this guidance is not always possible because some computer-drawn pie charts appear counterclockwise.)
3. If you shade the wedges, do so clockwise and from light to dark.
4. Keep all labels horizontal and, most important, give the percentage value of each wedge.
5. Finally, check to see that all wedges, as well as the percentage values given for them, add up to 100 percent.

Although pie graphs have a strong visual impact, they also have drawbacks. If more than five or six items of information are presented, the graph looks cluttered. And unless percentages are labeled on each section, the reader cannot compare the values of the sections as accurately as on a bar graph.

Picture Graphs

Picture graphs are modified bar graphs that use picture symbols to represent the item for which data are presented. Each symbol corresponds to a specified quantity of the item, as shown in Figure 16. Note that precise figures are also included, since the picture symbol can indicate only approximate figures. Here are some tips on preparing picture graphs:

1. Make the symbol self-explanatory.
2. Have each symbol represent a specific number of units.
3. Show larger quantities by increasing the number of symbols rather than by creating a larger symbol, because if the latter is done it is difficult to judge relative size correctly.

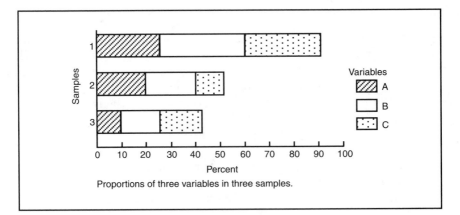

FIGURE 14 Bar graph in which not all parts of the whole have been included

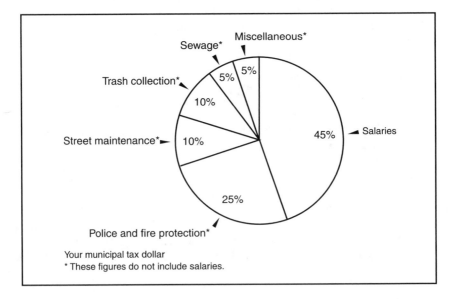

FIGURE 15 Pie graph

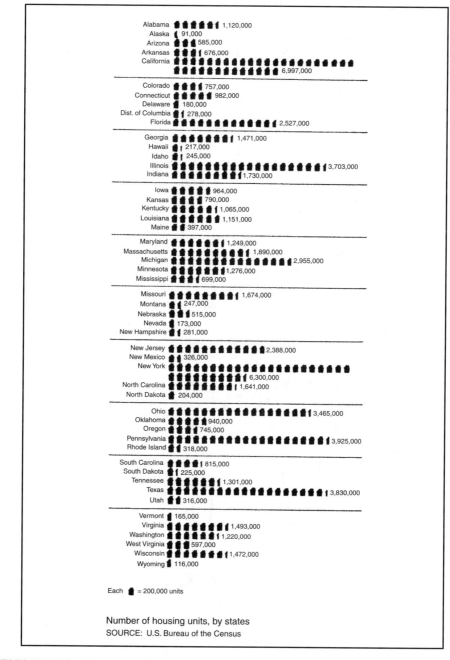

Each 🏠 = 200,000 units

Number of housing units, by states
SOURCE: U.S. Bureau of the Census

FIGURE 16 Picture graph

DRAWINGS

A drawing is useful when you wish to focus on details or relationships that a photograph cannot capture (see Figures 17 and 18). It can emphasize the significant piece of a mechanism, or its function, and omit what is not significant. However, if the precise details of the actual appearance of an object are necessary to your report or document, a photograph is essential.

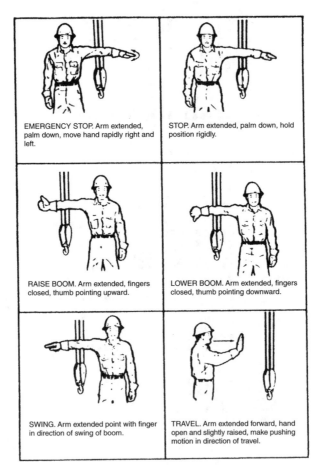

FIGURE 17 Drawing showing relationships among hand signals for crane operation
Source: Harnischfeger Corporation

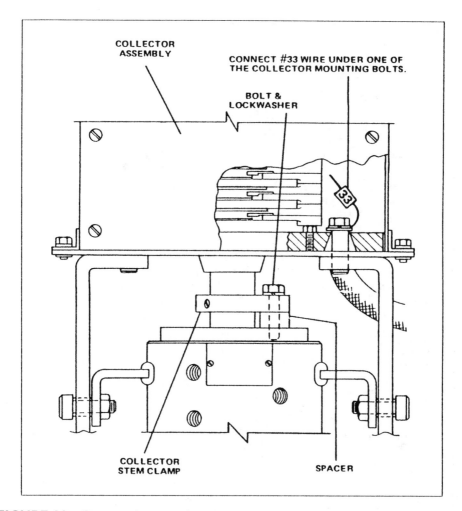

FIGURE 18   Cutaway drawing of a collector
Source: **Harnischfeger Corporation**

To show the proper sequence in which parts fit together, or when it is essential to show the details of each individual part, use an *exploded-view drawing.* (See Figure 19.)

Many organizations have their own format specifications for drawings. In the absence of such specifications, the following tips should be helpful:

1. Give the drawing a clear title and a figure number, both of which should be centered or flush left below the drawing.
2. Place the source line, if there is one, aligned beneath the title.

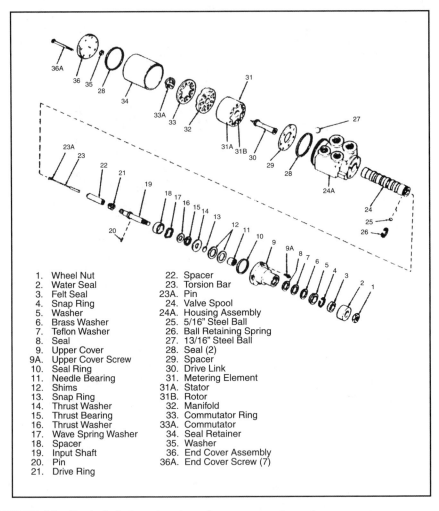

FIGURE 19   Exploded-view drawing of power steering valve

Source: **Harnischfeger Corporation**

3. Show the equipment from the point of view of the person who will use it.
4. When illustrating a subsystem, show its relationship to the larger system of which it is a part.
5. Draw the different parts of an object in proportion to one another, unless you indicate that certain parts are enlarged.
6. Where a sequence of drawings is used to illustrate a process, arrange them from left to right and from top to bottom.
7. Label parts in the drawing so that text references to them are clear.

8. Depending on the complexity of what is shown, labels may be placed on the parts themselves, or the parts may be given letter or number symbols, with an accompanying key. (See Figure 19.) . . .

FLOWCHARTS

A *flowchart* is a diagram of a process that involves stages, with the sequence of stages shown from beginning to end The flowchart presents an overview of the process that allows the readers to grasp the essential steps quickly and easily. The process being illustrated could range from the stages by which bauxite ore is refined into aluminum ingots for fabrication to the steps involved in preparing a manuscript for publication.

Flowcharts can take several forms to represent the steps in a process. They can consist of labeled blocks (Figure 20), pictorial representations (Figure 21), or standardized symbols (Figure 22). The items in any flowchart are always connected according to the sequence in which the steps occur. The normal direction of flow in a chart is left to right or top to bottom. When the flow is otherwise, be sure to indicate it with arrows.

Flowcharts that document computer programs and other information-processing procedures use standardized symbols. The standards are set forth in *U.S.A. Standard Flowchart Symbols and Their Usage in Information Processing,* published by the American National Standards Institute, publication X3.5. When creating a flowchart, follow these guidelines:

1. Label the flowchart clearly and concisely.
2. Assign the chart a figure number if it is being used in a document that contains five or more illustrations.
3. With labeled blocks and standardized symbols, use arrows to show the direction of flow only if the flow is opposite to the normal direction. With pictorial representations, use arrows to show the direction of all flow.
4. Label each step in the process, or identify it with a conventional symbol. Steps can also be represented pictorially or by captioned blocks.
5. Include a key if the flowchart contains symbols that your reader may not understand.
6. Leave adequate white space on the page. Do not crowd your steps and directional arrows too close together.
7. As with other illustrations, place the flowchart as near as possible to that portion of the text that refers to it.

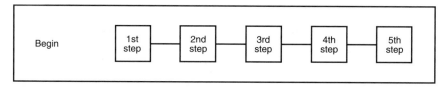

FIGURE 20   Simple block flowchart

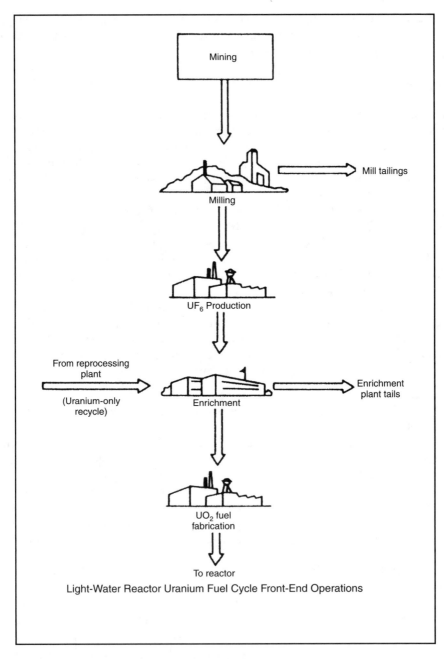

FIGURE 21   Flowchart using pictorial symbols

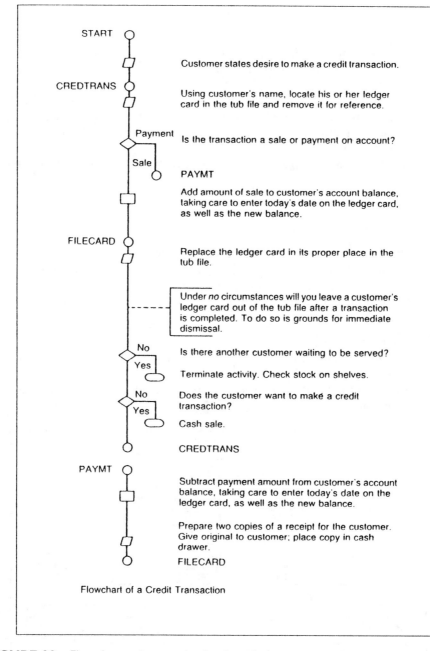

START ⭘

Customer states desire to make a credit transaction.

CREDTRANS ⭘

Using customer's name, locate his or her ledger card in the tub file and remove it for reference.

Payment

Is the transaction a sale or payment on account?

Sale

⭘ PAYMT

Add amount of sale to customer's account balance, taking care to enter today's date on the ledger card, as well as the new balance.

FILECARD ⭘

Replace the ledger card in its proper place in the tub file.

Under *no* circumstances will you leave a customer's ledger card out of the tub file after a transaction is completed. To do so is grounds for immediate dismissal.

No

Is there another customer waiting to be served?

Yes

Terminate activity. Check stock on shelves.

No

Does the customer want to make a credit transaction?

Yes

Cash sale.

⭘ CREDTRANS

PAYMT ⭘

Subtract payment amount from customer's account balance, taking care to enter today's date on the ledger card, as well as the new balance.

Prepare two copies of a receipt for the customer. Give original to customer; place copy in cash drawer.

⭘ FILECARD

Flowchart of a Credit Transaction

FIGURE 22 Flowchart using standardized symbols

ORGANIZATIONAL CHARTS

An organizational chart shows how the various components of an organization are related to one another. Such an illustration is useful when you want to give readers an overview of an organization or indicate the lines of authority within the organization. (See Figure 23.)

The title of each organizational component (office, section, division) is placed in a separate box. These boxes are then linked to a central authority. If your readers need the information, include the name of the person occupying the position identified in each box.

As with all illustrations, place the organizational chart as close as possible to the text that refers to it.

MAPS

Maps can be used to show the specific geographic features of an area (roads, mountains, rivers) or to show information according to geographic distribution (population, housing, manufacturing centers, and so forth).

Bear these points in mind as you create maps for use with your text (see Figure 24):

1. Label the map clearly.
2. Assign the map a figure number if you are using enough illustrations (five) to justify the use of figure numbers.
3. Make sure all boundaries within the map are clearly identified. Eliminate unnecessary boundaries.

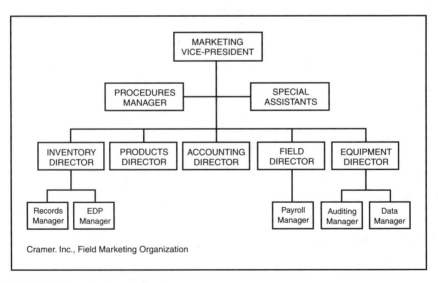

FIGURE 23   Organizational chart

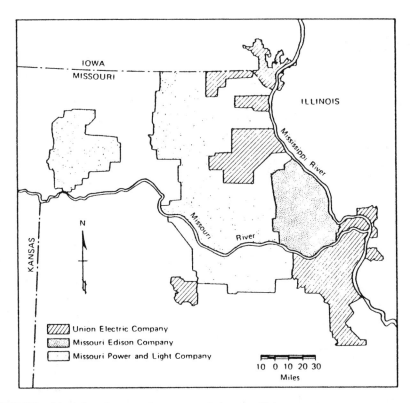

FIGURE 24 Map showing service areas of three utilities
Source: U.S. Nuclear Regulatory Commission

4. Eliminate unnecessary information from your map. For example, if population is the focal point, do not include mountains, roads, rivers, and so on.
5. Include a scale of miles or feet, or of kilometers or meters, to give your reader an indication of the map's proportions.
6. Indicate which direction is north.
7. Emphasize key features by using shading, dots, crosshatching, or appropriate symbols, and include a key telling what the different colors, shadings, or symbols represent.
8. Place maps as close as possible to the portion of the text that refers to them, but not preceding the first text reference.

COMPUTER GRAPHICS

The widespread availability of graphics capabilities on personal computers has made their use increasingly common. With access to the proper hardware, software, and printer, you can create tables, graphs, charts, drawings, even maps, with near-professional quality. Once an image has been created on the computer screen, you can save it in a file, print it out again and again, or recall it for subsequent modification. As you create such graphics materials, be sure to apply the principles specific to each type presented in this chapter, recognizing that software packages often have inflexible settings for some features, such as the sequence in which segments appear on a pie chart, or the size and kind of typeface used for words and numbers on the graphic. After creating a computer graphic, use the guidelines at the beginning of this chapter for integrating it into your document.

Computer graphics for personal computers fall into two broad categories: *business graphics* and *free-form graphics*.

Business Graphics

Business graphics programs, also called charting programs, produce preformatted charts, graphs, and tables based on numerical data taken either from a computer spreadsheet program, as in Figure 25, or from the text and data entered from the computer keyboard. Figure 26 shows a range of typical bar and pie charts created by business graphics software.

Free-Form Graphics

In free-form graphics, as the name implies, images are not automatically preformatted to become charts, graphs, and tables that display numerical data. Instead, the user controls the image by manipulating a variety of strokes, lines, shapes, and symbols, much as an artist wields pencil, brush, charcoal, templates, and other tools to create images on paper and canvas. Many such programs also contain "clip art" libraries of ready-to-use electronic images of various symbols, shapes, and pictures, samples of which are shown in Figure 27.

Free-form programs are further subdivided into *paint programs* and *draw programs,* categories defined by the software technology each uses. Paint programs allow users to create freehand images on screen, usually with an input device called a *mouse.* (A mouse is a hand-held device that the user pushes across a desktop pad to move the cursor on the screen and to draw and modify images.) You can control images created with paint programs down to the individual dot, "bit," or picture element (known as a pixel) on the screen. Because of this capability, these programs are often referred to as pixel- or bit-mapped. The pixels are similar to the dots that make up the images on a television screen. Figure 28 (p. 261) shows a typical paint program image. Note how the pixels create a "fuzzy" image.

Draw programs differ from paint programs in that the user manipulates predefined changes, called *graphics primitives* (boxes, triangles, arcs, lines, etc.), rather than pixels. That is, the user creates images by manipulating existing "building block" configurations, such as those shown in Figure 29. The images

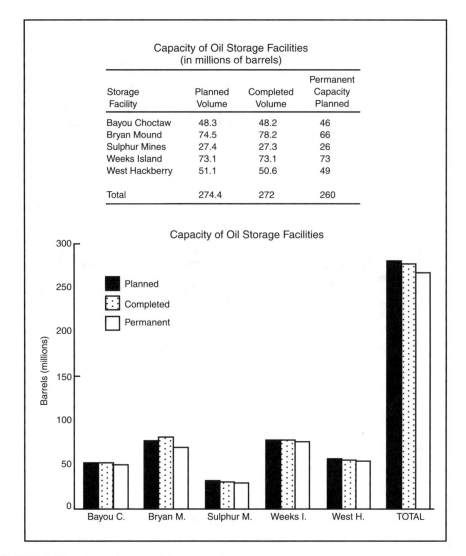

Capacity of Oil Storage Facilities
(in millions of barrels)

Storage Facility	Planned Volume	Completed Volume	Permanent Capacity Planned
Bayou Choctaw	48.3	48.2	46
Bryan Mound	74.5	78.2	66
Sulphur Mines	27.4	27.3	26
Weeks Island	73.1	73.1	73
West Hackberry	51.1	50.6	49
Total	274.4	272	260

FIGURE 25   Spreadsheet table-to-graph image

are manipulated and saved to memory by the coordinates, or vectors, of the beginning and ending points of the lines that make them up. For this reason they are sometimes called vector- or object-oriented programs. These images can be combined to form graphs, flowcharts, and organizational charts, as well as line drawings. Figure 30 shows an image rendered with draw-program software. Note the crispness of this image compared with the same subject created with a paint program in Figure 28.

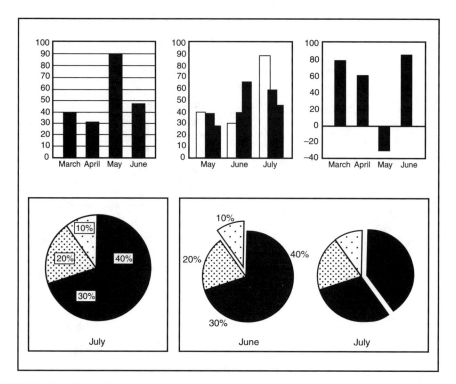

FIGURE 26   Typical computer-generated bar and pie charts

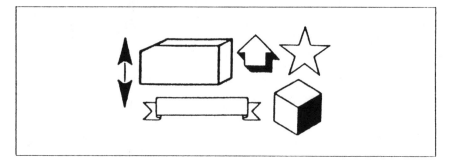

FIGURE 27   Sample computer clip art

FIGURE 28   Typical paint program image

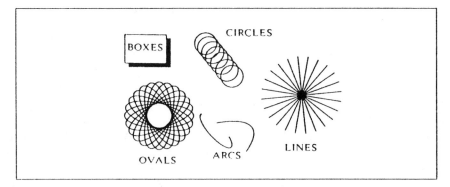

FIGURE 29   Graphics primitives

FIGURE 30   Image created from draw program

Specialized Graphics Programs

Other graphics programs are available for creating complex maps, statistical analyses, animation, and computer-aided design (CAD) images used in engineering and manufacturing. They offer specialized applications for the design, engineering, architectural, or scientific professional that are beyond the scope of this . . . [discussion].

SUMMARY

Tables and illustrations—graphs, drawings, photographs, charts, and maps—can increase the reader's understanding of what you are saying: they can express ideas or convey information in ways that words alone cannot. To ensure that visuals are effective, create or select those that are appropriate to your purpose and suitable to your reader's needs.

Visuals perform the following functions:

- *Tables* show comparisons between figures by arranging them in rows and columns.
- *Line graphs* show the relationship over time between two or more sets of figures.
- *Bar graphs* show
  - Varying quantities of the same item during a fixed period of time.
  - Quantities of the same item at different points in time.

- Quantities of different items during a fixed period of time.
- Quantities of the different parts that make up a whole.

- *Pie graphs* show data values as wedge-shaped sections of a circle, each wedge representing a percentage of the whole.
- *Picture graphs,* which are modified bar graphs, show picture symbols to correspond to a specific quantity of the item depicted.
- *Drawings* depict the details of an object or relationships between objects.
    - *Cutaway drawings* reveal the inner parts of a machine or other object.
    - *Exploded-view drawings* illustrate how parts fit together or present the details of each separate part. . . .
    - *Flowcharts* depict the various stages of a process.
    - *Organizational charts* show how the various jobs in an organization relate to one another and indicate lines of authority in an organization.
    - *Maps* show specific geographical features of an area or information according to geographic distribution.

Darrell Huff

How to Lie with Statistics

Darrell Huff, a freelance writer, expanded this article into a book with the same title (Norton, 1954).

"The average Yaleman, Class of '24," *Time* magazine reported last year after reading something in the New York *Sun,* a newspaper published in those days, "makes $25,111 a year."

Well, good for him!

But, come to think of it, what does this improbably precise and salubrious figure mean? Is it, as it appears to be, evidence that if you send your boy to Yale you won't have to work in your old age and neither will he? Is this average a mean or is it a median? What kind of sample is it based on? You could lump one Texas oil-man with two hundred hungry freelance writers and report *their* average income as $25,000-odd a year. The arithmetic is impeccable, the figure is convincingly precise, and the amount of meaning there is in it you could put in your eye.

In just such ways is the secret language of statistics, so appealing in a fact-minded culture, being used to sensationalize, inflate, confuse, and oversimplify. Statistical terms are necessary in reporting the mass data of social and economic trends, business conditions, "opinion" polls, this year's census. But without writers who use the words with honesty and understanding and readers who know what they mean, the result can only be semantic nonsense.

In popular writing on scientific research, the abused statistic is almost crowding out the picture of the white-jacketed hero laboring overtime without time-and-a-half in an ill-lit laboratory. Like the "little dash of powder, little pot of paint," statistics are making many an important fact "look like what she ain't." Here are some of the ways it is done.

The sample with the built-in bias. Our Yale men—or Yalemen, as they say in the Time-Life building—belong to this flourishing group. The exaggerated estimate of their income is not based on all members of the class nor on a random or representative sample of them. At least two interesting categories of 1924-model Yale men have been excluded.

First there are those whose present addresses are unknown to their classmates. Wouldn't you bet that these lost sheep are earning less than the boys from prominent families and the others who can be handily reached from a Wall Street office?

There are those who chucked the questionnaire into the nearest wastebasket. Maybe they didn't answer because they were not making enough money to brag about. Like the fellow who found a note clipped to his first pay check suggesting that he consider the amount of his salary confidential: "Don't worry," he told the boss. "I'm just as ashamed of it as you are."

Omitted from our sample then are just the two groups most likely to depress the average. The $25,111 figure is beginning to account for itself. It may indeed be a true figure for those of the Class of '24 whose addresses are known and who are willing to stand up and tell how much they earn. But even that requires a possibly dangerous assumption that the gentlemen are telling the truth.

To be dependable to any useful degree at all, a sampling study must use a representative sample (which can lead to trouble too) or a truly random one. If *all* the Class of '24 is included, that's all right. If every tenth name on a complete list is used, that is all right too, and so is drawing an adequate number of names out of a hat. The test is this: Does every name in the group have an equal chance to be in the sample?

You'll recall that ignoring this requirement was what produced the *Literary Digest*'s famed fiasco.* When names for polling were taken only from telephone books and subscription lists, people who did not have telephones or *Literary Digest* subscriptions had no chance to be in the sample. They possibly did not mind this underprivilege a bit, but their absence was in the end very hard on the magazine that relied on the figures.

This leads to a moral: You can prove about anything you want to by letting your sample bias itself. As a consumer of statistical data—a reader, for example, of a news magazine—remember that no statistical conclusion can rise above the quality of the sample it is based upon. In the absence of information about the procedures behind it, you are not warranted in giving any credence at all to the result.

The truncated, or gee-whiz, graph. If you want to show some statistical information quickly and clearly, draw a picture of it. Graphic presentation is the thing today. If you don't mind misleading the hasty looker, or if you quite clearly *want* to deceive him, you can save some space by chopping the bottom off many kinds of graph.

---

*Editor's note: The *Literary Digest* predicted that Alfred Landon would defeat Franklin Roosevelt in the 1936 presidential election. Landon carried only two states.

Suppose you are showing the upward trend of national income month by month for a year. The total rise, as in one recent year, is 7 percent. It looks like this:

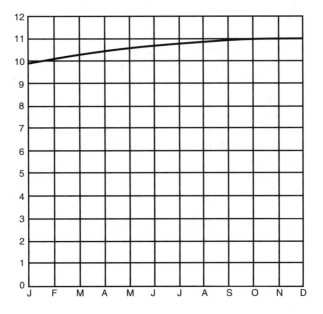

That is clear enough. Anybody can see that the trend is slightly upward. You are showing a 7 percent increase, and that is exactly what it looks like.

But it lacks schmaltz. So you chop off the bottom, this way:

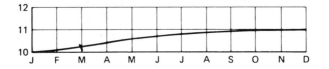

The figures are the same. It is the same graph and nothing has been falsified—except the impression that it gives. Anyone looking at it can just feel prosperity throbbing in the arteries of the country. It is a subtler equivalent of editing "National income rose 7 percent" into " . . . climbed a whopping 7 percent."

It is vastly more effective, however, because of that illusion of objectivity.

The souped-up graph. Sometimes truncating is not enough. The trifling rise in something or other still looks almost as insignificant as it is. You can make that 7 percent look livelier than 100 percent ordinarily does. Simply change the proportion between the ordinate and the abscissa. There's no rule against it, and it does give your graph a prettier shape.

But it exaggerates, to say the least, something awful:

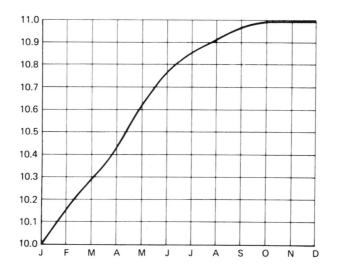

The well-chosen average. I live near a country neighborhood for which I can report an average income of $15,000. I could also report it as $3,500.

If I should want to sell real estate hereabouts to people having a high snobbery content, the first figure would be handy. The second figure, however, is the one to use in an argument against raising taxes, or the local bus fare.

Both are legitimate averages, legally arrived at. Yet it is obvious that at least one of them must be as misleading as an out-and-out lie. The $15,000-figure is a mean, the arithmetic average of the incomes of all the families in the community. The smaller figure is a median; it might be called the income of the average family in the group. It indicates that half the families have less than $3,500 a year and half have more.

Here is where some of the confusion about averages comes from. Many human characteristics have the grace to fall into what is called the "normal" distribution. If you draw a picture of it, you get a curve that is shaped like a bell. Mean and median fall at about the same point, so it doesn't make very much difference which you use.

But some things refuse to follow this neat curve. Income is one of them. Incomes for most large areas will range from under $1,000 a year to upward of $50,000. Almost everybody will be under $10,000, way over on the lefthand side of that curve.

One of the things that made the income figure for the "average Yaleman" meaningless is that we are not told whether it is a mean or a median. It is not that one type of average is invariably better than the other; it depends upon what you are talking about. But neither gives you any real information—and either may be highly misleading—unless you know which of those two kinds of average it is.

In the country neighborhood I mentioned, almost everyone has less than the average—the mean, that is—of $10,500. These people are all small farmers, except for a trio of millionaire week-enders who bring up the mean enormously.

You can be pretty sure that when an income average is given in the form of a mean nearly everybody has less than that.

The insignificant difference or the elusive error. Your two children Peter and Linda (we might as well give them modish names while we're about it) take intelligence tests. Peter's IQ, you learn, is 98 and Linda's is 101. Aha! Linda is your brighter child.

Is she? An intelligence test is, or purports to be, a sampling of intellect. An IQ, like other products of sampling, is a figure with a statistical error, which expresses the precison or reliability of the figure. The size of this probable error can be calculated. For their test the makers of the much-used Revised Stanford-Binet have found it to be about 3 percent. So Peter's indicated IQ of 98 really means only that there is an even chance that it falls between 95 and 101. There is an equal probability that it falls somewhere else—below 95 or above 101. Similarly, Linda's has no better than a fifty-fifty chance of being within the fairly sizeable range of 98 to 104.

You can work out some comparisons from that. One is that there is rather better than one chance in four that Peter, with his lower IQ rating, is really at least three points smarter than Linda. A statistician doesn't like to consider a difference significant unless you can hand him odds a lot longer than that.

Ignoring the error in a sampling study leads to all kinds of silly conclusions. There are magazine editors to whom readership surveys are gospel; with a 40 percent readership reported for one article and a 35 percent for another, they demand more like the first. I've seen even smaller differences given tremendous weight, because statistics are a mystery and numbers are impressive. The same thing goes for market surveys and so-called public opinion polls. The rule is that you cannot make a valid comparison between two such figures unless you know the deviations. And unless the difference between the figures is many times greater than the probable error of each, you have only a guess that the one appearing greater really is.

Otherwise you are like the man choosing a camp site from a report of mean temperature alone. One place in California with a mean annual temperature of 61 is San Nicolas Island on the south coast, where it always stays in the comfortable range between 47 and 87. Another with a mean of 61 is in the inland desert, where the thermometer hops around from 15 to 104. The deviation from the mean marks the difference, and you can freeze or roast if you ignore it.

The one-dimensional picture. Suppose you have just two or three figures to compare—say the average weekly wage of carpenters in the United States and another country. The sums might be $60 and $30. An ordinary bar chart makes the difference graphic.

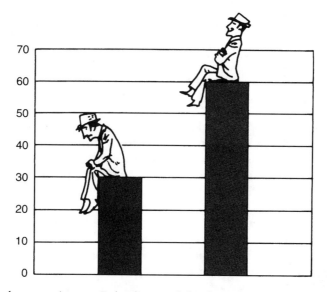

That is an honest picture. It looks good for American carpenters, but perhaps it does not have quite the oomph you are after. Can't you make that difference appear overwhelming and at the same time give it what I am afraid is known as eye-appeal? Of course you can. Following tradition, you represent these sums by pictures of money bags. If the $30 bag is one inch high, you draw the $60 bag two inches high. That's in proportion, isn't it?

The catch is, of course, that the American's money bag, being twice as tall as that of the $30 man, covers an area on your page four times as great. And since your two-dimensional picture represents an object that would in fact have three dimensions, the money bags actually would differ much more than that. The volumes of any two similar solids vary as the cubes of their heights. If the unfortunate foreigner's bag holds $30 worth of dimes, the American's would hold not $60 but a neat $240.

You didn't say that, though, did you? And you can't be blamed, you're only doing it the way practically everybody else does.

The ever-impressive decimal. For a spurious air of precision that will lend all kinds of weight to the must disreputable statistics, consider the decimal.

Ask a hundred citizens how many hours they slept last night. Come out with a total of, say, 781.3. Your data are far from precise to begin with. Most people will miss their guess by fifteen minutes or more and some will recall five sleepless minutes as half a night of tossing insomnia.

But go ahead, do your arithmetic, announce that people sleep an average of 7.813 hours a night. You will sound as if you knew precisely what you are talking about. If you were foolish enough to say 7.8 (or "almost" 8) hours it would sound like what it was—an approximation.

The semiattached figure. If you can't prove what you want to prove, demonstrate something else and pretend that they are the same thing. In the daze that follows the collision of statistics with the human mind, hardly anybody will notice the difference. The semiattached figure is a durable device guaranteed to stand you in good stead. It always has.

If you can't prove that your nostrum cures colds, publish a sworn laboratory report that the stuff killed 31,108 germs in a test tube in eleven seconds. There may be no connection at all between assorted germs in a test tube and the whatever-it-is that produces colds, but people aren't going to reason that sharply, especially while sniffling.

Maybe that one is too obvious and people are beginning to catch on. Here is a trickier version.

Let us say that in a period when race prejudice is growing it is to your advantage to "prove" otherwise. You will not find it a difficult assignment.

Ask that usual cross section of the population if they think . . . [Blacks] have as good a chance as white people to get jobs. Ask again a few months later. As Princeton's Office of Public Opinion Research has found out, people who are most unsympathetic to . . . [Blacks] are the ones most likely to answer yes to this question.

As prejudice increases in a country, the percentage of affirmative answers you will get to this question will become larger. What looks on the face of it like growing opportunity for . . . [Blacks] actually is mounting prejudice and nothing else. You have achieved something rather remarkable: the worse things get, the better your survey makes them look.

The unwarranted assumption, or **post** hoc *rides again.* The interrelation of cause and effect, so often obscure anyway, can be most neatly hidden in statistical data.

Somebody once went to a good deal of trouble to find out if cigarette smokers make lower college grades than non-smokers. They did. This naturally pleased many people, and they made much of it.

The unwarranted assumption, of course, was that smoking had produced dull minds. It seemed vaguely reasonable on the face of it, so it was quite widely accepted. But it really proved nothing of the sort, any more than it proved that poor grades drive students to the solace of tobacco. Maybe the relationship worked in one direction, maybe in the other. And maybe all this is only an indication that the sociable sort of fellow who is likely to take his books less than seriously is also likely to sit around and smoke many cigarettes.

Permitting statistical treatment to befog casual relationships is little better than superstition. It is like the conviction among the people of Hebrides that body lice produce good health. Observation over the centuries had taught them that people in good health had lice and sick people often did not. *Ergo,* lice made a man healthy. Everybody should have them.

Scantier evidence, treated statistically at the expense of common sense, has made many a medical fortune and many a medical article in magazines, including professional ones. More sophisticated observers finally got things straightened out in the Hebrides. As it turned out, almost everybody in those circles had lice most of the time. But when a man took a fever (quite possibly carried to him by those same lice) and his body became hot, the lice left.

Here you have cause and effect not only reversed, but intermingled.

There you have a primer in some ways to use statistics to deceive. A well-wrapped statistic is better than Hitler's "big lie": it misleads, yet it can't be pinned onto you.

Is this little list altogether too much like a manual for swindlers? Perhaps I can justify it in the manner of the retired burglar whose published reminiscences amounted to a graduate course in how to pick a lock and muffle a footfall: The crooks already know these tricks. Honest men must learn them in self-defense.

David W. Ewing

Strategies of Persuasion

David W. Ewing was Executive Editor-Planning of the Harvard Business Review *and a member of the faculty of the Harvard Business School.*

When we review reports, letters, and memoranda that get the intended results, we find a fascinating diversity of approaches. Some are gentle in approach, taking readers by the hand and leading them to a certain finding or recommendation. Others are brisk and abrupt. Some are objective in approach, carefully examining both sides of an idea, like a judge writing a difficult decision—at least until the end. Others burst with impatience to explain one side, and only one side, of a proposal or argument. Some flow swimmingly; others erupt like Mount Vesuvius.

How is it possible that communications taking such different approaches can all be effective? It is not sufficient to answer glibly, "It depends on the situation," or "Communications should mirror the personal style of the communicator." How does a written communication depend on the situation? Which style of the communicator should be reflected in a given communication? Examine the writing of most business executives, professionals, and public leaders, and you will find not one but *many* styles of exposition.

RULES EVERY PERSUADER SHOULD KNOW

The explanation lies in a set of relationships among the communicator, the reader, the message, and the time-space environment. These relationships work in predictable ways and are an important part of the knowledge of every good business and professional writer. They come into play in the planning stages of writing, when the writer is considering how he or she will proceed, and in the main body of the presentation, to accomplish what he or she has promised in the opening paragraphs. The relationships affect some of the most important decisions a writer makes—the choice of ideas to use, the comparative emphasis to be given various arguments pro and con, the types of reasons and supporting material used, the establishment of credibility, and other matters.

It is convenient but self-defeating to follow fixed prescriptions for persuasion, such as to put your strongest arguments first or last or to identify with the readers. Such nostrums were fine for the age of patent medicines and snake-oil peddlers, but not for the age of diagnostic medicine. What is more, they are belittling. They assume that writers are witless. Good writers vary their approaches in response to their readings of different situations. Just as a good golfer plays an approach to a green differently depending on the wind, so a good writer uses different strategies depending on the crosscurrents of mood and feeling.

How should you choose your approach to a group of readers? What elements of the approach should be tailored to the situation? These are the topics of this . . . [selection]. Let us assume that the substance of the intended message is clear in your mind. . . .

1. Consider Whether Your Views Will Make Problems for Readers.

J. C. Mathes and Dwight W. Stevenson tell of the student engineer who was asked to evaluate the efficiency of the employer plant's waste-treatment process.[1] He found that, by making a simple change, the company could save more than $200,000 a year. Anticipating an enthusiastic response, he wrote up and delivered his report. Although he waited with great expectations, no accolades came. Why? What he hadn't counted on was that now his supervisors would have to explain to their bosses why they had allowed a waste of $200,000 a year. They were far from elated to read his report.

"If you want to make a man your enemy," Henry C. Link once said, "tell him simply, 'You are wrong.' This method works every time." Under the illusion that their sciences are "hard," physicists, biologists, and others may assume that it is necessary only to worry about setting the facts forth accurately. From posterity's standpoint, perhaps yes—but not from the standpoint of writing for current results. As Kenneth Boulding, head of the American Association for the Advancement of Science, has pointed out, the so-called "hard" sciences are in many ways "soft," and vice versa. "You can knock forever on a deaf man's door," said Zorba the Greek, and the deaf man can be a physicist as well as a marketing manager or public official.

If your views are bad news for readers, you proceed to report them, but with empathy and tact and an effort to put yourself in the readers' shoes. You work as carefully as if you were licking honey off a thorn.

---

[1] J. C. Mathes and Dwight W. Stevenson, *Designing Technical Reports* (Indianapolis: Bobbs-Merrill, 1976), pp. 18–19.

2. Don't Offer New Ideas, Directives, or Recommendations for Change until Your Readers Are Prepared for Them.

"Should I state my surprising findings at the very beginning of my memorandum?" a writer asks. "Should I go slow with my heretical proposal and hold the reader's hand?" asks another.

Generally speaking, the answer to all such questions depends on the extent of your audience's resistance to change, the amount of change you are asking for, the uncertainty in readers' minds as to your understanding of their situation, and what psychologists call the "perceived threat" of your communication, that is, how much it seems (to readers) to upset their values and interests. The more change, uncertainty, and/or threat, the slower you should proceed, the more carefully you should prepare your readers.

For instance, if your boss is enthusiastic about a new promotion scheme that he (or she) has paid a consultant $25,000 to devise, naturally you will want to go slow in shooting it down (at least if you want to stay in his good graces). In fact, any written criticism of the scheme is probably out of order until you have had a chance to talk with him and get a feel for the proper timing of any forthcoming criticism. When you do commit yourself to writing, you should probably review the arguments for the new scheme as fairly as possible, making it crystal clear to him that you understand them. Only then does it become timely to turn to the facts or conditions that, in your opinion, raise serious questions about the plan.

On the other hand, suppose the faulty promotion scheme is of little personal interest to your boss—it is not his (or her) "baby." Now the situation is different. You can launch right into the shortcomings, throwing your heaviest objections first. The fact is that the boss may not even want to know about the lesser objections, much less the supporting arguments once advanced for the plan; the main things he should know are that (a) the plan is in trouble and (b) the major, most compelling reasons why.

Clearly, this strategy is plain, everyday common sense—what you would normally do in communicating orally instead of in writing. Only in writing you must be more explicit and thorough, because a document lacks the expressiveness and visual advantages of a spoken dialogue.

Now consider another type of situation. Suppose it is your unhappy task to write a department manager that the extra appropriations he (or she) was promised have been cancelled. First of all, how would you handle it if you saw him often and could talk with him personally? . . . You would not (we hope) pussyfoot around trying to withhold the bad news from him. Also, once you had indicated the main message, you would probably backtrack a little and make it clear that the step is being taken with reluctance.

"Joe, it looks as if we're not going to be able to give you the extra budget we promised," you might say, getting down to business. That is the message in a nutshell—now for the review of common ground. "We know how much you have counted on getting those people and funds. There's no doubt you could manage them well and put them to good use. And we know that the morale of your people is involved in this, too. But the fact is that the sales we counted on are not coming in. We've got to cut somewhere, and, frankly, we feel it's got to be your department because. . . ."

If you are communicating by letter or memorandum, the strategy is exactly the same (only "Joe" may now be "Mr. Wyncoop"). After your lead, you review the main needs as he and you understand them, perhaps spelling them out more than you would have in a face-to-face meeting but choosing the same ones. Only then do you turn to the new conditions that make it necessary to do an about-face.

3. Your Credibitility with Readers Affects Your Strategy.

In general, communication research indicates that the chances of opinion change vary with the communicator's authority with his or her readers. In their succinct summary of the field, *Persuasion,* social scientists Marvin Karlins and Herbert I. Abelson point out that credibility itself is a variable; that is, it can be influenced by the words of the communicator.[2] Above all, as psychologists repeatedly emphasize, credibility lies in the eye of the beholder. . . .[3]

In written communications there are two types of credibility. It may be given or it may be acquired. Between the two lies a world of practical difference.

Given credibility may result from your position in an organization. If, let us say, you are the boss writing directions to a subordinate, your credibility is likely to be high. Given credibility also may result from reputation—a well-known chemist has more credibility in communications about polymers than a good industrial engineer has, but the latter would possess more credibility in communications about time-and-motion studies. It may result from the individuals and groups the writer is associated with—if he or she is a member of the same trade union the reader belongs to, a union held in high esteem by both, he or she has more credibility in a memorandum on grievance procedures than a member of the board of directors would have.

Though you may be high in given credibility, you may yet need to remind some readers of the fact. In the case of an obvious credential, such as a position in an important organization, a letterhead may be enough to do the trick. Another device is to insert a few lines of biographical data at the head of a report or brochure. If you have had experience or associations that carry weight with the reader, perhaps you can interject them early in the message. "During a visit I had last week with Zach Jarvis," you might say, knowing that Dr. Zachary P. Jarvis is a magic name with your reader, or "The Executive Committee of the Aberjona Basin Association asked me to join their meeting on Monday . . ." knowing that group carries a triple-A rating in the mind of the reader.

---

[2]Marvin Karlins and Herbert I. Abelson. *Persuasion* (New York: Springer, 1970); see pp. 107–132.
[3]See, for example, Ralph L. Rosnow and Edward J. Robinson, Eds. *Experiments in Persuasion* (New York: Academic, 1967).

Of course, you do not want to overplay your hand at such name dropping. A report to a fairly diverse audience by a famous black organization began simply:

> The National Association for the Advancement of Colored People has for many years been dedicated to the task of defending the economic, social and political rights and interests of black Americans. The growing national debate about energy has led us to examine the question to ascertain the implications for black Americans.[4]

Again, an attorney of the American Civil Liberties Union, in a letter to members of the organization soliciting donations, began:

> My dear friend:
>
> I am the ACLU lawyer who went into court last April to defend freedom of speech in Skokie, Illinois, for a handful of people calling themselves "nazis."
>
> The case has had an enormous impact on my life.
>
> It has also gravely injured the ACLU financially. . . .[5]

Acquired credibility, on the other hand, is earned by thoughts and facts in the written message. I may not know you from Adam. Yet if you send me a letter or report that carefully, helpfully describes something I am interested in, you gain credibility in my estimation.

Some studies suggest that, if you are low in given credibility and seek to acquire it with an audience, a useful technique is to cite ideas or evidence that support the reader's existing views.[6] As Disraeli once said, "My idea of an agreeable person is a person who agrees with me." The very fact that you feel confident and knowledgeable enough to articulate these views is likely to lift you several notches in the reader's estimation.

Still another approach is that old standby of persuaders—identifying yourself, in an early section, with the goals and interests of the audience. Possibly the most famous example of this strategy is the opening of Marc Antony's funeral oration, in Shakespeare's *Julius Caesar:* "I come to bury Caesar, not to praise him. . . ."

Finally, you can acquire credibility by citing authorities who rate highly with your intended audience, or by exhibiting documentary evidence that, because of its source, lends prestige and authority to your proposals or ideas.

"For success in negotiation," says C. Northcote Parkinson and Nigel Rowe, "it is vitally important that people will believe what you say and assume that any promise you make will be kept. But it is no good saying: 'Trust me. Rely on my word.' Only politicians say that."[7]

---

[4] *The Wall Street Journal,* January 12, 1978.
[5] David Goldberger, letter dated March 20, 1978.
[6] Karlins and Abelson, op. cit., pp. 115–119.
[7] C. Northcote Parkinson and Nigel Rowe, "Better Communcation: Business's Best Defense," *The McKinsey Quarterly,* Winter 1978, p. 26.

Even if you have prestigious credentials, you cannot take too much for granted. In an age of television sets, radios, cassettes, and record players in every home, credibility—at least with the public—may come quicker for the singer or comedian than for the judge, business executive, or medical researcher. In fact, because of an association with an organization or profession, you may be stereotyped as a member of "them" or "the establishment." It may behoove you to establish that you are a person with a name, a personality, certain interests, certain experiences—not just a nameless representative.

4. If Your Audience Disagrees with Your Ideas or is Uncertain about Them, Present Both Sides of the Argument.

Behavioral scientists generally find that, if an audience is friendly to a persuader, or has no contrary views on the topic and will get none in the future, a one-sided presentation of a controversial question is most effective.[8] For instance, if your point is that sales of product X in the St. Louis territory could be doubled and you are writing to enthusiastic salespeople of product X, your best course is to concentrate on facts and examples showing the enormous potential of product X. There is no shortage of evidence showing that people generally prefer reading material that confirms their beliefs, and that they develop resistance to material that repudiates their beliefs. (As we shall see presently, however, this does not mean you cannot change their minds.)

But suppose your audience has not made up its mind, so far as you know? In this case you would do well to deal with *both* sides of the argument (or all sides, if there are more than two). Follow the same approach if the reader disagrees with you at the outset. For one thing, a two-sided presentation suggests to an uncertain or hostile audience that you possess objectivity. For another, it helps the reader remember your view by putting the pros and cons in relationship to one another. Also, it meets the reader's need to be treated as a mature, informed individual. As Karlins and Abelson point out:

> Conspicuously underlying your presentation is the assumption that the audience would be on your side if they only knew the truth. The other points of view should be presented with the attitude "it would be natural for you to have this idea if you don't know all the facts, but when you know all the facts, you will be convinced."[9]

Karlins and Abelson tested the reactions of audiences in postwar Germany to Voice of America broadcasts. They found that the most persuasive programs were those that included admissions of shortcomings in United States living conditions.[10]

---

[8]Experiments supporting this conclusion are reported by Carl I. Hovland, Arthur A. Lumsdaine, and F. Sheffield in *Experiments on Mass Communiation* (Princeton: Princeton University Press, 1949). Cited by Karlins and Abelson, op. cit., p. 22.

[9]Karlins and Abelson, op. cit., p. 26.

[10]See *Factors Affecting Credibility in Psychological Warfare Communications* (Washington, D.C.: Human Resources Research Office, George Washington University, 1956).

Again, observation of businessmen's reactions to scores of *Harvard Business Review* articles advocating controversial measures convinces me that the most influential articles have been those that have acknowledged the shortcomings, weaknesses, and limitations of their arguments. When an author wants to sell a new idea to a sophisticated audience, he or she should be candid about the soft spots in his argument.

5. Win Respect by Making Your Opinion or Recommendation Clear.

Although strategy may call for a two-sided argument, this does not mean you should be timid in setting forth your conclusions or proposals at the end. We assume here that you have definite views and seek to persuade your audience to adopt them. The two-sided approach is a *means* to that end; it does not imply compromising or obfuscating your conclusions. An official at Armour & Company once criticized many reports from subordinates to bosses on the ground that, after presenting much data, they concluded, in effect, "Here is what I found out and maybe we should do this or maybe we should do that." The typical response of a boss to such a memorandum, he noted, was to do nothing. Hence, the time taken both in writing and reading was wasted.[11]

6. Put Your Strongest Points Last If the Audience Is Very Interested in the Argument, First If It Is Not So Interested.

This question is referred to by social scientists as the "primacy-recency" issue in persuasion. The argument presented first is said to have primacy; the argument presented last, recency. Although studies of the question have produced inconsistent findings and no firm rules can be drawn, it appears that if your audience is deeply concerned with your subject you can afford to lead it along from the weakest points to the strongest. The audience's great interest will keep it reading, and putting the weaker points at the start tends to create rising reader expectations about what is coming. When you end with your strongest punch, therefore, you do not let readers down.

If your audience is not so concerned with the topic, on the other hand, it may be best to use the opposite approach. Now you cannot risk leading readers along a winding path. They may drop out before you reach the end. So grab their attention right at the beginning with your strongest argument or idea.

In any case, put the recommendation, facts, or arguments you most want the reader to *remember* first or last. Although experiments by social scientists on the primacy-recency issue are inconclusive, there is a firm pattern on the question of recall. The ideas you state first or last have a better chance of being remembered than the ideas stated in the middle of your appeal or case.

---

[11]John Ball and Cecil B. Williams, *Report Writing* (New York: Ronald, 1955).

7. Don't Count on Changing Attitudes by Offering Information Alone.

"People are hostile to big business because they don't know enough facts about it," businesspeople are heard to say. Or, "If customers knew the truth about our costs, they would not object to our prices." Companies have poured large sums into advertising and public relations campaigns on this assumption; civic organizations have often based their hopes on it.

"The trouble with the assumption," states Karlins and Abelson, "is that it is almost never valid. There is a substantial body of research findings indicating that cognition—knowing something new—increasing information—is effective as an attitude change agent only under very specialized conditions."[12]

Social scientists do concede, however, that presentations of facts alone may strengthen the opinions of people who already agree with the persuader. The information reassures them and helps them defend themselves in discussions with others.

8. "Testimonials" Are Most Likely to be Persuasive If Drawn from People with Whom Readers Associate.

It is well known that a person's attitudes and opinions are strongly influenced by the groups to which he or she belongs or wants to belong—work units in a company, labor unions, bowling teams, social clubs, church associations, ethnic associations and so on. To muster third-party support for your proposal or idea, therefore, you would do well to cite the behavior, findings, or beliefs of groups to which your readers belong. In so doing, you allay any feelings of isolation readers might have if tempted to follow your ideas. You suggest that they are not alone with you, that there is group support for the points being made.

As every school child learns, the predominant attitudes of a group toward individuals or regarding standards of behavior, performance, or status influence an individual member's perceptions. For instance, a study of boys at a camp demonstrated that their ratings of various individuals' performances at shooting and canoeing were biased by their knowledge of the status of the rated individuals in the camp society. Thus a boy generally regarded as a leader was seen as performing better with the rifle or canoe than was a boy generally regarded as a follower, even though the first boy's performance was not actually superior.[13]

In addition, it seems fair to say that as modern television, radio, records, and cassettes have brought national celebrities into the home and automobile, these people, too, have been stamped with approval or disapproval by millions of groups across the country.

Accordingly, if your readers are young, dissident, or "long hairs," refer to a Richard Dreyfuss or a Joan Baez for supporting statements, not to a Gerald Ford

---

[12]Karlins and Abelson, op. cit., p. 33.
[13]Ibid., p. 50.

or an Arnold Palmer. If your readers are electrical engineers, quote well-regarded scientific sources as your authority, not star salespeople or public relations people. Take into account also that the more deeply attached your readers are to a group, the greater the influence of the group norms on them. For instance, one experiment by social scientists showed that the opinions of Catholic students who took their religion seriously were less influenced by the answers of nonserious Catholics than were the opinions of Catholic students who placed little value on their church membership.[14]

9. Be Wary of Using Extreme or "Sensational" Claims and Facts.

Both research in behavioral science and common sense confirm this rule.[15] Do not be misled by the fact that flashy journalists make successful use of extreme and bizarre cases to dramatize a story. The situation in business and professional writing is different from that in journalism.

When you seek the confidence and cooperation of your readers—and typically you do in the kinds of communications we deal with in this book—it is best to write in terms of the real world as you and they perceive it. Observable, believable, realistic statements carry more weight than any other kind. Although you want reader attention, you do not want to shock your audience with outlandish examples or arguments. These may help you to succeed in making the reader sit up—but they will also provoke distrust and suspicion.

Examples are common in the letters sections of newspapers. A writer who identified himself as a former vice-president of a well-known bank opposed a large power company's plan to build a new plant in a rural area near his town. His letter began as follows: "A great many of us . . . are profoundly disturbed by the proposal now being considered to disrupt and destroy the marvelous little valley southwest of [name of town], in order to build bigger and better power plants. This would be a devastating blow to the last unspoiled bit of country left in Connecticut. . . ."[16]

Like a batter who hits the first two pitches foul and quickly gets two strikes against him, this writer managed to distort the first two sentences he wrote. The proposed plant, though a very large one, would not "disrupt and destroy" the valley—only a small section of the valley area would be affected. Moreover, the valley was not "the last unspoiled bit of country" in the state—it was only a small parcel of the state's beautiful countryside. These exaggerations might have drawn cheers from rabid foes of the project, but the writer wasn't intersted in appealing to them; he wanted to win uncommitted readers. At the very beginning, however, he antagonized them with hyperbole.

---

[14]See H. Kelley, "Salience of Membership and Resistance to Change of Group-Anchored Attitudes," *Human Relations,* August 1955, pp. 255–289. Cited in Karlins and Abelson, op. cit., p. 58.

[15]See, for example, *Building Opposition to the Excess Profits Tax* (Princeton: Opinion Research Corporation, August 1952), and R. Weiss, "Conscious Technique for the Variation of Source Credibility," *Psychological Reports,* Vol. 20, 1969, p. 1159. Both cited in Karlins and Abelson, op. cit., pp. 36–37.

[16]*Lakeville Journal,* April 2, 1970, p. 11.

10. Tailor Your Presentation to the Reasons for Readers' Attitudes, If You Know Them.

Your chances of persuading readers are better if you can plan your appeal or argument to meet the main feelings, prejudices, or reasons for their beliefs. For instance, if reader beliefs are the result of their wanting to go along with certain groups they like or associate with, your best bet (as indicated earlier) is to show the acceptability of your point to these groups. If their attitudes reflect personal biases, such as an old grudge against someone in power, it is best to tailor your presentation to that prejudice. And so on.

Summarizing the implications of several behavioral studies, Karlins and Abelson present the example of three people who say they are against private ownership of industry. How should their reasons for this position influence one's choice of strategy or persuasion? The authors explain:

> One of them feels that way because he has only been exposed to one side of the story and has nothing else on which to base his opinions. The way to change this man's opinion may be to expose him to facts, take him to visit some factories, meet some workers and supervisors. A second person is against private ownership because that is the prevailing norm or social climate in the circles in which he finds himself. His attitudes are caused by his being a part of a group and conforming to its standards. You cannot change this fellow just by showing him facts. The facts must be presented in an atmosphere which suggests a social reward for changing his opinion. Some kind of status appeal might be a start in that direction. A third person may have negative attitudes toward private industry because by making business the scapegoat for all his troubles, he can unload his pent-up feelings of bitterness and disappointment at the world for not giving him a better break. . . . Trying to change this third person with facts may actually do more harm than good. The more evidence shows how wrong he is, the more he looks for reasons to support his beliefs. This kind of person can sometimes be influenced by helping him to understand why he has a particular attitude.[17]

11. Never Mention Other People without Considering Their Possible Effect on the Reader.

Other people may, as we saw earlier, be introduced for the sake of "testimonials." More commonly, however, other people's names are mentioned in the course of explaining a situation, narrating an event, or completing the format of a message. This use of names, too, may affect the power of your message.

A reference to the actions of another person—however simple and unobtrusive it may seem to you the writer—may alter your relationship with readers. If readers consider that person a friend or enemy, their natural reaction is to begin thinking of the possible bearing of your communication on their friendship or antagonism. This reaction can have significant implications for your approach.

---

[17]Karlins and Abelson, op. cit., p. 92.

To illustrate, a doctoral student who had failed to meet his school's program requirements tried to muster faculty opinion in support of his petition for re-admission by appearing daily at the entrance to the dining hall and handing out leaflets to faculty members. One such leaflet contained these words: "I am very unhappy about the strain my case has created for Professor [name of the program director]. I am distressed if last Friday's handout . . . created the impression that I was harping on his mistakes. I have told him and I tell you that I could under-stand his actions and decisions. . . ." The leaflet went on at some length to explain the doctoral student's feelings about the problem.

What this writer did not realize was the impact of the professor's name on his communication strategy. Almost everyone who received the leaflet was a col-league of the professor in question. Therefore the leaflet made it necessary for them to think of their relationship with the professor when they made up their minds about the petition. And their relationship with the professor was more important to them than their relationship with the doctoral student.

If the doctoral student considered it essential to mention the professor, he could have elected to: (1) try to win readers over while convincing them that their relations with the professor would not be affected, or (2) show that the pro-fessor was so far off base that readers were morally bound to risk their relation-ship with him. In the latter case, the leaflet should have contained ready-to-use arguments that readers could draw on in explaining to the professor why they sympathized with the doctoral student. Since the leaflet did neither of these things, it was a failure in persuasion.

Don't overlook the possible effect of distribution. Letters often go to third parties, with "cc" typed at the bottom followed by the names of those people. A memorandum often contains the names of several addressees in the "To" line at the top. Covering letters with reports may indicate several groups of readers. All this may affect your strategy. The background information that you could omit if writing only to Jones may be quite necessary if Brown, too, is an important reader; and the rather offhand treatment you give to a certain test or episode if writing to Jones and Brown might not be fitting at all if Larabee also is an intended reader. Many times the wise manager or professional rewrites part of a letter or memo after deciding to send a copy of it to an additional person who was not considered when the first draft was made.

Many people have strong feelings about "blind copies," that is, copies sent to persons other than those indicated after "cc" at the end of a letter or in the "To" line of a memorandum. Some people feel that blind copies never should be sent. Others feel that since a letter or memo is the property of the writer, he or she can distribute it at will. Although the latter view is legally correct, only an obtuse writer will distribute copies thoughtlessley if the content is in any way confiden-tial, personal, or politically sensitive.

SIZING UP YOUR READERS

We have a tendency to abstract written communications from real life, to act as if the customary ground rules of influence and persuasion don't apply to a message that is in writing. We act with a naiveté almost unheard of in our face-to-face relationships. Not seeing readers, we act as if they weren't real people. "If we write the information clearly, accurately, and correctly," we think wishfully to ourselves, "surely that satisfies the requirements of a piece of paper." But Josh Billings's puckish maxim, "As scarce as truth is, the supply has always been in excess of the demand," applies to truth on paper as well as truth in conversation.

Think of your intended readers as the real people they will be when they take your letter or report out of the "in-box." Only then can you decide intelligently what information and ideas to emphasize and in what order to present them. To help you think of readers as three-dimensional people, ask yourself some questions about their situation and relationships with you. Are they:

- Deeply or only mildly interested in the subject of your communication?
- Familiar or unfamiliar with your views, competence, and feelings about them?
- Knowledgeable or ignorant of your authority in the area discussed, your status, and your associations of possible importance to them?
- Committed or uncommitted to a viewpoint, opinion, or course of action other than the one you favor in your letter, report, or other document?
- Likely or unlikely to find your proposal, idea, finding, or conclusion threatening or requiring considerable change in their thought or behavior?
- Inclined or uninclined to think and feel the way they do about the subject because of identifiable reasons, prejudices, or experiences?
- Associated formally or informally with groups or organizations involved in some way with the idea or proposal you deal with?

With answers to questions like these in mind, you will not see your readers as shadows on the wall. They will sit across from you. You can write as if talking *with* them, not talking to them.

Philip C. Kolin

Proposals

Philip C. Kolin, who has published widely in the fields of literature and communications, is professor of English at the University of Southern Mississippi.

GUIDELINES FOR WRITING A SUCCESSFUL PROPOSAL

Regardless of the type of proposal you are called on to write, the following guidelines will help you to persuade your audience to approve your plan. Refer to these guidelines both before and while you formulate your plan.

1. **Approach writing a proposal as a problem-solving activity.** Your goal is to solve a problem that affects the reader. Do not lose sight of the problem as you plan and write your proposal. Everything in your proposal should relate to the problem, and the organization of your proposal should reflect your ability as a problem-solver. Psychologically, make the reader feel confident that you can solve the problem.

2. **Regard your audience as skeptical readers.** Even though you offer a plan that you think will benefit readers, do not be overconfident that they will automatically accept it as the best and only way to proceed. To determine the feasibility of your plan, readers will question everything you say. They will withhold their approval if your proposal contains errors, omissions, or inconsistencies. Consequently, try to examine your proposal from the readers' point of view. . . .

3. **Research your proposal carefully.** A winning proposal is not based only on a few well-meaning, general suggestions. All your good intentions and enthusiasm will not substitute for the hard facts readers will demand. Concrete examples per-

suade readers; unsupported generalizations do not. To make your proposal complete and accurate, you will have to do a lot of homework; for example, reading previous correspondence or research about the problem, doing comparative shopping for the best prices, verifying schedules and timetables, interviewing customers and/or employees, making site visits. . . .

4. **Prove that your proposal is workable.** The bottom-line question from your reader is "Will this plan work?" Your proposal should be well thought out. It should contain no statements that say "Let's see what happens if we do X or Y." By analyzing and, when possible, by testing each part of your proposal in advance, you can eliminate any quirks and revise the proposal appropriately before readers evaluate it. What you propose should be consistent with the organization and capabilities of the company. It would be foolish to recommend, for example, that a small company (fifty employees) triple its workforce to accomplish your plan.

5. **Be sure that your proposal is financially realistic.** This point is closely associated with and follows from guideline 4. "Is it worth the money?" is another bottom-line question you can expect from readers. Do not submit a proposal that would require an unnecessarily large amount of money to implement. For example, it would be unrealistic to recommend that your company spend $20,000 to solve a $2,000 problem that might not ever recur. Study the ecomonic climate, too—are you in an economic slump or in a boom time?

6. **Package your proposal attractively.** Make sure your proposal is letter-perfect, inviting, and easy to read (e.g., use plenty of headings and other visual devices). . . . The appearance as well as the content of your proposal can determine whether it is accepted or rejected. Remember that readers, especially those unfamiliar with your work, will evaluate your proposal as evidence of the type of work you want to do for them. Take advantage of any software programs dealing with desktop publishing . . . that may be available to you. This software will allow you to prepare a proposal that looks as if it were done by a professional printer.

INTERNAL PROPOSALS

The primary purpose of an internal proposal, such as the one included in Figure 1, is to offer a realistic and constructive plan to help your company run its business more efficiently and economically. On your job you may discover a better way of doing something or a more efficient way to correct a problem. In the world of work, typical problems for which proposals are written focus on money, personnel, outdated technology, health concerns, and organizational communications. You believe that your proposed change will save your employer time, money, or further trouble. (Tina Escobar and Oliver Jabur in Figure 1 have identified and researched a more effective and less costly way for Community Federal Bank to do business and to satisfy its customers.) You decide to notify your department head, manager, or supervisor, or your employer may ask you for specific suggestions to solve a problem he or she has already identified. . . .

EQUAL HOUSING
LENDER

FDIC/DIFM

COMMUNITY FEDERAL BANK

Powell	*Monroe*	*Langston*
584-5200	*413-6000*	*796-3009*

TO: Michael L. Sappington, Executive Vice President
        Dorothy Woo, Langston Regional Manager

FROM: Tina Escobar, Oliver Jabur, ATM Services

DATE: June 2, 1994

RE: A Proposal to Install an ATM at the Mayfield Park Branch

PURPOSE

Clearly states why proposal is being sent

We are writing to propose a cost-effective solution to what is a growing problem at the Mayfield Park branch in Langston: inefficient servicing of customer needs and rising personnel costs. We recommend that you approve the purchase and installation, within the next three to four months, of an ATM at Mayfield. Such action is consistent with Community's goals of expanding electronic banking services and promoting our image as a self-serve yet customer-oriented institution.

THE PROBLEM WITH CURRENT SERVICES AT MAYFIELD PARK

Identifies problem by giving reader necessary background information

Currently, we employ four tellers at Mayfield. However, we are spending too much on personnel/salary for routine customer transactions. In fact, as determined by teller activity reports, nearly 25 percent of the 4 tellers' time each week is devoted to routine activities easily accommodated by ATMs. Here is a breakdown of teller activity for the month of May:

Teller #	Total Transactions	Routine Transactions
1	6,205	1,551
2	5,989	1,383
3	6,345	1,522
4	6,072	1,518
	24,611	5,974

FIGURE 1   An internal unsolicited proposal

Divides problem into parts— volume, financial, personnel, customer service

Clearly, we are not fully using our tellers' sales abilities when they are kept busy with routine activities. To compound the problem, we expect business to increase by at least 25 percent at Mayfield in the next few months, as projected by this year's market survey. If we do not install an ATM, we will need to hire a fifth teller, at an annual cost of $20,800 ($15,500 base pay plus approximately 30 percent for fringes), for the additional 6,000 transactions we project.

Verifies that problem is widespread

Most important, customer needs are not being met efficiently at Mayfield. Recent surveys done for Community Federal by Watson-Perry demonstrate that customers are dissatisfied about not having the convenience of an ATM at Mayfield. Seventy-seven percent of the respondents to the Watson-Perry questionnaire pointed to the lack of an ATM as Mayfield's biggest drawback. Customers are unhappy about long waits in line to do simple banking business, such as deposits, withdrawals, and loan payments, and about having to drive to other branches to do their after-hours banking. Conversations we had with manager Rachael Harris-Ignara at Mayfield confirm that customers regularly complain to tellers and loan officers about not having an ATM.

Ultimately, the lack of an ATM at Mayfield Park hurts Community's image. With ATMs available to Mayfield residents at local stores and other banks, our institution risks having customers and potential customers disassociate Community from their banking needs. We not only miss the opportunity of selling them on our other services but also risk losing their business entirely.

A SOLUTION TO THE PROBLEM

Relates solution to individual parts of the problem

The purchase and installation of an ATM at Mayfield Park will initially result in significant savings in personnel costs and time. We will not have to hire a fifth teller at $20,800 for the anticipated increase in transactions. Thanks to an ATM, we will also be able to allocate more effectively the talents of the four existing tellers at Mayfield. These tellers will then be able to assist customers with questions and transactions that cannot be handled through an ATM, such as purchase of savings bonds, traveler's checks, or foreign currency. Mayfield tellers, therefore, will have more opportunities for greater involvement in customer services and can spend more time cross-selling our services. As a result, we will get additional personnel achievements without reducing the level of customer service. In fact, we will be improving that service.

The increase in teller availability will inevitably lead to greater customer satisfaction. An ATM will allow customers the option of meeting their banking needs electronically or through a teller. Customers in a hurry can

FIGURE I *(Continued)*

easily make a withdrawal with their ATM cards, while those who need more personal attention can take advantage of window service. As customer frustration is eased, so too will be the stress on tellers because of shorter lines and fewer complaints.

Shows problem can be solved and stresses how

It is feasible to install an ATM at Mayfield. This location does not pose the difficulties we faced at some older branches. Mayfield offers ample room to install a drive-up ATM in the stubbed-out fourth drive-up lane. This location is away from the heavily congested area in front of the bank, yet it is easily accessible from the main driveway and the side drive facing Cornith Avenue.

Judging from the ATM vendor's past work, the ATM could be installed and operational within three to four months. That is the amount of time it took to install ATMs at the first two locations in Powell and for Archer Avenue in Langston. Moreover, by authorizing the expenditure at Mayfield within the next month, you will ensure that ATM service is available before the Christmas season.

COSTS

Itemizes costs

The costs of implementing our proposal are as follows:

Diebold Drive-up ATM	$28,000.00
Installation fee	2,000.00
Maintenance (1 year)	1,500.00
	$31,500.00

Interprets costs for reader

This $31,500.00, however, does not truly reflect into our annual costs. We would be able to amortize, for tax purposes, the cost of installation of the ATM over five years. Our annual expenses would, therefore, look like this:

$30,000 (28,000 + 2,000) ÷ 5 years =
$6,000.00 + 1,500 (maintenance) or $7,500 per year.

Compared with the $20,800 a year the bank would have to expend for an additional fifth teller at Mayfield, the annual depreciated cost for the ATM ($7,500) reduces by nearly two-thirds the amount of money the bank will have to spend for much more efficient customer service.

FIGURE I   *(Continued)*

CONCLUSION

Stresses benefits
for reader and
bank as a whole

Authorizing an ATM for the Mayfield Park branch is both feasible and cost-effective. Adoption of this proposal will save our bank more than $13,000.00 in teller services annually, reduce customer complaints, and increase customer satisfaction and approval. We will be happy to discuss this proposal with you anytime at your convenience.

FIGURE 1   *(Continued)*

Generally speaking, your proposal will be an informal, in-house message, so a brief (usually one- or two-page) memo should be appropriate.

An internal proposal can be written about a variety of topics, such as:

- purchasing new or more advanced equipment: word processors, transducers, automobiles
- hiring new employees or training current ones to learn a new technique or process
- eliminating a dangerous condition or reducing an environmental risk to prevent accidents—for employees, customers, or the community at large
- improving communication within or between departments of a company or agency
- revising a policy to improve customer relations (eliminating an inconvenience, speeding up a delivery) or employees' morale (offering vanpooling; adding more options for a schedule).

As this list shows, internal proposals cover almost every activity or policy that affects the day-to-day operation of a company or agency.

Your Audience and Office Politics

Writing an internal proposal requires you to be aware of and sensitive to office politics. To be successful, your internal proposal should be written with the needs and likes of your audience in mind. Remember that your boss will expect you to be very convincing about both the problem you say exists and the solution you propose to correct it. You cannot assume that your reader will automatically agree with you that there is a problem or that your plan is the only way to solve it.

Your reader in fact may feel threatened by your plan. After all, you are advocating a change. Some managers regard such changes as a challenge to their administration of an office or department. Or your reader may be indifferent, not even wanting to give your work serious consideration. Or your manager-reader may have certain "pet" projects or ways of doing things that you must take into account. To surmount these and other obstacles, show that the change you propose is in everyone's best interest. Do not overlook the possibility that your boss may have to take your proposal up the organizational ladder for commentary and, eventually, approval. . . . Writing a proposal may mean working as a team with your boss—putting his or her name on the document, too, to get notice and credit.

Before you write an internal proposal, consider the implications of your plan for your boss and for other offices or sections in your company. A change you propose for your department or office (transfers; new budgets or schedules) may have sweeping and potentially disruptive implications for another office or division within your company. It is wise to discuss your plan with your boss before you put it in writing. Then you might provide your boss with a draft, asking for his or her revisions or feedback. . . .

Never submit an internal proposal that offers an idea you think will work but relies on someone else to supply the specific details on *how* it will work. For example, do not write an internal proposal that says the payroll, community relations, maintenance, or advertising department can give the reader the details and costs he or she needs about your proposal. That pushes the responsibility onto someone else, and your proposal could be rejected for lack of concrete evidence.

The Organization of an Internal Proposal

A short internal proposal follows a relatively straightforward plan of organization, from identifying the problem to solving it. Internal proposals usually contain four parts, as shown in Figure 1. Refer to this proposal as you read the following discussion.

The Introduction

Begin your proposal with a brief statement of why you are writing to your boss: "I propose that. . . ." State why you think a specific change is necessary now. Then succinctly define the problem and emphasize that your plan, if approved by the reader, will solve that problem. Where necessary, stress the urgency to act—within the next week? month?

Background of the Problem

In this section prove that a problem does exist by documenting its importance for your boss and your company. As a matter of fact, the more you show how the problem affects the boss's work (and area of supervision), the more likely you are to persuade him or her to act. And the more concrete evidence you cite, the easier it will be to convince the reader that the problem is significant and that action needs to be taken now.

Avoid vague (and unsupported) generalizations such as "we're losing money each day with this procedure (or piece of equipment)"; "costs continue to escalate";

"the trouble occurs frequently in a number of places"; "numerous complaints have come in"; "if something isn't done soon, more difficulty will result."

Instead, provide readers with quantifiable details about the number of dollars a company is actually losing per day, week, or month. Emphasize the financial trouble so that you can show how your plan (described in the next section) offers an efficient and workable solution. Indicate how many employees (or work-hours) are involved or how many customers are inconvenienced or endangered by a procedure or condition. Notice how Escobar and Jabur include this information in their proposal in Figure 1. Verify how widespread a problem is or how frequently it occurs by citing specific occasions. Rather than just saying that a new word processor would save the company "a lot of money," document how many work-hours are lost using other equipment for the routine jobs your company now has employees perform.

The Solution or Plan

In this section describe the change you want approved. Tie your solution (the change) directly to the problem you have just documented. Each part of your plan should help eliminate the problem or should help increase the productivity, efficiency, or safety you think is possible.

Your reader will again expect to find factual evidence. Do not give merely the outline of a plan or say that details can be worked out later. Supply details that answer the following questions: (1) Is the plan workable—can it be accomplished here in our office or plant? and (2) Is it cost-effective—will it really save us money in the long run or will it lead to even greater expenses?

To get the boss to say yes to both questions, supply the facts that you have gathered as a result of your research. For example, if you propose that your firm buy a new piece of equipment, do the necessary homework to locate the most efficient and cost-effective model available, as Tina Escobar and Oliver Jabur do in Figure 1. Supply the dealer's name, the costs, major conditions of service and training contracts, and warranties. Describe how your firm could use this equipment to obtain better results in the future. Cite specific tasks the new equipment can perform more efficiently at a lower cost than the equipment now in use.

It is also wise to raise alternative solutions, before the reader does, and to discuss their disadvantages. Notice how Tina Escobar and Oliver Jabur do this in Figure 1, in their discussion of the solution to the bank's customer-srvice problem. They show why installing an ATM is more necessary than hiring a fifth teller. Their discussion is based on a strong persuasive strategy, demonstrating to their readers that they have examined all alternatives and chosen the best.

If you are proposing that your company hire or reassign employees, indicate where these employees will come from, when they will start, what they will be paid, what skills they must have, where they will work and for how long. If you propose to assign current employees to a job, keep in mind that their salary will still have to be paid by your company. Just because they are coworkers does not mean they will work for nothing. Note again in Figure 1 how Tina Escobar and

Oliver Jabur raise and resolve the problem of new and profitable responsibilities for the tellers at Community Federal Bank.

A proposal to change a procedure must include the following details:

1. How the new (or revised) procedure will work
2. How many employees or customers will be affected by it
3. When it will go into operation
4. How much it will cost the employer to change procedures
5. What delays or losses in business might be expected while the company switches from one procedure to another
6. What employees, equipment, or locations are available to accomplish this change.

As these questions indicate, your boss will be concerned about schedules, working conditions, employees, methods, locations, equipment, and the costs involved in your plan for change. The costs, in fact, will be of utmost importance. Make sure that you supply a careful and accurate budget so that your reader will know what the change is going to cost. Moreover, make those costs attractive by emphasizing how inexpensive they are as compared to the cost of not making the change, as Escobar and Jabur do in the section labeled "Costs." Double-check your math.

The Conclusion

The conclusion to your internal proposal should be short—a paragraph or two at the most. Your intention is to remind the reader that the problem is serious, that the reason for change is justified, and that you think the reader needs to take action. Select the most important benefits and emphasize them again. In Figure 1, Escobar and Jabur again emphasize the savings that the bank will see by following their plan as well as the increase in customer satisfaction. Also indicate that you are willing to discuss your plan with your reader.

SALES PROPOSALS

A sales proposal is the most common type of external proposal. Its purpose is to sell your company's products or services for a set fee. Whether short or long, a sales proposal is a marketing tool that includes a sales pitch as well as a detailed description of the work you propose to do. Figures 2 and 3 contain sample sales proposals.

The Audience and Its Needs

Your audience will usually be one or more business executives who have the power to approve or reject a proposal. Unlike readers of an internal proposal, your audience for a sales proposal may be even more skeptical since they may not know you or your work. Your proposal may also be evaluated by experts in

REYNOLDS INTERIORS
250 Commerce Avenue, S.W.
Portland, OR 97204–2129

503–555–8733   FAX 503–555–1629

January 21, 1994

Mr. Floyd Tompkins, Manager
General Purpose Appliances
Highway 41, South
Portland, OR 97222

Dear Mr. Tompkins:

In response to your Request 7521 for bids for an appropriate floor covering at your new showroom, Reynolds Interiors is pleased to submit the following proposal. After carefully reviewing your specifications for a floor covering and inspecting your new facility, we believe that the Armstrong Classic Corlon 900 is the most suitable choice. I am enclosing a sample of the Corlon 900 so that you can see how it looks.

Corlon's Advantages

Guaranteed against defects for a full three years, Corlon is one of the finest and most durable floor coverings manufactured by Armstrong. It is a heavy-duty commercial floors 0.085-inch thick for protection. Twenty-five percent of each tile consists of interface backing; the other 75 percent is an inlaid wear layer that offers exceptionally high resistance to everyday traffic. Traffic tests conducted by the Independent Floor Covering Institute repeatedly proved the superiority of Corlon's construction and resistance.

Another important feature of Corlon is the size of its rolls. Unlike other leading brands of similar commercial flooring—Remington or Treadmaster—Corlon comes in 12-foot-wide rather than 6-foot-wide rolls. This extra width will significantly reduce the number of seams in your floor, thus increasing its attractiveness and reducing the danger of tile split.

Installation Procedures

The Classic Corlon requires that we use the inlaid seaming process, a technical procedure requiring the services of a trained floor mechanic. Herman Goshen, our floor mechanic, has over fifteen years of experience working with the inlaid seam process. His professional work has been consistently praised by our customers.

Installation Schedule

We can install the Classic Corlon on your showroom floor during the first week of March, which fits the timetable specified in your request. The tile will take three and one-half days to install and will be ready to walk on immediately. We recommend, though, that you not move equipment onto the floor for twenty-four hours after installation.

FIGURE 2   An external solicited proposal

Costs

The following costs include the Classic Corlon tile, labor, equipment, and tax:

750 sq. yards of Classic Corlon at $12.50/yard	$ 9,375.00
Labor (28 hr. @ $15.00/hr.)	420.00
Sealing fluid (10 gal. @ 10.00/gal.)	100.00
Total	9,895.00
Tax (5 percent)	494.75
GRAND TOTAL	**$10,389.75**

Our costs are $250.00 under those you specified in your request.

Reynolds' Qualifications

Reynolds Interiors has been in business for more than twenty-eight years. In that time, we have installed many commercial floors in Portland and its suburbs. In the last year, we have served more than sixty customers, including the new multi-purpose Tradex plant in Portland.

Conclusion

Thank you for the opportunity to submit this proposal. We believe you will have a great deal of success with an Armstrong floor. If we can provide you with any further information, please call us.

Sincerely yours,

Sharon Scovill

Sharon Scovill
Sales Manager

Jack Rosen

Jack Rosen
Installation Supervisor

FIGURE 2   *(Continued)*

National Business Equipment

NBE

470 Rodgers Rd. ■ Camden, NJ 08104–0826 ■ (201) 555–1100

September 20, 1994

Ms. Denise Taylor
Business Manager
Madison Tool and Die Company
3400 Veterans Boulevard
Camden, NJ 08104

Dear Ms. Taylor:

While we were servicing your Piko 2500 copier last week, it occurred to me that you might be interested in purchasing a newer model that will give you state-of-the-art features to make your copying work more efficient and reliable. In the past three years since you purchased your 2500, copier technology has advanced tremendously, giving users many benefits at surprisingly low cost.

Based on our assessment of Madison Tool and Die Company's needs for the most reliable office equipment available, we recommend that you purchase the Piko 4000. We believe that this copier will satisfactorily meet all of your copying needs for the present and well into the future.

ADVANTAGES OF THE PIKO 4000

With the Piko 4000, your cost per single copy will be reduced to only 1 1/2 cents, compared with approximately 3 cents per copy with your present machine. The 4000 model is designed to make as many as **75,000 copies per month,** almost double the recommended work load of your current copier. The following description of the main features of the Piko 4000 will show you its other advantages over your current copier.

Printing Technology

The Piko 4000's **printing rate of 80 copies per minute** will save you valuable time. The automatic duplexing capability allows you to turn single-sided documents into double-sided ones, thus reducing paper costs. In addition to the standard paper sizes—8 1/2 × 11. 8 1/2 × 14, and 11 × 17—**the 4000 functions effectively with other paper sizes and shapes,** including heavyweight sheet stock suitable for specifications and drawings. The Piko 4000 has original-size sensing, which can copy varying sizes automatically.

FIGURE 3 An external unsolicited proposal

Copier Control Security System

The Piko 4000's **electronic ID keypad** allows only authorized users, with individual entry codes, to run the copier. This keypad can control access, limit copies, and provide tabulated records of use for your accounting purposes.

Feeding/Sorting Capacity

The 400 includes as standard equipment both a recirculating automatic document feeder and an automatic sorter/stapler. These features **can reduce the time spent on large copying jobs by as much as 50 percent.**

Printing Quality

The new Piko 4000 offers top-quality, high resolution reproduction **in four colors** in addition to black and white. Since seeing is believing, I am enclosing a copy of this proposal made on the 4000 as well as a copy of a blueprint to show you how both copied documents look.

Reduction/Enlargement Function

The Piko 4000's zoom capability allows you to increase or decrease the size of your copy within a range of 50–200 percent. Since these changes are made in increments of 1 percent, your copy is always the exact size you want.

Size

The Piko 4000 measures 4' × 3' × 4', **occupying one-quarter less space** than your present copier. Once the machine is installed, you will find this extra space useful for your storage needs.

INSTALLATION, TRAINING, AND SERVICE

We will deliver and install your new Piko 4000 within one week of receiving your purchase order. The installation requires approximately forty-five minutes.

Our local sales representative, Darlene Simpson, is available to instruct your employees in the operation and routine maintenance of the 4000. A phone call will allow us to set up a mutually convenient date for such instruction. If a problem occurs, **we offer customers the latest in remote diagnostics.** A modem installed in the Piko 4000 allows us to diagnose specific problems (and needed parts) before dispatching a service representative. We save time, so you save time. In addition, our "hot line" is available to you on business days from 8:00 A.M. to

FIGURE 3 *(Continued)*

5:00 P.M. We guarantee that a service representative will arrive at your office **within two business hours** from the time of your call.

COSTS

Below is the price of the Piko 4000 and appropriate initial supplies:

Piko 4000	$18,295.00
Toner (tube)	23.00
Paper (5,000 sheets of medium grade 8 1/2 × 11)	25.00
Service contract—optional	
(per year, parts and labor included)	2,300.00
Total	$20,643.00

We install your new Piko 4000 **free of charge.**

NATIONAL'S REPUTATION

Over the past 22 years, National has supplied copiers to more than 90 companies in the greater Camden area, including Biscayne Industries, Northeast Manufacturing, and Teunissen Accounting, Inc. In addition to providing quality products, we are **dedicated to giving our customers fast and efficient service long after a sale.**

We appreciate your using **National Business Equipment** for your repair needs and hope that you will decide to purchase the new Piko 4000. Please call me if you have any questions about the Piko 4000 or National.

If this proposal is acceptable, please sign and return a copy of this letter.

Sincerely yours,

Marion Copely

Marion Copely

Encls. Copier samples

I accept this proposal made by National Business Equipment.

_____

for Madison Tool and Die Company

FIGURE 3 *(Continued)*

other fields employed by your prospective customer. Readers of a sales proposal will evaluate your work according to (1) how well it meets their needs and (2) how well it compares with the proposals submitted by your competitors. Your proposal must convince readers that you can provide the most appropriate work or service and that your company is more reliable and efficient than any other firm.

The key to success is incorporating the "you attitude" . . . throughout your proposal. Relate your product, service, or personnel to the reader's exact needs as stated in the RFP [request for proposal] for a solicited proposal or through your own investigations for an unsolicited proposal. You cannot submit the same proposal for every job you want to win and expect to be awarded a contract. Different firms have different needs. The most important question the reader will raise, therefore, about your work is "How does this proposal meet our company's special requirements?" Some other fairly common questions readers will have as they evaluate your work include the following:

- Does the writer's firm understand our problem?
- Can the writer's firm deliver what it promises?
- Can the job be completed on time?
- What assurances does the writer offer that the job will be done exactly as proposed?

Answer each of these questions by demonstrating how your product or service is tailored to the customer's needs.

Organizing a Sales Proposal

A sales proposal can have the following parts: introduction, description of the proposed product or service, timetable, costs, qualifications of your company, and conclusion.

Introduction

The introduction to your sales proposal can be a single paragraph in a short sales proposal or several pages in a more complex one. Basically, the introduction should prepare readers for everything that follows in your proposal. The introduction itself may contain the following sections, which sometimes may be combined.

1. **Statement of purpose and subject of proposal.** Tell readers why you are writing and identify the specific subject of your work. If you are responding to an RFP, use specific code numbers or cite application dates, as the proposal in Figure 2 does. If your proposal is unsolicited, indicate how you learned of the problem, as Figure 3 does. Briefly define the solution you propose.
2. **Background of the problem you propose to solve.** Show readers that you are familiar with their problem and that you have a firm grasp of the importance and implications of the problem. In a solicited proposal, such as the one shown in

Figure 2, a section outlining the problem is usually unnecessary, because the potential client has already identified the problem and just wants to know how you would handle it. In that case, just point out how your company would solve the problem, mentioning your superiority over your competitors (see the third paragraph of Figure 2). In an unsolicited proposal, you need to describe the problem in convincing detail, identifying the specific trouble areas. However, if it is an external proposal to a current customer, such as the one in Figure 3, it would be unwise to point out past problems your client may have had with your company's services or products. In Figure 3, note how the problems of the Piko 2500 model are described mainly as a way to sell the advantages of the Piko 4000.

Description of the Proposed Product or Service

This section is the heart of your proposal. Before spending their money, customers will demand hard, factual evidence of what you claim can and should be done. Here are some points that you should cover.

1. **Carefully show your potential customers that your product or service is right for them.** Stress specific benefits of your product or service most relevant to your reader. Blend sales talk with descriptions of hardware.
2. **Describe your work in suitable detail—what it looks like, what it does, and how consistently and well it will perform in the readers' office, plant, hospital, or agency.** You might include a brochure, picture, or, as the writers of the proposals in Figures 2 and 3 do, a sample of your product for customers to study. Convince readers that your product is the most up-to-date and efficient one they could select. Note how the proposals in Figures 2 and 3 emphasize this.
3. **Stress any special features, maintenance advantages, warranties, or service benefits.** Highlight features that show the quality, consistency, or security of your work. For a service, emphasize the procedures you use, the terms of that service, and even the kinds of tools you use, especially any "state-of-the-art" equipment. Be sure to provide a step-by-step outline of what will happen and why each step is beneficial for readers.

Timetable

A carefully planned timetable shows readers that you know your job and that you can accomplish it in the right amount of time. Your dates should match any listed in an RFP. Provide specific dates when the work will begin, how long it will continue, and when you will be finished installing equipment, testing equipment, or training employees to use equipment. For proposals offering a service, specify how many times—by the hour, week, month—customers can expect to receive your help; for example, spraying three times a month for an exterminating service or making deliveries by 10:00 A.M. for a trucking company. Indicate whether follow-up visits or service calls will be provided.

Costs

Make your budget accurate, complete, and convincing. Accepted by both parties, a proposal is a binding legal agreement. Don't underestimate costs in the hope that a low bid will win you the job. You may get the job but lose money doing it, for the customer will rightfully hold you to your unrealistic figures. Neither should you inflate prices; competitors will beat you in the bidding.

Give customers more than the bottom-line cost. Show exactly what readers are getting for their money so they can determine if everything they need is included. Itemize costs for specific services, equipment, labor (by the hour or by the job), transportation, travel, or training you propose to supply. If something is not included or is considered optional, say so—additional hours of training, replacement of parts, and the like. If you anticipate a price increase, let the customer know how long current prices will stay in effect. That information may spur them to act favorably now.

Qualifications of Your Company

Emphasize your company's accomplishments and expertise in using relevant services and equipment. You might list previous work you have done that is identical or similar to the type of work you are proposing to do for the customer. You may even want to mention the names of a few local firms for whom you have worked that would be pleased to recommend you. Never misrepresent your qualifications or those of the individuals who work with or for you. Your prospective client may verify if you have in fact worked on similar jobs for the last five to six years.

Conclusion

This is the "call to action" section of your proposal. As with the conclusion in an internal proposal, encourage your reader to approve your plan. Stress the major benefits your plan has for the customer. Notice how the last paragraphs of Figures 2 and 3 do this effectively. Offer to answer any questions the reader may have. Some proposals end by asking readers to sign and return a copy of the proposal thus indicating their acceptance of it, as the proposal in Figure 3 does.

Part 5

Resumes and Other Written Materials for a Job Search

The written materials that are part of an application for a job face one of the most difficult audiences imaginable: experienced recruiting managers. Because the selections that follow in this section of *Strategies* offer extensive advice on how to prepare and write resumes and cover letters, I will limit my remarks here to some general comments.

First, there is no *one* way to write a resume or a cover letter. Slavishly following some model that an applicant mistakenly thinks represents the ideal can defeat the purpose of writing a resume and a cover letter: to get an interview. The shelves of bookstores and libraries across the country are filled with guides offering advice on how to write resumes and cover letters. These guides offer advice, not commandments. Job applicants should consult as many of these guides as they want, read the selections that follow in *Strategies,* but, in the end, let common sense be their guide when they write what are essentially advertisements for themselves.

The best advice on how to write effective resumes and cover letters quite naturally comes from the recruiting managers who read them. Several years ago, my own university's placement bureau sponsored a panel featuring the recruiting managers from the companies that traditionally have hired the greatest number of our graduates. All the managers agreed that they looked at resumes and cover letters to determine as quickly as possible—sometime merely by scanning candidate's application materials—what preparation and experience candidates had in the following skills and areas:

- written and oral communication skills
- MIS or PC skills
- interpersonal skills, as demonstrated by the ability to work as a member of a team
- self-reliance and initiative, as demonstrated by the ability to work alone
- a sense of what the world of work demands in terms of professionalism and deadlines
- specific skills in at least one business or technical area supplemented by secondary skills in a variety of related areas
- a sense of business and personal ethics
- the ability to manage time, set priorities, and work under stress.

While recruiting managers do not expect job candidates to excel in all these areas or possess all these skills from the start, the list does provide some general guidelines for communicating with recruiting managers by resumes and cover letters—or, for that matter, in person during an interview.

This section of *Strategies* begins with John L. Munschauer's detailed discussion of how to write a resume and cover letter. Based on his many years of experience working at the Cornell University Career Center, Munschauer places the writing of resumes and cover letters within the context of how they are read by busy recruiting managers. Blythe Camenson offers additional advice on how to begin a job search by putting together a self-assessment designed to allow applicants to present themselves in terms of their strengths and job expectations. Marcia R. Fox follows with some practical suggestions about the mechanics of a good cover letter, and Richard H. Beatty discusses the advantages and disadvantages of using a functional rather than a more traditional chronological resume in applying for a job.

John L. Munschauer

Writing Resumes and Letters in the Language of Employers

John L. Munschauer is Director Emeritus of the Cornell University Career Center.

"I am sorry, Father O'Mega, I can't let you in."

"But, St. Peter, I did everything I was supposed to. I changed the Mass from Latin to English. I had the communicants hold their hands the new way when they took communion. Everyone, so far as I knew, genuflected properly, and even Mrs. O'Reilley's Protestant husband stood when I read the gospel. I can't think of a thing I did that was wrong."

"The message, Father O'Mega, what about the message?"

"The message? What message?"

"The message of the Lord, Father. Don't you remember? The Ten Commandments? The Beatitudes? The Golden Rule? It's the *message* that gets people in here, not the ritual. The ritual was supposed to help deliver the message, but you made the ritual the message and the meaning got lost."

Meanwhile, back on earth . . . in Tucumcari, New Mexico, and all over the United States, job hunters are at their typewriters and computers pushing words around to make their resumes look like the ones they have seen in books. Whether to have it professionally printed or not, that is the question. Does it have enough action verbs?

In Peoria, Illinois, and elsewhere, employers are scanning resumes. Some are beautiful to look at; a few are even printed commercially on expensive paper. But the resumes are laid aside. Employers are reading them with one thought in mind: What can the candidates do for us? The message they are looking for isn't there.

WHY USE A RESUME?

The purpose of a resume is to convey a message, a purpose easily forgotten in the ritual of preparing it. At every turn, you will get conflicting advice about how to conduct the ritual:

- You must have a resume to get a job.
- The purpose of a resume is to get an interview.
- Every resume must have a job objective.
- A resume should never, ever be longer than one page.
- On a resume, list experience chronologically.

On the other hand, others advise:

- Don't use a resume if you are looking for an executive position or merely seeking information.
- Resumes typecast you and narrow your options.
- Interview for information first, determine what an employer wants, and then decide whether offering a resume will suit your purposes.
- Two or more pages present no problem if your resume follows a logical, easy-to-read outline.
- List your experience by function, not chronologically.

The more opinions you get, the more confused you become, but you finally work up something. You send it out. You get little or no response. You change your resume from one page to two pages—or from two pages to one. The results are no better. Somebody says to try pink paper; *that* will get attention. You decide you should have used blue. You fiddle and fiddle with your resume, trying to find the magic formula that will get you what you want. You are caught up in the ritual, forgetting that the purpose of a resume is to send a message.

It is hard not to be distracted from the message. You concentrate on developing a format, forgetting the message, instead of concentrating on the message and *then* working on a format that will convey it best. But what is your message? To find the answer, use your imagination to step out of yourself and become an employer. Think about what the employer is trying to accomplish and the talent required to get it done. Now, from your imaginary employer's chair, look back at yourself and ponder the answer. If you can't come up with one, you will have difficulty writing a resume that will say to the employer, "I have something to offer you."

Resumes are not tickets to a job. They are just one of several ways to court employers. And, as in any courtship, sometimes it's just as well not to put everything in print so the other party can't draw the wrong conclusion. Take the case of Cheryl Fender, an alto who learned that the Springfield Opera Company was auditioning for an alto to sing *Carmen*. Cheryl was well-qualified, and her voice teacher was well-known as a coach of only the most gifted singers. On the other hand, her resume, incomplete and with gaps in her history, showed extensive experience as a secretary. Wisely, she did not send the resume, which might have established her as a secretary who wanted to sing rather than as a singer who had supported herself by being a secretary. Instead, she asked herself what was important to the director of an opera company. Voice, of course, so she outlined her training in a letter. But dramatic ability also mattered, so she included pictures of herself on stage in roles that beautifully illustrated her dramatic ability. She got an audition. Cheryl followed a good marketing rule: Don't confuse customers by flaunting things that don't speak to their needs.

GIVING YOUR MESSAGE

While the language of employment for you is "I want" and for employers it is "I need," you can create resumes and letters in your language that will be read by employers in theirs. I don't mean statements such as "I have analyzed my qualifications and feel confident that they fit your needs." If you were one of thousands of employers who read this kind of thing, would you ask what qualifications and what needs? If the answer isn't clear—and it rarely is—there won't be a message.

It isn't all that difficult to create a statement that is effective. Start with the written word, with prose. Even if you are going to be approaching employers in person, go through the exercise of writing letters. Otherwise, like most people, you can get caught up in the resume ritual and neglect to develop the words that can be more effective in telling your story. People sweat over resumes, then dash off letters without much thought. Perfect resumes arriving in the mail won't even be read if the cover letters don't impress and engage employers.

I recall being asked to review a proposed application letter for a job on the staff of a yachting magazine. The letter was beautifully written, but it left the impression of being all "I": "I want to write so very much, and I am sure I can learn if you give me a chance. . . . I got straight A's in English. . . . I love boats. . . . I am a sailor. . . . I was captain of the sailing team in college. . . ."

What do you say to a person who has written a letter like that? He had obviously spent hours composing it. In its appearance and use of language, it met the highest standards. I could not squelch the young man's hopes, so I had him read *Book Publishing*, a pamphlet Daniel Melcher wrote when he was president of R. R. Bowker Company. Although the pamphlet is concerned only with book publishing, I thought its message would be useful for someone interested in magazines as well.

Let me paraphrase a portion of Melcher's pamphlet:

I like publishing and you might like it too. It is only fair to warn you, however, that publishing attracts a great many more people than the industry can possibly absorb. Sometimes it seems as though half of the English majors in the country besiege publishing offices for jobs each year. While we hope that you will have something to offer us, you might as well face it. Publishers are experts in the art of the gentle brush-off.

The interviewer hopes that you have what he needs, but it turns out that you have never looked into any of the industry's trade journals nor read any books about the industry. You haven't even acquainted yourself with the work of your university press. You tell the interviewer that you are willing to start anywhere, but it develops that a file clerk's job would not interest you, you do not want to type, you don't think you can sell, and you know nothing about printing.

The fact is, all you have thought about is what you want, but it is his needs that create jobs, and you must address yourself to needs.

Your problem, therefore, is to learn as much as possible about the industry before you go looking for a job. Only in this way will you be able to put yourself in the publisher's place and talk to him about his needs rather than about your wants.

After reading Melcher's advice, the fellow said he got the point, thanked me, and left, returning about three weeks later to show me his revised letter. It began in much the same way as his earlier effort—with "I's"—"I majored in English. . . . I have done considerable sailing. . . ."—but after a few short sentences, he suddenly changed his tack. A new three-sentence paragraph began like this: "With my interests, naturally I want to work for you. But more to the point is not what I want but what you need." He followed this with three words that many women refuse to use and that men almost never think to use: "I can type." Right away, he showed that he knew he must be useful.

So much for the easy part of the letter. Next came the difficult part. Although he had had no experience and had never submitted an article to a magazine or even written for a student newspaper, he still had to come up with something that would interest the editor. He found the solution. The letter continued in this manner:

> . . . In looking into the field of journalism, I visited the editor of our alumni magazine, and I talked to magazine space salesmen and to executives responsible for placing ads in magazines. I also visited a printer who has contracts with magazines. In addition, I have been reading trade journals and several books on the industry. As I looked into publishing, it occurred to me that, of all the things I have done, the one I could most closely relate to the field was, strangely enough, an experience I had as a babysitter.

Immediately, he had the editor's full attention. How could babysitting fit in with publishing? During the summer of his junior year in college, he had taken a job as a sailing instructor, tutor, and companion to the children of a wealthy family that summered on the coast of Maine. The parents often went away for a week or more at a time, leaving a governess in charge, with a cook, chauffeur, maid, and gardener to do the chores and the student to keep the children busy. While the parents were on a cruise, the governess suffered a stroke, sending the cook into a tizzy, the maid into tears, and the chauffeur and gardener to the local bar. Only the student could cope, and he took charge and managed the estate for the rest of the summer.

In his letter of application, he described the crisis and its subsequent problems and told how he had met them. Then he related those experiences to the problems that he had learned editors, advertisers, printers, and others encounter in the publishing industry. Reading his letter, you could picture the student working for a publisher. There would be no slipups with the printers. Advertisers and authors would be handled with tact, yet he would get them to turn in their copy on time. He came through as someone with ingenuity, energy, and reliability; and the letter itself testified to his writing ability.

Did he get the job? Yes and no. He got an offer, but because of the publisher's urgent need, the job had to be filled immediately, and he could not accept it. He was teaching school at the time and felt that in fairness to his pupils he should finish the school year. However, a while later, the letter surfaced again when a group of editors at a meeting chatted during lunch. The subject of good editorial help came up. It followed the usual theme: "They don't make 'em like they used to." The editor of another sailing magazine complained that he had been looking for an assistant, but despite hundreds of applicants he had found no one suitable. At this point, the editor of the yachting magazine described the young man's letter and agreed to share it. The upshot was, again, a job offer. This time the timing was right, and the young man took the job. There is nothing quite like the staying power of a well-written letter. It is remembered.

Later, I complimented the young man, telling him I had never read a better letter. "It was easy," he said. "The first letter was the tough one. I didn't have anything to say other than 'I have a good record. Please give me a chance,' but I knew everyone else was saying the same thing, so I would have to say it better. I struggled with every word, trying to make an ordinary message extraordinary, but even elegant words can't make something out of fluff. I didn't know anything about publishing, so I didn't have anything to say to publishers; nor could I be really convincing without knowing enough about the work to decide whether I wanted that kind of job or not. But publishing sounded exciting, so I thought I would give it a fling."

The Importance of Knowing What the Job Is All About

"When I looked into the field in depth," the young man continued, "I became confident that I had something to offer. Thanks to a few good high school teachers and college professors, I knew where to place a comma and a colon, so I had a technical skill to offer publishers. And I knew sailing. But the big need I saw was one I discovered when I was teaching school and again when I was an assistant manager at a McDonald's—a need for people who can get things done. Such a quality is hard to describe without an analogy. I could have alluded to any number of jobs I had held, but I chose the baby-sitting job because I thought it would be different and would introduce an element of surprise. Apparently, the analogy worked. I got the job. More important, I wanted it. If I had received a job offer after sending the first letter, before I had really investigated publishing, I would have taken the job with an attitude of 'teach me.' That's a passive role, an observer's role, and observers tend to be critical. The chances are fifty-fifty that I would not have known how or where to contribute and would have quit after a while. Instead of offering my employer solutions to his problems, I would have become one of his problems."

I asked the young man if he had used any other supporting documents, such as a resume, to help him get the job. "I had a resume," he replied, "but I held it back, because my task was to transfer the qualities I had demonstrated as a baby-sitter to the needs of a publisher, and I couldn't seem to do that in a resume. A resume is a good way to outline facts, but I had to use prose to develop the analogy.

"I did think about including a writing sample, but when I studied the magazine I realized that most of the articles were written by contributors rather than staff. Their job was to select and edit articles. If I presented myself as a writer, I wouldn't have received an offer; the magazine didn't hire authors. When a friend of mine saw an advertisement for an editing job, he made a list of the required qualifications, then presented his case point by point, right on target. Then the dope attached a resume that said loud and clear, 'I want to be a writer.' He either should have skipped the resume or prepared a new one."

Are these examples intended to be good arguments for not using resumes? No. They simply emphasize that it is important to determine the best way to get a message across. There are times when there is no substitute for a resume. When employers advertise and list the qualifications they seek, there is no better way to respond than to send a resume outlining qualifications.

LETTERS OF APPLICATION

Sometimes, however, it is hard to figure out how to make the letter of application you send with your resume more effective than those of the hundreds of other people who are responding to the ad. Imagine being on the receiving end of applications at General Motors or Exxon.

To find out about the effectiveness of letters and resumes, I visited corporations and asked employment managers for their comments. "Here," said one employment manager, as he picked up an 18-inch stack of letters and handed it to me. "This is my morning's mail. Read these letters and you'll have your answer."

"I can't," I protested. "I have only 2 hours, and there's a day's reading here."

"Yes, you can," he replied. "Unfortunately, you'll get through the pack in half an hour, because a glance will tell you that most are not worth reading."

It was hard to believe that the letters could be that bad, but he was right. The typical letter was an insult. Among the letters that I did not finish reading was one that was obviously a copy—the machine must have run out of toner thirty copies back. It began:

Dear Sir:

    I am writing to the top companies in each industry and yours is certainly that. I want to turn my outstanding qualities of leadership and my can-do abilities to. . . .

Enough of that. Also, the applicant hadn't even bothered to type in the employment manager's name, which he could easily have found in reference books such as the CPC Annual, published by the College Placement Council, and Peterson's Job Opportunities, published by Peterson's Guides.

The next letter was written in pencil on notepaper. There may have been an Einstein behind that one, but I can't imagine anyone taking the time to find it out. Many other letters were smudged and messy. Some applicants tried to attract attention with stunts, such as putting cute cartoons on their letters. One piece of mail contained a walnut and a note that read, "Every business has a tough nut to crack. If you have a tough nut to crack and need someone to do it, crack this nut." Inside the nut, all wadded up, was a resume. Cute tricks and cleverness don't work at the General Mammoth Corporation.

At the same time, the good letters stood out like gold. Five letters, only five letters in that pile of hundreds, were worth reading. They had this in common:

- They looked like business letters. Their paragraphing, their neatness, and their crisp white 8 1/2″ × 11″ stationery attracted attention like good-looking clothing and good grooming.
- They were succinct.
- There were no misspellings or grammatical errors.

As I read them, I heard a voice—the voice of a fusty old high school English teacher—commanding out of the past:

- If you can't spell a word, look it up in a dictionary.
- Use a typewriter. Pen and ink are for love letters.
- Clean your typewriter. Avoid fuzzy type. You wouldn't interview in a dirty shirt, so don't send a dirty resume or letter.
- For format, use a secretarial manual. If you don't have one, borrow one from the library. What do you think libraries are for?

How that English teacher would have loved the following advice given by the late Malcolm Forbes when he was editor-in-chief of *Forbes* magazine:

Edit ruthlessly. Somebody ~~has~~ said that words are ~~a lot~~ like inflated money—the more ~~of them that~~ you use, the less each one ~~of them~~ is worth. ~~Right on.~~ Go through your entire letter ~~just~~ as many times as it takes. ~~Search out and~~ Annihilate all unnecessary words ~~and~~ sentences—even ~~entire~~ *paragraphs.*

The following letter is typical of those I saw that day. Give it the Forbes treatment, and see what you can do with it. You may need a scissors as well as a pencil.

Dear Mr. Employer,

I am writing to you because I am going to be looking for employment after I graduate which will be from Michigan where I studied Chemical Engineering and I will be getting a Bachelor of Science degree in June. The field in which I am interested and hope to pursue is process design and that is why I am writing your company to see if you have openings like that. I think I have excellent qualifications and you will find them described in the resume which I have attached to this letter.

The five letters that stood out favorably were characterized by their simplicity. Here is a letter, fictitious of course, but enough like the letter I remember to give you an idea of the ones that created a favorable impression:

February 1, 1991

Mr. Paul Boynton
Manager of Employment
The United States Oil Company
1 Chicago Plaza
Chicago, Illinois 60607

Dear Mr. Boynton:

This June I will receive a Bachelor of Science in Chemical Engineering from the University of Michigan, and I hope to work in process design or instrumentation. I saw your description in *Peterson's Job Opportunities for Engineering, Science, and Computer Graduates* soliciting applicants with my interests. Enclosed is a resume to help you evaluate my qualifications.

While I find all aspects of refining interesting, my special interest in process design and instrumentation developed while working as a laboratory assistant to Professor Juliard Smith, who teaches process design. I wrote my senior thesis on the subject of instrumentation under him, and part of what I wrote will be used in a textbook he is writing and editing

Would it be possible to have an interview with you in Chicago during the week of March 1? To be even more specific, could it be arranged for 10 A.M. on Tuesday, March 3? I am going to be in Chicago that week, and this time and date would be best for me, but of course I would work out another time more convenient for you. In any event, I will call your office the week before to determine whether an interview is possible.

Sincerely yours,

Charles C. Thompson

Thompson's effective letter, and three of the four others, followed a similar pattern:

1. The first paragraph stated who the writer was and what he wanted.
2. The second paragraph, sometimes the third, and in one case a fourth paragraph, indicated why the writer wrote to the employer and mentioned areas of mutual interest, special talents that might be of interest to employers, or other factors relating to qualifications that could be better described in a letter than in a resume.
3. A final paragraph suggested a course of action.

The fifth letter covered the same points in a different order. I remember it because it complemented the good but not outstanding resume shown in Figure 1.

LANCE ZAROTE

Campus Address
101 Morril Hall
University of Puget Sound
Tacoma, Washington 95840
(206) 555–6206

Permanent Address
25 The Byway
Provincetown, Massachusetts 05840
(617) 555–6026

GOAL: A SALES CAREER

EDUCATION: UNIVERSITY OF PUGET SOUND
               Bachelor of Arts 1991
               Philosophy Major

               Business-Related Courses

Statistics	Introduction to Computers
Economics	Calculus
Accounting	English

EMPLOYMENT: UNIVERSITY OF PUGET SOUND (Work-Study Program)

Dining hall supervisor	1990–91
Kitchen helper	1987–89

               OTHER WORK while in college

Ma and Pa Motel, Tacoma—night clerk	1990
Joe's Bar and Grill—weekend waiter & bartender	1989–1991
Baby-sitting, gardening, house cleaning	1985–1991

               WATCHEE OUTEY SUMMER CAMP, Nome, Alaska

Sailing coach and waterfront director	1990
Counselor	1986–1989

ACTIVITIES: CAMPUS
               Chair, Campus Chest Drive
               Intramural hockey, tennis, and volleyball
               LIVING UNIT
               House Manager, $25,000 budget
               Secretary
               Membership Committee
               CIVIC
               Reader-companion in nursing home
               Big Brother-Sister Program, Southside Youth Center
               CHURCH
               Choir
               Youth leader
               Sunday school teacher

INTERESTS:
Skiing	Music
Chess	Dancing

FIGURE I   An ordinary resume can become effective when attached to a powerful letter.

In that resume, a perceptive employer could see a person he might like, someone who was energetic and personable. Yet it didn't quite hang together, because the work history and activities didn't seem to support what the writer wanted to do. But look at the letter that "made" the resume:

Mr. Paul Boynton, Manager of Employment
The United States Oil Company
1 Chicago Plaza
Chicago, Illinois 60607

March 10, 1991

Dear Mr. Boynton:

This June, following my graduation from the University of Puget Sound, I want to pursue a career in sales. Between April 10 and 23, I plan to call on leading companies whose products I would like to sell. The purpose of this letter is to determine whether you would like to have me include you in my itinerary.

Let me tell you why I believe I can sell. It seems to me that I am always selling. As a camp counselor, I persuaded the director to buy a fleet of small sailboats so I could start a sailing program. When our college housing co-op needed painting, I persuaded the members to give up a vacation to do the job. In thinking about how I enjoyed selling these and other projects, I decided to look into a sales career. I persuaded several sales representatives to let me spend a day or more traveling with them to see what it was like.

While with them I realized something more about myself that further convinced me I belong in sales. The best sales representatives were well organized, had high energy levels, and used their time efficiently, qualities I feel I have. As evidence, I have enclosed a resume that outlines my accomplishments in college and during vacations.

I hope to hear from you. United States Oil is in the top group of employers on my list.

Sincerely,

Lance Zarote
Encl: Resume

Hard Work and Attention to Detail Make for a Good Letter

While only the five letters I have mentioned were effective, the rest of the correspondents could have done as well. The point is they didn't. Most people won't. Therein lies your opportunity, because, like Thompson and Zarote, you can write letters that set you apart. You don't have to create a literary masterpiece; just don't knock off a letter hastily with thoughts that wander all over the page. Write it and rewrite it, following Forbes's advice. Unless you are an exceptional typist, you are not good enough to type it yourself. Hire a professional or use a word processor, but be sure the print is letter quality. Also, get an English teacher or someone in the word business to check your spelling, punctuation, and grammar.

Don't Delegate the Job of Letter Writing

More important than style, however, is the thought process used in preparing letters and resumes. Don't shortchange yourself by delegating your thinking to someone else. Be sure it is *your* letter. Somehow, a ghost-written letter always has a phony ring to it. When you write to employers, think about their needs; then think about yourself and what you offer, and relate this to what you would like to do. Putting your thoughts on paper—thoughtfully—will make you sort out your ideas and interrelate them. When you see them on paper they will talk back to you, at times to suggest better ideas, at other times to tell you that you are off the mark. To organize your ideas, create an outline. In other words, prepare a resume even if you decide not to use it. *The value of a resume is frequently more in its preparation than in its use.*

RESUME PREPARATION

When you do give an employer your resume, make it a testimony to your ability to organize your thoughts. Remember, too, it must look sufficiently attractive to get an employer to read it. Unfortunately, most of the resumes I have seen on employers' desks were just as unattractive as the letters; they had sloppy, crowded margins, were poorly organized, and were badly reproduced. At least 30 percent of the resumes had been put aside with hardly a glance because their physical appearance was so awful. The rest got a 20-second scan to see if they were worth studying.

Following are two resumes that pass the appearance test with flying colors. Let's see how they fare during a 20-second scan and beyond.

Nancy Jones—A Good Resume Made Better

Nancy Jones's resume has arrived at the desk of a laboratory director who needs an assistant to help run a quality-control laboratory in a pharmaceutical company (Figure 2). With candidates far outnumbering openings in biology, the advertisement for the job has brought in hundreds of applications, and the director is wearily scanning them one by one to find the few that will be of interest to him. Conditions are not favorable for Nancy. She has to catch his eye with impressive qualifications, or she is not going to get anywhere.

The director picks up Nancy's resume. Immediately, he is impressed, because it looks attractive. He thinks the resume reflects an orderly mind. Most resumes he has looked at just do not put it all together.

He begins to read. The job objective annoys him; it strikes him as being long-winded. Why couldn't she simply say she is interested in applied biology? What is this business about working with people? Is that there because she has doubts about biology?

NANCY O. JONES

Present Address
105 Belleville Place
Ames, Iowa 50011
Phone: 515-555-6674

After June 1, 1991
1212 Centerline Road
Old Westbury, New York 11568
Phone: 516-555-7664

Born January 6, 1969   5'7"   135 lbs.   Single   Excellent health

Career
Objective

Research and development in most areas of applied biology, with an opportunity to work with people as well

Education

Iowa State University, Ames, Iowa
Bachelor of Science, June 1991
Major: Biology   Concentration: Physiology

GPA: 3.3 on a 4.0 scale

Major Subjects	Minor Subjects
Mammalian Physiology	Qualitative Analysis
Vertebrate Anatomy	Quantitative Analysis
Histology	Organic Chemistry
Genetics	Biochemistry

Scholarships
and Honors

University Scholarship: $2850/year
Iowa State Science and Research Award

Dean's List two semesters

Activities

Volunteer Probation Officer, 1988–89
Probation Department, Ames, Iowa

Tutor, Chemistry and Math, 1988–91
Central High School, Ames, Iowa

Member, Kappa Zeta social sorority

Women's Intercollegiate Hockey Team

Special Skills

Familiarity with Spanish; PL/C and FORTRAN computer languages; typing

Work Experience

Teaching Assistant and Laboratory Instructor
Freshman Biology   School year, 1990–91

Waitress, Four Seasons Restaurant
Catalina Island   Summers, 1989–90

FIGURE 2   The resume of Nancy O. Jones does not communicate her career-related experience.

If resumes are supposed to say only what needs to be said, what about this line?

Born January 6, 1969   5'7"   135 lbs.   Single   Excellent health

Does it say anything about her ability to do the job? The biologist doesn't think so.

His eyes move down the page:

Education   Iowa State University, Ames, Iowa
                 Bachelor of Science, June 1991
                 Major: Biology   Concentration: Physiology

                 GPA: 3.3 on a 4.0 scale

                 Major Subjects   Minor Subjects
                 Mammalian Physiology   Qualitative Analysis
                 Vertebrate Anatomy   Quantitative Analysis
                 Histology   Organic Chemistry
                 Genetics   Biochemistry

She has used the outline form well, so he is able to take in a great deal of information in one look. Her education impresses him.

Double spacing above and below her grade point average makes it stand out. However, if her average had not been quite as good, and if she had not wanted to feature it, she could have used single spacing to make it less conspicuous, like this:

Education   Iowa State University, Ames, Iowa
                 Bachelor of Science, June 1991
                 Major: Biology   Concentration: Physiology
                 GPA: 2.7 on a 4.0 scale

Now that she has told him about her grades and her courses she wants to drive home the point that she was no ordinary student. She flags him down with the headline "Scholarships and Honors," and he sees her financial aid and science awards, which make a favorable impression. Two times on the dean's list may not be important enough to set apart by double spacing, but it makes a modest impression.

Up until now, Nancy has made a favorable impression overall. The director is ready to take in the next batch of information:

Activities   Volunteer Probation Officer, 1988–89
                 Probation Department, Ames, Iowa

                 Tutor, Chemistry and Math, 1988–91
                 Central High School, Ames, Iowa

                 Member, Kappa Zeta social sorority

                 Women's Intercollegiate Hockey Team

She almost loses him by featuring her work as a probation officer. That is not of primary interest to him, but through good spacing and placement, she draws his eye to the next item, which states that she has tutored chemistry and math. Unfortunately, Nancy now loses him permanently by ranking tutoring along with the sorority and the hockey team. He guesses she has made her major statement about biology, and so he turns to the next resume.

Good as it is, Nancy Jones's resume could be improved by reorganization. Her tutoring chemistry and her two years as a teaching assistant and laboratory instructor in biology reveal a more-than-academic interest in biology. They should be featured. Everything related to biology and any other information of possible use to the employer should be put under a new marginal headline and statement, as follows:

Career-Related Experience	Biology Lab Instructor and Teaching Assistant Freshman Biology (1990–91)
	Chemistry and Mathematics Tutor Central High School, Ames, Iowa (1988–91)

Skills and Interests	Microscopy	Computer Languages:
	Electron Microscopy	FORTRAN, PL/1,
	Histology	COBOL
	Spectrum Analysis	Statistics
	Small-Animal Surgery	

Now, the director is able to see the things she can do. "Good," he says to himself, "she can use an electron microscope. We need someone with that skill."

By no means should Nancy eliminate mention of the hockey team, but since she is applying for a professional job, the first bait to throw out is credentials; they testify to her ability to do the work. After that, what may sink the hook is how the employer sees her as a person. He may have picked out bits here and there that testify to her diligence, but the picking out depends on chance reading; it would be best not to leave anything to chance. With a slightly different presentation, she might ensure that he receives an impression of her diligence.

There are other areas she could strengthen, things she has underplayed or totally neglected to mention. Sticking the waitress job in down at the bottom of the page is almost an apology for it. She may also have had other jobs, such as baby-sitting, household work, or door-to-door sales, that she belittles in her mind and has not even mentioned. Mention of such things might make a good impression on an employer looking for somebody who is not afraid to work and who is mature for her years.

If we quizzed Nancy, we might find that she could put something like this on her resume:

Scholarships and Financial Support	90% self-supporting through college as follows: University Scholarship: $2850/year Iowa State Science and Research Award

> Waitress, Four Seasons Restaurant
> Catalina Island (Summers, 1989–90)
>
> Teaching and instructing, baby-sitting, home
> maintenance, selling

And this, because her activities tell something about her as a person:

> Activities
>
> Volunteer Probation Officer (1988–89)
> Kappa Zeta social sorority
> Women's Intercollegiate Hockey Team
> Skiing, sailing, singing, tennis

Double spacing has been cut down so as not to overemphasize the less important items, yet a string of other things not terribly important in themselves has been inserted to support the impression of an active, interesting person. Sometimes you want to leave an impression, at other times you want to emphasize a qualification. For example, Nancy wanted to feature her studies, and the way she brought them out by listing them in a column was good. If she had listed them like this, they would not have stood out:

> Major Subjects: Mammalian Physiology, Vertebrate
> Anatomy, Histology, Genetics
> Minor Subjects: Quantitative Analysis, Qualitative
> Analysis, Organic Chemistry, Biochemistry

In Figure 3, you will see the Nancy Jones resume as she might have revised it to emphasize the strong points that would have been of interest to that director who was looking for a resourceful assistant. The added emphasis might have made the difference that would have landed the job for her.

Nancy's revised resume is pure outline, devoid of prose. It works well for her. When she states that she has studied quantitative and qualitative analysis and knows PL/1 and FORTRAN, a scientist reading her resume knows what this means.

Janet Smith—The Proper Use of Headlines

Janet Smith, whose resume is shown in Figure 4, has a different problem in presenting her qualifications. She needs to *describe* what she did in order to tell an employer about their qualifications, so her resume calls for a mixture of key words and prose to get her message across. Yet prose can destroy the effect of an outline. The solution lies in imitating newspaper editors, who use headlines and subheadlines to attract readers. Like a newspaper, a resume should lend itself to skimming so the reader can quickly pick up a good overview of what is important. Then the reader can select specific things of interest and read further. There is an art to using headlines, but Janet Smith hasn't mastered it, at least not in the resume she used to apply for a job with Hermann Langfelder, a hardbitten old

---

NANCY O. JONES

Present Address
105 Belleville Place
Ames, Iowa 50011
Phone: 515-555-6674

After June 1, 1991
1212 Centerline Road
Old Westbury, New York 11568
Phone: 516-555-7664

Career Objective   Research and development

Education   Iowa State University, Ames, Iowa
Bachelor of Science, June 1991
Major: Biology   Concentration: Physiology

GPA: 3.3 on a 4.0 scale

Major Subjects	Minor subjects
Mammalian Physiology	Qualitative Analysis
Vertebrate Anatomy	Quantitative Analysis
Histology	Organic Chemistry
Genetics	Biochemistry

Career-Related   Biology Lab Instructor and Teaching Assistant
Experience   Freshman Biology (1990–91)

Chemistry and Mathematics Tutor
Central High School, Ames, Iowa (1988–91)

Skills and Interests   Microscopy   Computer Languages:
Electron Microscopy   FORTRAN, PL/1,
Histology   COBOL
Spectrum Analysis   Statistics
Small-Animal Surgery

Scholarships   90% self-supporting through college as follows:
and Financial
Support   University Scholarship: $2850/year
Iowa State Science and Research Award

Waitress, Four Seasons Restaurant
Catalina Island (Summers, 1989–90)

Teaching and instructing, baby-sitting, home
maintenance, selling

Activities   Volunteer Probation Officer (1988–89)
Kappa Zeta social sorority
Women's Intercollegiate Hockey Team
Skiing, sailing, singing, tennis

FIGURE 3 The revised resume of Nancy O. Jones effectively features her career-related experience in a separate section.

JANET V. SMITH
111 Main Street
North Hero, Vermont 05073
802-555-1234

CAREER
OBJECTIVE

A challenging position in personnel administration requiring organizational ability and an understanding of how people function in business and industry.

EDUCATION

PURDUE UNIVERSITY
Hammond, Indiana
Master of Industrial Relations   June 1991

SMITH COLLEGE
Northampton, Massachusetts
Bachelor of Arts, magna cum laude   June 1986

WORK
EXPERIENCE

UNIVERSAL METHODIST CHURCH
1 Central Square
New York, New York 10027
Assistant Personnel Officer. 1988–90
Responsible for interviewing applicants for clerical positions within the organization and for placing those who demonstrated appropriate skills, for accepting and dealing with employees' grievances, and for developing programs on career advancement.

CORTEN STEEL COMPANY
10 Lake Street
Akron, Ohio 44309
Assistant, Personnel Office. 1986–88
Responsible for all correspondence of Personnel Director and for interviewing some custodial applicants and referring them to appropriate supervisors for further interviewing.

BORG-WARNER, INC.
Ithaca, New York 14850
Assembly-line worker. Summer 1985
Assembled parts of specialized drive chains in company with thirty other men and women.

COMMUNITY
SERVICE

PLANNED PARENTHOOD
North Hero, Vermont 05073
Counselor. Summers, 1983–84
Explained various aspects of family planning and provided birth control information to clients of Planned Parenthood. Made referrals to other counselors and physicians where appropriate.

AUXILIARY
SKILLS

French: Fluent. Knowledge of office procedures.
Experience with mainframe computes and PCs.

FIGURE 4   The resume for Janet V. Smith misleads the reader with irrelevant headings.

hand with thirty years in personnel and labor relations in the machinery business. He picked up her resume and read:

| CAREER OBJECTIVE | A challenging position in personnel administration requiring organizational ability and an understanding of how people function in business and industry. |

That was pure baloney, and he choked on it. He thought of the hours he had spent in meetings, listening to a lot of hot air. "Challenging, my foot!" he muttered. Then he read the bit about her organizational ability and her understanding of how people function in business and industry. "I've been at this business for thirty years," he groused to himself, "and I still can't figure out how people function in industry. But *she* knows all about it."

Beware of Misleading Headlines

"Well, let's see what she's done," he said to himself, and then his eyes fell on "UNIVERSAL METHODIST CHURCH." That did it! He didn't want to bring any do-gooder into his factory to preach. He rejected her. And the fault was hers, in using the headline.

Janet's job at the church was administrative, not ministerial, and the church didn't care whether she was Jewish, Catholic, or agnostic. Only in its ministerial work does the Methodist Church need Methodists. But people—Langfelder and the rest of us—respond to symbols and make snap judgments on the basis of symbols. Janet had put the symbol of the church—its name—in a heading, when she could have done something better.

When you lay out your resume, think of symbols. Imagine you are a newspaper editor who wants to put a story across. As an editor planning headlines, you must imagine yourself in the position of the reader. Ask yourself, "What words will catch the reader's eye? What word will put the reader off?"

Use words that fit the job in question, and play down those that can lead an employer to think of you in terms that don't relate to the job. Ask yourself, "Does this say something to the employer?" Janet Smith missed Langfelder, and some of her duties would have yielded key words to send appropriate messages. The following arrangement would have been more effective for her:

| WORK EXPERIENCE | ASSISTANT PERSONNEL OFFICER, 1988–90
Universal Methodist Church
1 Central Square
New York, New York 10027
<u>Interviewing, placement, grievances, and training</u> of applicants and employees in the clerical and support services of the church organization. Developed programs for the career advancement of employees. |

ASSISTANT, PERSONNEL OFFICE, 1986–88
Corten Steel Company
10 Lake Street
Akron, Ohio 44309
Interviewing and referring as an assistant to Personnel Director. Responsible for all correspondence and for interviewing some custodial applicants and referring them to supervisors for further interviews.

Let's suppose that Janet has a chance to send Langfelder the revised resume shown in Figure 5. This time, she uses headlines that pinpoint the ideas she most wants to get across, so that he makes it past the Universal Methodist Church and gets down to the assembly line. "Hey, now, look at that!" he thinks. "She's worked out there on the floor. That means she's heard all the language and knows the gripes and the tedium. We have a lot of women in this factory, and it might be a good thing to have a down-to-earth, smart woman on my staff." (His sexism may have been showing, but this kind of employer is alive and kicking somewhere out there, and you may have to deal with him.)

What if Janet Smith wanted a job in the computer industry and had had ten years' experience with IBM? Employers are impressed by "graduates" of companies like IBM that are known as leaders in their fields. Ten years with them is significant. It might make sense to present her experience this way:

EXPERIENCE   1981–91
IBM Corporation
Binghamton, New York
Assistant Personnel Officer

Were she an engineer after a technical job that could use her IBM experience, highlighting the name of the company would have been a good idea. Imagine employers giving the resume a 20-second scan. What words should you use and how should they be placed to catch the eye and make employers want to read further?

Mark Meyers—The Functional Resume

Janet Smith and Nancy Jones were lucky. Their training and experience translated into satisfactory headlines to highlight their experience, but that doesn't always work. Mark Meyers, whose resume appears in Figure 6, adopted a different technique to help him get a job in community recreation. He got his message across by creating a resume based on functions.

When he began to write his resume, he tried time and time again to get his message across in a conventional form in which he first listed his education, then his experiences in chronological order, and finally his activities, hobbies, and interests. But writing it conventionally raised all sorts of problems.

JANET V. SMITH
111 Main Street
North Hero, Vermont 05073
802-555-1234

CAREER
INTERESTS   Personnel Administration and Labor Relations

EDUCATION   PURDUE UNIVERSITY   June 1991
                           Hammond, Indiana
                           Master of Industrial Relations

                           SMITH COLLEGE June 1986
                           Northampton, Massachusetts
                           Bachelor of Arts, magna cum laude

WORK   ASSISTANT PERSONNEL OFFICER. 1988–90
EXPERIENCE   Universal Methodist Church
                           1 Central Square
                           New York, New York 10027
                           Interviewing, placement, grievances, and training of
                           applicants and employees in the clerical and support services
                           of the church organization. Developed programs for the
                           career advancement of employees.

                           ASSISTANT, PERSONNEL OFFICE. 1986–88
                           Corten Steel Company
                           10 Lake Street
                           Akron, Ohio 44309
                           Interviewing and referring as an assistant to Personnel
                           Director. Responsible for all correspondence and for
                           interviewing some custodial applicants and referring them to
                           supervisors for further interviews.

                           ASSEMBLY-LINE WORKER. Summer 1985
                           Borg-Warner, Inc.
                           Ithaca, New York 14850
                           Factory work experience on an assembly line. Worked with a
                           team of thirty other men and women.

COMMUNITY   COUNSELOR. Summers, 1983–84
SERVICE   Planned Parenthood
                           North Hero, Vermont 05073
                           Explained various aspects of family planning and provided
                           birth control information to clients of Planned Parenthood.
                           Made referrals to other counselors and physicians where
                           appropriate.

AUXILIARY   French: Fluent. Knowledge of office procedures.
SKILLS   Experience with mainframe computers and PCs.

FIGURE 5   The revised resume for Janet V. Smith stresses what she did rather than the less important point of where she did it.

He wanted to highlight his public relations and promotion experience, some of which he had been paid for and some not. Some of it had also been secondary to a primary assignment. Dividing up this experience and placing bits and pieces of it in various parts of the resume to make it conform to a conventional style diluted its impact. Also, his athletic ability and experience would mean a great deal to an employer in his field, but how could he show it effectively? Some of it had been gained as a participant, some through training, and some as a coach. Could he expect an employer to sift through the various sections of the resume to find out all he had done in athletics? (Remember, you can only count on an employer giving a resume a quick scan before deciding whether or not to study it more fully.)

His solution, as seen in Figure 6, was to feature the functions of the job he wanted and then describe things he had done that pertained to each area. Thus, under each function he developed the equivalent of a mini-resume.

Preparing a Resume for a Specific Job

Mark stated his case well, but you can't get blood from a stone. He found he had to look outside his field, because jobs in it were virtually nonexistent due to cutbacks in government funding. In his search, he ran across the following job listing from the publisher of a magazine for parents:

EDITORIAL SECRETARY

BA. in Liberal Arts

Interested in childhood training. Well organized, outstanding language skills. Typing and clerical skills, potential to use electronic text-editing equipment. Reporting to Coordinating Editor, Happy Days magazine. Assist in all editorial functions. Evidence of creativity essential. Entry-level position with career potential.

Can you put yourself in his shoes, analyze the job, and devise a resume that speaks to the stated needs of this employer? The clue to doing it is to go through the job description and step by step take your cue from the employer. Right off, you will hit a bit of a snag because the employer has specified a B.A., while Mark has a B.S. His drama minor might give him appropriate credentials, however. If he shows his education as in the following example it might reflect the liberal background the employer apparently prefers:

EDUCATION   HOBOKEN UNIVERSITY, B.S., 1987
                        Major: Recreation   Minor: Drama

                        Humanities courses:
                        Introduction to Dramatic Literature
                        British Drama to 1700

MARK MEYERS

1414 South Harp Road
Dover, Delaware 19901
302-555-4444

CAREER INTEREST
Community recreation.

RECREATION PROGRAMMING EXPERIENCE
Planned and implemented programs in stagecraft and drama; assisted with pro-
gramming in ceramics, photography, and physical fitness for Dover Youth
Bureau summer program. Summer 1986.

Lectured and led tours at Atlantic County (Delaware) Park nature trail and visi-
tors' center. Prepared slides to illustrate lecture; helped in construction of
nature exhibits. Summer 1985.

Coordinated men's intramural sports competitions for Hoboken University.
Had responsibility for equipment and scheduling. 1984–86.

PUBLIC RELATIONS AND PROMOTION EXPERIENCE
Directed publicity efforts of University Drama Club for several productions.
Developed innovative techniques—such as a costumed cast parade to
arouse interest in an avant-garde staging of "Alice in Wonderland", which
gained campuswide attention. 1984–85.

Advertised in various media and became familiar with advertising methods,
including writing news releases, taping radio announcements, designing
graphics for posters and fliers. Drama Club, 1985–87; men's sports,
1984–87.

LEADERSHIP AND ATHLETIC ABILITIES
Trained in outdoor leadership and survival skills by Outward Bound.
Summer 1983.

Coached hockey and basketball. Dover Youth Bureau, 1986.

Participate in hockey, swimming, backpacking.

RESEARCH AND EVALUATION ABILITIES
Prepared college research project on the recreational needs of residents of a
Hoboken neighborhood. Designed questionnaire to solicit residents' own
perceptions of their needs; interviewed residents and local officials. Fall
semester 1986.

Reported on effectiveness of Dover Youth Bureau programming. Summer
1986.

OFFICE SKILLS
Dealt with unhappy and irate customers at McGary's Department Store, Dover.
Worked at service desk, tracking down problems and rectifying errors.
Summer 1982.

Typing and use of office machinery.

EDUCATION
Bachelor of Science, June 1987   Major: Recreation
Hoboken University, Hoboken, New Jersey   Minor: Drama

FIGURE 6   The resume of Mark Meyers illustrates the functional style.

History of Theater
Playwriting
Introduction to Poetry
Shakespeare
English History

Next, the job calls for an interest in childhood training, then language skills, and so on. Each specification suggests a headline for a resume. Mark faces another stumbling block when it comes to demonstrating an interest in childhood training, since his experience has been with older youths only. Since a resume makes points by stating facts, he cannot demonstrate an interest in childhood training in his resume because he lacks the appropriate experience. However, he *can* describe his interest in the letter that usually goes hand in glove with a resume. He has a good basis for doing so, because the field of recreation certainly has much to do with the entire range of human development from childhood on. He should be able to point out correlatives in his education and experience with the work being done in childhood training, and he could check a library for information about childhood training to help develop the correlatives. Above all, he should read the magazine and try to tie in as many of his experiences as possible with the purpose of the magazine.

With his interest in childhood training brought out in a letter to complement his resume, he might then proceed to develop his outline as follows:

EXPERIENCE

HUMAN DEVELOPMENT

<u>Dover Youth Bureau</u>. Planned and implemented programs in drama, photography, athletics, health. Summer 1986

<u>Research Project</u>. Studied recreational needs of a Hoboken neighborhood. Interviewed residents, developed questionnaire. Project provided an insight into the family life and problems of parents in a neighborhood setting. Fall semester 1986

<u>Outdoor Leadership Training</u>. Practical experience in human development through a knowledge of nature and survival skills. Summer 1983.

LANGUAGE SKILLS AND CREATIVITY

<u>Writing</u>. Wrote releases, developed advertising, and prepared radio announcements for Drama Club, men's sports programs, and other events. Wrote report on Hoboken research project. 1984–87

<u>Lecturing</u>. Gave talks and led tours at Atlantic County (Delaware) Park nature trail and visitors' center. Summer 1985

<u>Audiovisual</u>. Prepared slides for audiovisual presentation for visitors' center. 1985.

Promotional. Created innovative techniques—such as costume parade of cast—to arouse interest in avant-garde staging of "Alice in Wonderland," which gained campuswide attention. 1984–85.

Design. Designed posters, fliers, and other graphics for sporting events, plays, and other campus events. 1984–87.

Constructed nature exhibits at nature center. 1985.

CLERICAL AND ADMINISTRATIVE

Clerical. Typing and use of office machinery. McGary's Department Store. Summer 1982.

Administration. Responsible for equipment, scheduling of programs, and coordinating of competitions. Hoboken University, 1984–86.

The functional resume allows you to develop a different message for each job or type of job you wish to apply for. Different functions can be highlighted, depending on what the job requires, and your specific experiences rearranged under different headings. It gives you the flexibility you need if your experience has been diverse.

Almost every resume ought to have something of the functional resume about it. With computers making it so easy to change a text, there is no reason why each resume can't be slanted to appeal to the particular employer, even if it's the resume of a generalist like Bruce Gregory Robinson, who hopes to get into a training program.

Bruce Gregory Robertson—A Resume Reflecting an Active Mind and Body

Put yourself in the chair of a merit employer, someone who is interested in candidates not so much for what they know as for what they can learn. Imagine yourself as a banker, a merchant, or a manufacturer. The position you have to fill can be learned easily on the job. What interests you are candidates' traits—their energy, intelligence, leadership qualities—the things that tell you the candidate can grow in your employ and eventually become an executive in the company. You are looking for a resume that reflects an active person with an active mind. Bruce Robertson is such a person, and his resume shown in Figure 7 has been designed accordingly.

Now make yourself a textbook publisher. Again, you are a merit employer and promote from within. Your company hires people with a good liberal education and little or no postgraduate experience. The men and women you hire will travel from college to college either soliciting manuscripts from professors or showing them texts that might fit their courses. Impressive candidates should have a high energy level and be able to show that they have used time efficiently and worked independently. How does Bruce look to you?

BRUCE GREGORY ROBERTSON

Home Address	College Address
105 Comstock Drive	5 Erasmus Drive
Pierre, South Dakota 57501	St. Paul, Minnesota 55101
605-555-5445	612-555-4554

CAREER
INTERESTS
— The marketing of products or services in industries such as banking, publishing, and retailing.

EDUCATIONAL
BACKGROUND
— Macalester College, St. Paul, Minnesota   A.B., May 1991
Major: English   Minor: Economics
Honors: Dean's List, 1990–91

EMPLOYMENT
— Entrepreneurial—Organized a company to paint road markings in parking lots; rented necessary equipment; employed two additional workers; had personal contact with mall managers, store owners, etc., to promote the company's services: summers, 1989–90.

— Administrative and Clerical—Temporary Help, Inc. Typing and other business machine operations, bookkeeping, clerking, complaint adjustments; short-term assignments with auto dealers, banks, real estate operations, schools, and similar employers: summer 1988; part-time 1989.

— Miscellaneous—Camp counselor, stock clerk in grocery store, baby-sitter, newspaper boy. From high school on have earned money for clothes, travel, and purchase and maintenance of an automobile.

ACTIVITIES
— Vice President, Seven-Come-Eleven investment club
— Varsity basketball
— Coach and tutor, St. Paul Concordia Boys Club
— Debate Club
— Book and Bottle literary club

SKILLS
— Typing
— PL/C and FORTRAN computer languages
— German: fluent

HONORS
— Dakota Interstate Scholarship
— Hubert H. Humphrey First Prize, Minnesota Intercollegiate Debate

INTERESTS
— Public speaking and debate
— Investments
— Writing
— Parachute jumping

FIGURE 7   A resume designed more to reflect an active, energetic personality than specific experience.

Now change the scenario slightly. This time you are a publisher of texts for primary and secondary school. To get your texts adopted, your representatives will have to make presentations to state regents, school superintendents, and other educational groups. Remember that experience and college major don't matter, but do you think it might help Bruce's cause if he did a bit more to high-light his interest in public speaking and his honors as a debater? He has a computer. It would be a cinch for him to change his resume slightly. Can you think how he might do it?

Next, put yourself in the chair of an employer on Main Street. You need someone in customer relations, a person to follow up on complaints and misunderstandings about credit arrangements or bills. You want to hire someone who has had experience of this kind. Still, the work easily could be learned on the job . . . if the right person came along. In response to your classified advertisement, Bruce's resume arrives accompanied by a routine cover letter. He recounts people problems he has resolved and describes his administrative experience in a way that indicates he is a team worker. What's your decision? What if his letter is more inspired?

Michelle Trio—The Curriculum Vitae

A curriculum vitae (literally, "course of life" in Latin), sometimes called a C.V. or vita, is a resume for academic positions and as such does not need a statement of goals or interest. While there is merit in keeping nonacademic resumes brief by focusing on employers' needs, a faculty tends to select colleagues not just to teach but for the prestige they will bring to the department, especially in the long run. An eminent faculty attracts eminent associates. Publications, research, memberships, and honors all contribute to telling what a candidate is like; hence long vitae that reflect many achievements are traditional. The same candidate applying for an industrial position wouldn't list a raft of publications, for example, if those publications didn't relate to the job in question.

The C.V. that is shown in Figure 8 lays out in a logical way all the essentials of a good vita. Note the correct way to list publications. In academia, those who list them incorrectly are jeopardizing their chances of being hired.

The Job Objective

"Do I have to have a job objective?" According to my calculations, as of this writing I have been a career counselor for 2,288 weeks, and I have been asked this question at least six times a week, except for the 132 weeks when I was on vacation. When I answer, "No, I don't like the heading JOB OBJECTIVE," the sigh of relief is audible. The job seekers think I have let them off the hook for one of the most important parts of a resume. I haven't. With rare exceptions, a resume *should* open with an objective—it's the way it is stated that can be changed.

MICHELLE TRIO

Department of English   406 East Bates Street
Athens College   Athens, NY 14850
Athens, NY 14850   607-555-7654
607-555-4567

EDUCATION

Ph.D., 1984, Cornwall University.

Dissertation:	"The Anatomy of Sin: Violations of *Kynde* and *Trawpe* in *Cleanness*," directed by L. E. Cooper (DAI 40/09, p. 5046-A).
Major Subject:	Old and Middle English language and literature.
Minor Subject:	Medieval philology.
Courses:	Old English; *Beowulf;* seminars on the Junius Manuscript, the Exeter Book, and hagiography; Middle English literature; Chaucer; *Piers Plowman;* Medieval Latin; paleography; Old French; Middle High German; Old Icelandic; Dante.

A.M., 1979, with High Honors, Boston University.

B.A., 1977, *magna cum laude,* Honors in English, State University of New York at Stony Brook.

PROFESSIONAL EXPERIENCE

1985–present: Assistant Professor, Medieval and Renaissance Literature, Department of English, Athens College. Position includes: Engl 325 Chaucer, Engl 323 Triumph of English (a course that I instituted on the history of English), Engl 232 Medieval Literature, Engl 420 Shakespeare Seminar, Engl 231 Ancient Literature, Engl 107 Introduction to Literature: Myth, Legend, and Folktale.

1985: Assistant Professor, Department of English, Cornwall University Summer Program. One section of Practical Prose and Composition and training of a graduate teaching assistant.

1984–85: Lecturer, Department of English, Cornwall University. Two courses in Freshman Seminar Program: Shakespeare and Politics, Practical Prose Composition. I was Co-director of the latter, with responsibility for planning the syllabus, training and evaluating graduate teaching assistants, and leading staff meetings on problems and goals of teaching composition.

PUBLICATIONS

"Heroic Kingship and Just War in the Alliterative *Morte Arthure*," to be published in *Acta,* 11.

Articles on the Beatitudes, the Handwriting on the Wall, the Parable of the Marriage Feast, Sarah, and Sodom and Gomorrah in *Dictionary of Biblical Tradition in English Literature,* ed. David L. Jeffrey, to be published by Oxford University Press and W. B. Eerdmans.

"On Reading *Bede's Death Song:* Translation, Typology, and Penance in Symeon of Durham's Text of the *Epistola Cuthbert: de Obitu Bedae.*" *Neuphilologische Mitteilungen,* 84 (1983), 171–81.

FIGURE 8   A curriculum vitae

PROFESSIONAL SERVICE AND ACTIVITIES

1. At Athens College
    1989–90: English Department Library Representative.
    1989–90: Member, Athens College Faculty Enrichment Committee.
    1988–90: English Department Personnel Committee.
    1987–89: English Department Committee on the London Center.

2. Elsewhere
    1985: Organizer and Chair, Latin Section, Northeast Modern Language Association. Topic: "Eschatology and Apocalypticism."
    1980–82: Organizer and Chair, *Quodlibet:* The Cornwall Medieval Forum.
    Memberships in MLA, Medieval Academy of America, International Arthurian Society.

HONORS

Charles A. Dana Fellowship for Excellence in Teaching, Athens College.

Goethe Prize in German Literature, Cornwall University.

George Lincoln Fellowship in Medieval Studies, Cornwall University.

Teaching Assistantship, Medieval Studies program, Cornwall University ("Medieval Literature in Translation").

Teaching Fellowship, English Department, Boston University ("Freshman Rhetoric and Composition").

LANGUAGES

Reading knowledge of Latin, Old French, Spanish, Italian, Middle High German, and Old Icelandic, in addition to the usual French, German, and Old and Middle English.

CREDENTIALS

Dossier may be obtained from the Educational Placement Bureau, Barnes Hall, Cornwall University, Athens, NY 14853.

FIGURE 8   *(Continued)*

I show them the preceding Robertson, Meyers, and Smith resumes and tell them that I prefer the headline CAREER INTEREST, because it leads to a simple and direct way of stating the purpose of the resume. For example, I suggest they try rewriting the Robertson and revised Smith statements of interest as job objectives to see if they don't find it awkward. After a struggle they come up with the same kind of baloney found in the unrevised Jones and Smith resumes.

They listen politely. I may have helped them with a minor problem of phrasing, but I know that I really haven't dealt with their question. Then they tell me what I already know: They don't want to state a goal because they don't know what they want to do. I ask them to imagine themselves as the employers reading their resumes. If theirs is like the revised Nancy Jones resume, which is so obviously slanted toward biology, then employers can figure out what the resume is for without a stated objective. However, I wouldn't fool with a resume that didn't tell me at the outset what it was all about. When I have to start studying a resume to guess what the writer wants, I throw it in the wastebasket.

Next I point to the Zarote and Robertson resumes, which wouldn't make any sense without a statement of purpose. And I can't get up much enthusiasm for a letter as a substitute for an objective. A letter stating a purpose accompanied by a resume without purpose is a wasted letter.

I remember a young woman who, on short notice, got a chance to be interviewed by a recruiter from a large department store. Knowing very little about merchandising, she headed for the nearest department store, where several managers were kind enough to give her information and the loan of their trade journals. In three hours of investigation, her eyes were opened to an industry in which people pursued careers in training, employee relations, promotion, credit, public relations, merchandising, and other occupations. Several of these looked interesting, so thanks to word processing, she easily changed a few things on her resume, then listed her objective like this:

CAREER INTERESTS: Training, promotion, and public relations in a retail setting.

Like most of us, her range of aptitudes was wide, so, like a chameleon, she showed the recruiter only those that seemed to match retailing.

One Page or Two?

If a resume can be kept to one page, so much the better. The length depends on the message. In reading resumes in which everything is jammed on one page with none of the white space or headings that can make them attractive and readable, I have wearied of trying to find information and have given up. On the other hand, it has never been the least bit tiring to lift a piece of paper and turn to a second page when scanning an interesting resume. A resume is an outline. It needs a white space. It needs headings that stand out. Don't sacrifice them for some arbitrary notion about a one-page maximum.

Additional Advice About Resumes

No matter how you develop your message, test it before you send it to employers. Get friends to give you a critique of your resume or vita, especially if they are in an occupation in which you hope to find a job. A word of caution, however. Unless you guide them, their critique may relate more to the ritual than to the message. One job hunter had modeled a resume after Janet Smith's. The critic took a red pencil and put all the dates of employment in the left-hand margin. That would have been a good idea if it had indicated long years of experience with a company such as 3M, signifying considerable experience with one of the best-managed companies in the country. But the dates in question referred to short-term summer jobs that were of no consequence to the message and cluttered the margin with information that distracted from the headlines.

One way to get a resume criticized is to hold it up a few feet from the reader and ask for comments on its appearance. Does it look neat? Is the layout pleasing? Does it look easy to read? Is the print good-looking?

Next, give the critics the resume to read. Let them make all the comments they want. You may pick up valuable ideas for improving its style and layout, but be careful you don't get caught up in inconsequentials. What you really want is to have your critics look at the resume as if they didn't know you. You might even show them a resume with an alias, then ask:

- What qualifications does this person have?
- What do you see this person doing with these qualifications?
- What kind of an employer would want to hire this person?
- Does the resume project an image of a certain kind of person? What kind? Aggressive? Thoughtful? Energetic? What?

In other words, ask your critics the most important question about your resume: "What message do you get about me?"

Blythe Camenson

The Self-Assessment

Blythe Camenson is the Director of the Fiction Writers Connection and the author of more than twenty books on career-related topics.

Self-assessment is the process by which you begin to acknowledge your own particular blend of education, experiences, values, needs, and goals. It provides the foundation for career planning and the entire job search process. Self-assessment involves looking inward and asking yourself what can sometimes prove to be difficult questions. This self-examination should lead to an intimate understanding of your personal traits, your personal values, your consumption patterns and economic needs, your longer-term goals, your skill base, your preferred skills, and your underdeveloped skills.

You come to the self-assessment process knowing yourself well in some of these areas, but you may still be uncertain about other aspects. You may be well aware of your consumption patterns, but have you spent much time specifically identifying your longer-term goals, or your personal values as they relate to work? No matter what level of self-assessment you have undertaken to date, it is now time to clarify all of these issues and questions as they relate to the job search.

The knowledge you gain in the self-assessment process will guide . . . all of the following tasks:

- Writing resumes
- Exploring possible job titles
- Identifying employment sites
- Networking
- Interviewing

THE SELF-ASSESSMENT. By Blythe Camenson. Reprinted from *Great Jobs for Liberal Arts Majors* by Blythe Camenson, 1997. Used with permission of NTC/Contemporary Publishing Company.

- Following up
- Evaluating job offers.

In each of these steps, you will rely on and return often to the understanding gained through your self-assessment. Any individual seeking employment must be able and willing to express to potential recruiters and interviewers throughout the job search these facets of his or her personality. This communication allows you to show the world who you are so that together with employers you can determine whether there will be a workable match with a given job or career path.

HOW TO CONDUCT A SELF-ASSESSMENT

The self-assessment process goes on naturally all the time. People ask you to clarify what you mean, or you make a purchasing decision, or you begin a new relationship. You react to the world, and the world reacts to you. How you understand these interactions and any changes you might make because of them are part of the natural process of self-discovery. There is, however, a more comprehensive and efficient way to approach self-assessment with regard to employment.

Because self-assessment can become a complex exercise, we have distilled it into a seven-step process that provides an effective basis for undertaking a job search. The seven steps include the following:

1. Understanding your personal traits
2. Identifying your personal values
3. Calculating your economic needs
4. Exploring your longer-term goals
5. Enumerating your skill base
6. Recognizing your preferred skills
7. Assessing skills needing further development.

As you work through your self-assessment, you might want to create a worksheet similar to the one shown in Figure 1. Or you might want to keep a journal of the thoughts you have as you undergo this process. There will be many opportunities to revise your self-assessment as you start down the path of seeking a career.

Step 1   Understanding Your Personal Traits

Each person has a unique personality that he or she brings to the job search process. Gaining a better understanding of your personal traits can help you evaluate job and career choices. Identifying these traits, then finding employment that allows you to draw on at least some of them can create a rewarding and fulfilling work experience. If potential employment doesn't allow you to use these preferred traits, it is important to decide whether you can find other ways to

Self-Assessment Worksheet

STEP 1.   **Understand Your Personal Traits**
The personal traits that describe me are:
(Include all of the words that describe you.)

The ten personal traits that most accurately describe me are:
(List these ten traits.)

STEP 2.   **Identify Your Personal Values**
Working conditions that are important to me include:
(List working conditions that would have to exist for you to accept a position.)

The values that go along with my working conditions are:
(Write down the values that correspond to each working condition.)

Some additional values I've decided to include are:
(List those values you identify as you conduct this job search.)

STEP 3.   **Calculate Your Economic Needs**
My estimated minimum annual salary requirement is:
(Write the salary you have calculated based on your budget.)

Starting salaries for the positions I'm considering are:
(List the name of each job you are considering and the associated starting salary.)

STEP 4.   **Explore Your Longer-Term Goals**
My thoughts on longer-term goals right now are:
(Jot down some of your longer-term goals as you know them right now.)

(continued)

FIGURE I   Self-assessment worksheet

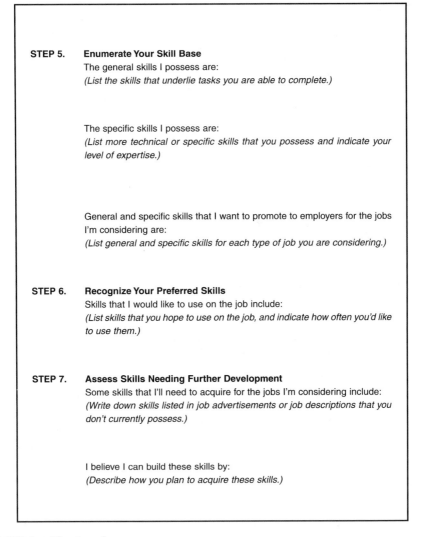

STEP 5. **Enumerate Your Skill Base**
The general skills I possess are:
(List the skills that underlie tasks you are able to complete.)

The specific skills I possess are:
(List more technical or specific skills that you possess and indicate your level of expertise.)

General and specific skills that I want to promote to employers for the jobs I'm considering are:
(List general and specific skills for each type of job you are considering.)

STEP 6. **Recognize Your Preferred Skills**
Skills that I would like to use on the job include:
(List skills that you hope to use on the job, and indicate how often you'd like to use them.)

STEP 7. **Assess Skills Needing Further Development**
Some skills that I'll need to acquire for the jobs I'm considering include:
(Write down skills listed in job advertisements or job descriptions that you don't currently possess.)

I believe I can build these skills by:
(Describe how you plan to acquire these skills.)

FIGURE I   *(Continued)*

express them or whether you would be better off not considering this type of job. Interests and hobbies pursued outside of work hours can be one way to use personal traits you don't have an opportunity to draw on in your work. For example, if you consider yourself an outgoing person and the kinds of jobs you are examining allow little contact with other people, you may be able to achieve the level of interaction that is comfortable for you outside of your work setting. If such a compromise seems impractical or otherwise unsatisfactory, you probably should explore only jobs that provide the interaction you want and need on the job.

Many young adults who are not very confident about their attractiveness to employers will downplay their need for income. They will say, "Money is not all that important if I love my work." But if you begin to document exactly what you need for housing, transportation, insurance, clothing, food, and utilities, you will begin to understand that some jobs cannot meet your financial needs and it doesn't matter how wonderful the job is. If you have to worry each payday about bills and other financial obligations, you won't be very effective on the job. Begin now to be honest with yourself about your needs.

Inventorying Your Personal Traits

Begin the self-assessment process by creating an inventory of your personal traits. Using the list in Figure 2, decide which of these personal traits describe you.

Active	Critical	Generous
Accurate	Curious	Gentle
Adaptable	Daring	Good-natured
Adventurous	Decisive	Helpful
Affectionate	Deliberate	Honest
Aggressive	Detail-oriented	Humorous
Ambitious	Determined	Idealistic
Analytical	Discreet	Imaginative
Appreciative	Dominant	Impersonal
Artistic	Eager	Independent
Brave	Easygoing	Individualistic
Businesslike	Efficient	Industrious
Calm	Emotional	Informal
Capable	Empathetic	Innovative
Caring	Energetic	Intellectual
Cautious	Excitable	Intelligent
Cheerful	Expressive	Introverted
Clean	Extroverted	Intuitive
Competent	Fair-minded	Inventive
Confident	Farsighted	Jovial
Conscientious	Feeling	Just
Conservative	Firm	Kind
Considerate	Flexible	Liberal
Cool	Formal	Likable
Cooperative	Friendly	Logical
Courageous	Future-oriented	Loyal

FIGURE 2 Personal traits

Mature	Principled	Sincere
Methodical	Private	Sociable
Meticulous	Productive	Spontaneous
Mistrustful	Progressive	Strong
Modest	Quick	Strong-minded
Motivated	Quiet	Structured
Objective	Rational	Subjective
Observant	Realistic	Tactful
Open-minded	Receptive	Thorough
Opportunistic	Reflective	Thoughtful
Optimistic	Relaxed	Tolerant
Organized	Reliable	Trusting
Original	Reserved	Trustworthy
Outgoing	Resourceful	Truthful
Patient	Responsible	Understanding
Peaceable	Reverent	Unexcitable
Personable	Sedentary	Uninhibited
Persuasive	Self-confident	Verbal
Pleasant	Self-controlled	Versatile
Poised	Self-disciplined	Wholesome
Polite	Sensible	Wise
Practical	Sensitive	
Precise	Serious	

FIGURE 2   *(Continued)*

Focusing on Selected Personal Traits

Of all the traits you identified from the list in Figure 2, select the ten you believe most accurately describe you. If you are having a difficult time deciding, think about which words people who know you well would use to describe you. Keep track of these ten traits.

Considering Your Personal Traits in the Job Search Process

As you begin exploring jobs and careers, watch for matches between your personal traits and the job descriptions you read. Some jobs will require many personal traits you know you possess, and others will not seem to match those traits.

An editor's job, for example, requires the ability to work as part of a team, often coordinating schedules and activities with writers, typesetters, proofreaders, printers, and advertising and promotion departments. Excellent organizational and interpersonal skills are essential qualities. Freelance writers, on the other hand, usually work alone, with limited opportunities to interact with others. Both have deadlines to meet, but the writer usually has far fewer people to answer to and must be able to work independently.

Your ability to respond to changing conditions, decision-making ability, productivity, creativity, and verbal skills all have a bearing on your success in and enjoyment of your work life. To better guarantee success, be sure to take the time needed to understand these traits in yourself.

Step 2 Identifying Your Personal Values

Your personal values affect every aspect of your life, including employment, and they develop and change as you move through life. Values can be defined as principles that we hold in high regard, qualities that are important and desirable to us. Some values aren't ordinarily connected to work (love, beauty, color, light, marriage, family, or religion), and others are (autonomy, cooperation, effectiveness, achievement, knowledge, and security). Our values determine, in part, the level of satisfaction we feel in a particular job.

Defining Acceptable Working Conditions

One facet of employment is the set of working conditions that must exist for someone to consider taking a job.

Each of us would probably create a unique list of acceptable working conditions, but items that might be included on many people's lists are the amount of money you would need to be paid, how far you are willing to drive or travel, the amount of freedom you want in determining your own schedule, whether you would be working with people or data or things, and the types of tasks you would be willing to do. Your conditions might include statements of working conditions you will *not* accept; for example, you might not be willing to work at night or on weekends or holidays.

If you were offered a job tomorrow, what conditions would have to exist for you to realistically consider accepting the position? Take some time and make a list of these conditions.

Realizing Associated Values

Your list of working conditions can be used to create an inventory of your values relating to jobs and careers you are exploring. For example, if one of your conditions stated that you wanted to earn at least $25,000 per year, the associated value would be financial gain. If another condition was that you wanted to work with a friendly group of people, the value that goes along with that might be belonging or interaction with people. Figure 3 provides a list of commonly held values that relate to the work environment: use it to create your own list of personal values.

Achievement	Development	Physical activity
Advancement	Effectiveness	Power
Adventure	Excitement	Precision
Attainment	Fast pace	Prestige
Authority	Financial gain	Privacy
Autonomy	Helping	Profit
Belonging	Humor	Recognition
Challenge	Improvisation	Risk
Change	Independence	Security
Communication	Influencing others	Self-expression
Community	Intellectual stimulation	Solitude
Competition	Interaction	Stability
Completion	Knowledge	Status
Contribution	Leading	Structure
Control	Mastery	Supervision
Cooperation	Mobility	Surroundings
Creativity	Moral fulfillment	Time freedom
Decision making	Organization	Variety

FIGURE 3   Work values

Relating Your Values to the World of Work

As you read the job descriptions in this . . . [selection], think about the values associated with that position.

··

> For example, the duties of a reporter would include researching, investigating, and conducting interviews; organizing the information in a logical format; and writing and editing articles and profiles. Associated values are intellectual stimulation, organization, communication, and creativity.

··

If you were thinking about a career in this field, or any other field you're exploring, at least some of the associated values should match those you extracted from your list of working conditions. Take a second look at any values that don't match up. How important are they to you? What will happen if they are not satisfied on the job? Can you incorporate those personal values elsewhere? Your answers need to be brutally honest. As you continue your exploration, be sure to add to your list any additional values that occur to you.

Step 3   Calculating Your Economic Needs

Each of us grew up in an environment that provided for certain basic needs, such as food and shelter, and, to varying degrees, other needs that we now consider basic, such as cable TV, reading materials, or an automobile. Needs such as privacy, space, and quiet, which at first glance may not appear to be monetary needs, may add to housing expenses and so should be considered as you examine your economic needs. For example, if you place a high value on a large, open living space for yourself, it would be difficult to satisfy that need without an associated high housing cost, especially in a densely populated city environment

As you prepare to move into the world of work and become responsible for meeting your own basic needs, it is important to consider the salary you will need to be able to afford a satisfying standard of living. The three-step process outlined here will help you plan a budget, which in turn will allow you to evaluate the various career choices and geographic locations you are considering. The steps include (1) developing a realistic budget, (2) examining starting salaries, and (3) using a cost-of-living index.

Developing a Realistic Budget

Each of us has certain expectations for the kind of life-style we want to maintain. In order to begin the process of defining your economic needs, it will be helpful to determine what you expect to spend on routine monthly expenses. These expenses include housing, food, transportation, entertainment, utilities, loan repayments, and revolving charge accounts. A worksheet that details many of these expenses is shown in Figure 4. You may not currently spend for certain items, but you probably will have to once you begin supporting yourself. As you develop this budget, be generous in your estimates, but keep in mind any items that could be reduced or eliminated. If you are not sure about the cost of a certain item, talk with family or friends who would be able to give you a realistic estimate.

If this is new or difficult for you, start to keep a log of expenses right now. You may be surprised at how much you actually spend each month for food or stamps or magazines. Household expenses and personal grooming items can often loom very large in a budget, as can auto repairs or home maintenance.

Income taxes must also be taken into consideration when examining salary requirements. State and local taxes vary by location, so it is difficult to calculate exactly the effect of taxes on the amount of income you need to generate. To roughly estimate the gross income necessary to generate your minimum and annual salary requirement, multiply the minimum salary you have calculated . . . by a factor of 1.35. The resulting figure will be an approximation of what your gross income would need to be, given your estimated expenses.

Examining Starting Salaries

Starting salaries . . . can be used in conjunction with the cost-of-living index . . . to determine whether you would be able to meet your basic economic needs in a given geographic location

		Could Reduce Spending? (Yes/No)
Cable	$ _____	_____
Child care	_____	_____
Clothing	_____	_____
Educational loan repayment	_____	_____
Entertainment	_____	_____
Food	_____	_____
At home	_____	_____
Meals out	_____	_____
Gifts	_____	_____
Housing		
Rent/mortgage	_____	_____
Insurance	_____	_____
Property taxes	_____	_____
Medical insurance	_____	_____
Reading materials		
Newspapers	_____	_____
Magazines	_____	_____
Books	_____	_____
Revolving loans/charges	_____	_____
Savings	_____	_____
Telephone	_____	_____

FIGURE 4 Estimated monthly expenses worksheet

Using a Cost-of-Living Index

If you are thinking about trying to get a job in a geographic region other than the one where you now live, understanding differences in the cost of living will help you come to a more informed decision about making a move. By using a cost-of-living index, you can compare salaries offered and the cost of living in different locations with what you know about the salaries offered and the cost of living in your present location.

Many variables are used to calculate the cost-of-living index, including housing expenses, groceries, utilities, transportation, health care, clothing, entertainment, local income taxes, and local sales taxes. Cost-of-living indices can be found in many resources, such as *Equal Employment Opportunity Bimonthly, Places Rated Almanac,* or *The Best Towns in America.* They are constantly being recalculated based on changes in costs.

If you lived in Cleveland, Ohio, for example, and you were interested in working as an assistant editor for the *Cleveland Plain Dealer,* you would earn, on average, $26,643 annually. But let's say you're also thinking about moving to

Transportation
   Auto payment   _____   _____
   Insurance   _____   _____
   Parking   _____   _____
—or
   Cab/train/bus fare   _____   _____
Utilities
   Electric   _____   _____
   Gas   _____   _____
   Water/sewer   _____   _____
Vacations   _____   _____
Miscellaneous expense 1 _____   _____
Expense: _____
Miscellaneous expense 2 _____   _____
Expense: _____
Miscellaneous expense 3 _____   _____
Expense: _____

TOTAL MONTHLY EXPENSES: _____

YEARLY EXPENSES (Monthly expenses 12):_____

INCREASE TO INCLUDE TAXES (Yearly expenses 1.35): _____ =
MINIMUM ANNUAL SALARY REQUIREMENT _____

FIGURE 4   *(Continued)*

either New York, Los Angeles, or Minneapolis. You know you can live on $26,643 in Cleveland, but you want to be able to equate that salary in the other locations you're considering. How much will you need to earn in those locations to do this? Figuring the cost of living for each city will show you.

Let's walk through this example. In any cost-of-living index, the number 100 represents the national average cost of living, and each city is assigned an index number based on current prices in that city for the items included in the index (housing, food, etc.). In the index we used, New York was assigned the number 213.3, Los Angeles's index was 124.6, Minneapolis's was 100.0, and

Job: Assistant Editor

CITY	INDEX	EQUIVALENT SALARY
$\dfrac{\text{New York}}{\text{Cleveland}}$	$\dfrac{213.3}{114.3}$	$\times\ \$26{,}643 = \$49{,}720$ in New York
$\dfrac{\text{Los Angeles}}{\text{Cleveland}}$	$\dfrac{124.6}{114.3}$	$\times\ \$26{,}643 = \$29{,}044$ in Los Angeles
$\dfrac{\text{Minneapolis}}{\text{Cleveland}}$	$\dfrac{100.0}{114.3}$	$\times\ \$26{,}643 = \$24{,}382$ in Minneapolis

Cleveland's index was 114.3 In other words, it costs more than twice as much to live in New York as it does in Minneapolis. We can set up a table to determine exactly how much you would have to earn in each of these cities to have the same buying power that you have in Cleveland. You would have to earn $49,720 in New York, $29,044 in Los Angeles, and $24,382 in Minneapolis to match the buying power of $26,643 in Cleveland.

If you would like to determine whether it's financially worthwhile to make any of these moves, one more piece of information is needed: the salaries of assistant editors in these other cities. The Dow Jones Newspaper Fund reports the following average salary information for assistant editors as of 1994:

Newspaper	Annual Salary	Salary Equivalent to Ohio	Change In Buying Power
New York Times	$60,266	$49,720	+$10,546
Los Angeles Daily News	$24,700	$29,044	–$ 4,344
Minneapolis Star Tribune	$23,244	$24,382	–$ 1,138
Cleveland Plain Dealer	$26,643	—	—

If you moved to New York City and secured employment as an assistant editor at the *New York Times,* you would be able to maintain a lifestyle similar to the one you led in Cleveland; in fact, you would even be able to enhance your lifestyle given the increase in buying power. The same would not be true for a move to Los Angeles or Minneapolis. You would decrease your buying power given the rate of pay and cost of living in these cities.

....

You can work through a similar exercise for any type of job you are considering and for many locations when current salary information is available. It will be worth your time to undertake this analysis if you are seriously considering a relocation. By doing so you will be able to make an informed choice.

Step 4   Exploring Your Longer-Term Goals

There is no question that when we first begin working, our goals are to use our skills and education in a job that will reward us with employment, income, and status relative to the preparation we brought with us to this position. If we are not being paid as much as we feel we should for our level of education, or if job demands don't provide the intellectual stimulation we had hoped for, we experience unhappiness and, as a result, often seek other employment.

Most jobs we consider "good" are those that fulfill our basic "lower-level" needs of security, food, clothing, shelter, income, and productive work. But even when our basic needs are met and our jobs are secure and productive, we as individuals are constantly changing. As we change, the demands and expectations we place on our jobs may change. Fortunately, some jobs grow and change with us, and this explains why some people are happy throughout many years in a job.

But more often people are bigger than the jobs they fill. We have more goals and needs than any job could fulfill. These are "higher-level" needs of self-esteem, companionship, affection, and an increasing desire to feel we are employing ourselves in the most effective way possible. Not all of these higher-level needs can be fulfilled through employment, but for as long as we are employed, we increasingly demand that our jobs play their part in moving us along the path to fulfillment

Another obvious but important fact is that we change as we mature. Although our jobs also have the potential for change, they may not change as frequently or as markedly as we do. There are increasingly fewer one-job, one-employer careers; we must think about a work future that may involve voluntary or forced moves from employer to employer. Because of that very real possibility, we need to take advantage of the opportunities in each position we hold to acquire skills and competencies that will keep us viable and attractive as employees in a job market that is not only increasingly technology/computer dependent, but also is populated with more and more small, self-transforming organizations rather than the large, seemingly stable organizations of the past.

It may be difficult in the early stages of the job search to determine whether the path you are considering can meet these longer-term goals. Reading about career paths and individual career histories in your field can be very helpful in this regard. Meeting and talking with individuals further along in their careers can be enlightening as well. Older workers can provide valuable guidance on "self-managing" your career, which will become an increasingly valuable skill in the future. Some of these ideas may seem remote as you read this now, but you should be able to appreciate the need to ensure that you are growing, developing valuable new skills, and researching other employers who might be interested in your particular skills package.

....

If you are considering a position as an editor at a newspaper, you would gain a better perspective on this career if you talked to an entry-level editorial assistant, a more experienced assistant or full editor, and finally a senior editor, managing editor, or editor-in-chief who has a considerable work history in the newspaper field. Each will have a different perspective, unique concerns, and an individual set of values and priorities.

....

Step 5   Enumerating Your Skill Base

In terms of the job search, skills can be thought of as capabilities that can be developed in school, at work, or by volunteering and then used in specific job settings. Many studies have documented the kinds of skills that employees seek in entry-level applicants. For example, some of the most desired skills for individuals interested in the teaching profession include the ability to interact effectively with students one on one, to manage a classroom, to adapt to varying situations as necessary, and to get involved in school activities. Business employers have also identified important qualities, including enthusiasm for the employer's product or service, a businesslike mind, the ability to follow written or verbal instructions, the ability to demonstrate self-control, the confidence to suggest new ideas, the ability to communicate with all members of a group, awareness of cultural differences, and loyalty, to name just a few. You will find that many of these skills are also in the repertoire of qualities demanded in your college major.

In order to be successful in obtaining any given job, you must be able to demonstrate that you possess a certain mix of skills that will allow you to carry out the duties required by that job. This skill mix will vary a great deal from job to job; to determine the skills necessary for the jobs you are seeking, you can read job advertisements or more generic job descriptions. . . . If you want to be effective in the job search, you must directly show employers that you possess the skills needed to be successful in filling the position. These skills will initially be described on your resume and then discussed again during the interview process.

Skills are either general or specific. General skills are those that are developed throughout the college years by taking classes, being employed, and getting involved in other related activities such as volunteer work or campus organizations. General skills include the ability to read and write, to perform computations, to think critically, and to communicate effectively. Specific skills are also acquired on the job and in the classroom, but they allow you to complete tasks that require specialized knowledge. Computer programming, drafting, language translating, and copy editing are just a few examples of specific skills that may relate to a given job.

In order to develop a list of skills relevant to employers, you must first identify the general skills you possess, then list specific skills you have to offer, and, finally, examine which of these skills employers are seeking.

Identifying Your General Skills

Because you possess or will possess a college degree, employers will assume that you can read and write, perform certain basic computations, think critically, and communicate effectively. Employers will want to see that you have acquired these skills, and they will want to know which additional general skills you possess.

One way to begin identifying skills is to write an experiential diary. An experiential diary lists all the tasks you were responsible for completing for each job you've held and then outlines the skills required to do those tasks. You may list several skills for any given task. This diary allows you to distinguish between the tasks you performed and the underlying skills required to complete those tasks. Here's an example:

Tasks	Skills
Answering telephone	Effective use of language, clear diction, ability to direct inquiries, ability to solve problems
Waiting on tables	Poise under conditions of time and pressure, speed, accuracy, good memory, simultaneous completion of tasks, sales skills

For each job or experience you have participated in, develop a worksheet based on the example shown here. On a resume, you may want to describe these skills rather than simply listing tasks. Skills are easier for the employer to appreciate, especially when your experience is very different from the employment you are seeking. In addition to helping you identify general skills, this experiential diary will prepare you to speak more effectively in an interview about the qualifications you possess.

Identifying Your Specific Skills

It may be easier to identify your specific skills, because you can definitely say whether you can speak other languages, program a computer, draft a map or diagram, or edit a document using appropriate symbols and terminology.

Using your experiential diary, identify the points in your history where you learned how to do something very specific, and decide whether you have a beginning, intermediate, or advanced knowledge of how to use that particular skill. Right now, be sure to list *every* specific skill you have, and don't consider whether you like using the skill. Write down a list of specific skills you have acquired and the level of competence you possess—beginning, intermediate, or advanced.

Relating Your Skills to Employers

You probably have thought about a couple of different jobs you might be interested in obtaining, and one way to begin relating the general and specific skills you possess to potential employer needs is to read actual advertisements for these types of positions. . . .

....

For example, you might be interested in a career as a senior editor for a magazine. A typical job listing might read, "Requires 2–5 years experience, organizational and interpersonal skills, imagination, drive, and the ability to work under pressure." If you then used any one of a number of general sources of information that describe the job of senior editor, you would find additional information. Senior editors also develop story ideas, make assignments, work with staff and freelance writers, edit articles, and coordinate tasks with other magazine departments.

Begin building a comprehensive list of required skills with the first job description you read. Exploring advertisements for and descriptions of several types of related positions will reveal an important core of skills necessary for obtaining the type of work you're interested in. In building this list, include both general and specific skills.

Following is a sample list of skills needed to be successful as a senior editor for a magazine. These items were extracted from general resource and actual job listings.

On separate sheets of paper, try to generate a comprehensive list of required skills for at least one job you are considering.

Job: Senior Editor

General Skills	**Specific Skills**
Disseminate information	Write editorials
Gather information	Take notes
Conduct research	Write letters

Job: Senior Editor (*Continued*)

General Skills	**Specific Skills**
Work in hectic environment	Write memos
Meeting deadlines	Use tape recorder
Work in noisy environment	Develop story ideas
Work long hours near deadline	Assign articles
Work well with other people	Edit articles
Exhibit creativity	Schedule articles
Exhibit drive	Proofread
Be able to work under pressure	Familiar with word
Be organized	processing
Be able to supervise the work of others	Layout pages
Have excellent written and verbal skills	Select illustrations
Be able to conduct meetings	

The list of general skills that you develop for a given career path would be valuable for any number of jobs you might apply for. Many of the specific skills would also be transferable to other types of positions. For example, developing story ideas is a required skill for senior editors working on a newspaper.

Now review the list of skills you developed and check off those skills that *you know you possess* and that are required for jobs you are considering. You should refer to these specific skills on the resume that you write for this type of job. . . .

Step 6   Recognizing Your Preferred Skills

In the previous section, you developed a comprehensive list of skills that relate to particular career paths that are of interest to you. You can now relate these to skills that you prefer to use. We all use a wide range of skills (some researchers say individuals have a repertoire of about 500 skills), but we may not be particularly interested in using all of them in our work. There may be some skills that come to us more naturally or that we use successfully time and time again and that we want to continue to use; these are best described as our preferred skills. For this exercise, use the list of skills that you developed for the previous section and decide which of them you are *most interested in using* in future work and how often you would like to use them. You might be interested in using some skills only occasionally, while others you would like to use more regularly. You probably also have skills that you hope you can use constantly.

As you examine job announcements, look for matches between this list of preferred skills and the qualifications described in the advertisements. These skills should be highlighted on your resume and discussed in job interviews.

Step 7 Assessing Skills Needing Further Development

Previously you developed a list of general and specific skills required for given positions. You already possess some of these skills; those that remain to be developed are your underdeveloped skills.

If you are just beginning the job search, there may gaps between the qualifications required for some of the jobs being considered and skills you possess. These are your underdeveloped skills. The thought of having to admit to and talk about these underdeveloped skills, especially in a job interview, is a frightening one. One way to put a healthy perspective on this subject is to target and relate your exploration of underdeveloped skills to the types of positions you are seeking. Recognizing these shortcomings and planning to overcome them with either on-the-job training or additional formal education can be a positive way to address the concept of underdeveloped skills.

On your worksheet or in your journal, make a list of up to five general or specific skills required for the positions you're interested in that you *don't currently possess*. For each item, list an idea you have for specific action you could take to acquire that skill. Do some brainstorming to come up with possible actions. If you have a hard time generating ideas, talk to people currently working in this type of position, professionals in your college career services office, trusted friends, family members, or members of related professional associations.

If, for example, you are interested in a job for which you don't have some specific required experience, you could locate training opportunities such as classes or workshops offered through a local college or university, community college, or club or association that would help you build the level of expertise you need for the job.

You might have noticed . . . that many excellent positions for your major demand computer skills. These computer skills were probably not part of your required academic preparation. While it is easy for the business world to see the direct link between oral and written communication and high technology, some college departments have been markedly reluctant to add this dimension to their curriculums. What can you do now? If you're still in college, take what computer courses you can before you graduate. If you've already graduated, look at evening programs, continuing education courses, or tutorial programs that may be available commercially. Developing a modest level of expertise will encourage you to be more confident in suggesting to potential employers that you can continue to add to your skill base on the job. . . .

Generally speaking, though, employers want genuine answers to these types of questions. They want you to reveal "the real you," and they also want to see how you answer difficult questions. In taking the positive, targeted approach discussed above, you show the employer that you are willing to continue to learn and that you have a plan for strengthening your job qualifications.

USING YOUR SELF-ASSESSMENT

Exploring entry-level career options can be an exciting experience if you have good resources available and will take the time to use them. Can you effectively complete the following tasks?

1. Understand and relate your personality traits to career choices
2. Define your personal values
3. Determine your economic needs
4. Explore longer-term goals
5. Understand your skill base
6. Recognize your preferred skills
7. Express a willingness to improve on your underdeveloped skills.

If so, then you can more meaningfully participate in the job search process by writing a more effective resume, finding job titles that represent work you are interested in doing, locating job sites that will provide the opportunity for you to use your strengths and skills, networking in an informed way, participating in focused interviews, getting the most out of follow-up contacts, and evaluating job offers to find those that create a good match between you and the employer. . . . The time you will need to put into your job search will depend on the type of job you want and the geographic location where you'd like to work. Think of your effort as a job in itself, requiring you to set aside time each week to complete the needed work. Carefully undertaken efforts may reduce the time you need for your job search.

Marcia R. Fox

The Style and Appearance of the Cover Letter

When she wrote the book from which this selection is taken, Marcia R. Fox directed career counseling and job placement at New York University's Graduate School of Public Administration.

To an employer, your cover letter is the sole clue to your proficiency with the English language. Not only must your letters be grammatically and typographically *perfect*, but you must also write them in appropriate style.

1. Aim for simple but precise language. For example: "I am a forthcoming master's in electrical engineering interested in an engineering position with your company." Such virtues of simplicity were ignored in another version of this applicant's cover letter.

   Having graduated from the University of Colorado with a master's in electrical engineering, I have decided to pursue a career in a company such as yours.

2. If there is a reason for your particular interest in a certain employer, say so directly and succinctly. For example: "Professor John Sears, my research adviser, has repeatedly told me of your leadership in the field of prenatal nutrition."

3. Strive for precise, specific language. Avoid trite clichés and empty phrases:

   Employment at a highly diversified company such as yours represents a *challenging and stimulating beginning* to a career in business as well as the opportunity to significantly augment one's education.

> Although I am young, I have learned a great deal about human nature. I am in search of a *meaningful growth position* where I can use my talents and skills in a *responsible, challenging way*. I am a *creative thinker*.

Avoid vague terms such as "meaningful growth opportunity" and "creative thinker." Substitute "management trainee" or another specific term for "meaningful growth opportunity" or a "responsible" or "challenging" position. It's not the employer's job to figure out what you might find challenging; he has specific jobs to fill. Tell him what you want and assess whether or not the opportunity is meaningful at the interview!

4. Be concise. The best antidote for excessive wordiness is to read over a draft of your letter with a red pen in hand. Be prepared to justify each word. Notice, for example, how only the last sentence of the following paragraph is pertinent or substantive:

> I understand that your personal obligations, in combination with the huge volume of applications through which you must wade, leave you with a limited amount of time to spend on each. With this in mind, thank you for your time and consideration.

In trying to identify with the employer's burden, the applicant unwittingly took up even more of the employer's time!

5. Use the active voice. The only way to sound energetic is to use the active voice. For example, state:

"I would welcome the opportunity for an interview," not "An interview at your convenience is requested." Write, "I acquired a skill in economic analysis from these courses," not "From these courses, a skill in economic analysis was acquired."

An applicant's discomfort at actually writing a cover letter or even applying for a job can often result in excessive abruptness.

To help you hear the tone of your letter, wait a day before mailing it out, or have someone else read your letters aloud or silently.

Many professionals equate a professional style with a cool tone. The ideal cover letter, however, conveys the person behind the applicant. To create the rapport that will win you interviews, you must reveal some human warmth. Read the following example aloud to hear the tone of commanding coldness to avoid:

> Contact me if you have a suitable opening.

Do not be afraid to sound enthusiastic for fear it will be misinterpreted as feigned interest by a cynical employer. Always convey your enthusiasm professionally, by anchoring it to a credential or relevant fact. For example: "As a result of my fascination with gerontology, I obtained an H.E.W. fellowship to write a dissertation on the problems of indigent elderly women."

Remember that the employer is the stranger who must be persuaded to interview you. Problems of inappropriate tone may occur if you confuse him with another group or person. For example, naïve job hunters often mistakenly perceive the employer as a career counselor or as a "buddy" with whom to share fantasies. Don't close a letter this way: "I would very much appreciate talking with you so as to obtain a clearer perspective on my career goals." It is not the employer's responsibility to counsel you. Get the advice elsewhere!

A second problem arises when applicants approach the employer in a narcissistic way. To win interviews you must convince the employer you can help him rather than vice versa, as in this unfortunate example:

Employment at a highly diversified company such as yours offers many attractive health and vacation benefits. Moreover, it gives a challenging start to one who hopes to open his own business some day.

Some applicants mistakenly regard the employer as an audience in search of entertainment. This can lead to the kind of self-conscious "cuteness" exhibited in the following example:

Is there something special about a person who graduated six months early from both high school and college and has never cut a class in her entire educational career?

If you are tempted to point to any perceived inadequacies, remind yourself that employers are not confessors. It is self-destructive to point out facts that can harm you and which might otherwise have gone unnoticed. For example:

I am a forthcoming graduate of M.I.T.'s program in electrical engineering. While I did not win any significant research prizes, I. . . .

Here is how a severe case of sour grapes was transmitted in a cover letter that manages to sound both confessional and hostile at the same time:

Academically, I stand well within the top half of my class. A failure to be glib on exams prevented me from qualifying for Phi Beta Kappa. However, this failure has not prevented me from succeeding in any other endeavor.

Sometimes it will be necessary to carefully stress the positive aspects of a situation that could easily have negative connotations to an employer. One woman reentered the job market using this bad opening sentence in her cover letter:

Seven years ago I resigned from a career in business to raise a family.

Her opening could easily have suggested an enterprising person had she thought to lead with her strengths:

Your organization may be interested in a seasoned personnel administrator who has started her own successful business and consulted with an executive search firm concurrently with starting a family.

Do not indulge in subjective self-praise:

The enclosed resume sketches my background and shows my creativity, outstanding leadership skills, and writing ability.

Let the employer draw inferences from factual statements! Or if you wish to point to the clear inference, be modest and less blunt. For example:

> While still a student, I have obtained useful managerial skills through various activities. For example, my work on the student newspaper helped to sharpen my business, editorial, and writing skills. During my second summer, an administrative-assistant position in a small office helped to train me in bookkeeping procedures, employee benefits, and general office systems. Recently, as a management intern assigned to the Deputy Mayor's Office, I obtained excellent exposure to public management problems.

Here is another good example:

> During my graduate school career, I have tried to combine a seriousness of intellectual purpose with the development of interpersonal leadership skills. As chairman of the Student-Faculty Committee on Student Life, president of the student body, and senator-at-large in the university senate, I have gained useful training for a career in educational administration.

Sometimes a personality trait, such as leadership, is an important criterion of the job. If it seems important to stress such a quality do so substantively, as in the following example:

> I have learned that an effective personality is an important asset in a successful public relations career. As a result, I have tried to polish and refine my interpersonal skills through a number of special workshops on leadership styles and interpersonal effectiveness. When I received the "Outstanding Student Leader" Award at Yale last year, I felt encouraged about my own future potential to work effectively with others.

Avoid the puffery of the alternative version:

> I believe that I have the kind of sparkling personality that is essential to a successful career in public relations.

A good cover letter should not exceed one page. Save the elaborate detail for the interview, when the employer is certain to be more interested. Keep the paragraphs in your letter short for the greater visual ease of your reader.

Center the letter, and allow two to four inches as a border under your signature. The letter should not look crowded. Top and side margins should be generous.

Use bond paper in 8 1/2 × 11″ size. Ideally, the paper should match that of your resume, but if that proves too difficult, don't worry about it. Avoid erasable paper unless you are such a poor typist that *not* using it will make it impossible for you to type letters. You might consider paying someone else to type the letter on bond for you. If at all possible, don't use an old college typewriter for your letters. As with your resume, an IBM Selectric or Executive typewriter will enhance your text.

Neatness is important. If your cover letters have obvious corrections, there may be adverse reactions. In any event you'll look unprofessional.

Proofread carefully for errors. A misspelled name, a run-on sentence, or a typographical error can easily cost you an interview! You cannot be too much of a perfectionist in this area. . . .

A Checklist

Want to write a cover letter? Sit down and fill in the blanks:

1. I am (*June graduate; M.B.A. from N.Y.U.*) _____
2. I want a job as (*a junior accountant*) _____
3. I can offer you (*skills, experience*) _____
4. I am enclosing a resume _____
5. I want to see you (*request an interview*) _____
6. Thank you!

Now polish the language, flesh out the concepts, and you have an effective cover letter.

Richard H. Beatty

The Functional Resume

Richard H. Beatty, a nationally recognized authority on job searches, is currently president of the Brandywine Consulting Group, a corporate outplacement consulting firm.

Although, as a seasoned employment professional, I must admit to not being a strong advocate . . . [the functional] resume, I will quickly add, however, that it does occasionally have its place in the employment process. As an indication of this fact, I would estimate that approximately 5 to 10 percent of all resumes received by corporations make use of the functional format. This makes the functional format the second most popular resume style (second only to the chronological format which is estimated to account for 90 to 95 percent of all resumes). . . .

ADVANTAGES OF THE FUNCTIONAL RESUME

Advocates of the functional resume are quick to point out that it makes use of an excellent marketing strategy. This resume format is designed to capture the interest of the resume reader right from the beginning. This is accomplished by positioning the candidates's most salable strengths at the beginning of the resume in the form of some brief statements that capsulize the candidate's major accomplishments and most salable professional experience.

The marketing premise upon which this resume design is based is that most employment professionals are not thorough in their reading of employment resumes. Further, advocates of this style of resume would tell you that many employment professionals in fact read only the first paragraph or two of a resume.

If these initial paragraphs do not capture their interest, the reader simply discards the resume and moves on to the next.

Although there may be some employment professionals who are not particularly thorough and conscientious in their resume reading, my experience suggests that this is certainly not the case with most. In my opinion, the majority of employment professionals discover early in their professional career that thorough resume reading is a must. In having read thousands of resumes as an employment manager, I have found that the specific experience and/or skills sought by my employer were sometimes not described until the second or third page of the applicant's resume. This experience has conditioned me to read resumes with a fair degree of thoroughness. Most professional employment managers are likewise sufficiently concerned with finding well-qualified employees for their organizations that they are equally thorough in their resume reading approach.

This is not to say that there are not those who are less diligent in their reading. To the contrary, I am sure there are some who are much less thorough. Such individuals may be superficially impressed with these opening paragraphs and may elect to invite the candidate in for an employment interview with a less than a complete screening of the candidate's qualifications. In the better companies, however, this superficial approach to resume screening will quickly attract attention, and the employment professional will soon be exposed as a nonprofessional. Poor resume screening will assuredly result in unnecessary interviews, wasting the valuable time of line managers and costing the company unnecessary money in the form of candidate travel reimbursement. Such inefficiencies will not long be tolerated, and careless employment managers may well find themselves in the unemployment line.

Of course there are well-qualified candidates who without professional guidance sometimes elect to use the functional employment resume. It is up to the employment manager, therefore, to thoroughly read the *entire* resume to determine whether or not the candidate possesses the prerequisite skills and qualifications sought by his or her company.

SOME CAUTIONS

Perhaps the biggest drawback of the functional resume is that it seems to be the format most frequently chosen by individuals who wish to disguise some flaw in their credentials. In fact, as an employment professional who has read several thousand resumes, I would estimate that in seven out of every ten cases functional resumes have intentionally been chosen by the employment candidate to camouflage such problems. Awareness of this history of deception causes most employment professionals to become suspicious when this particular format is used. The tendency is thus to read these resumes as if one were on a witch hunt. Energy and focus are directed toward finding out what is wrong with the candidate instead of being directed toward whether the candidate possesses the desired qualifications.

I have discussed my feelings concerning the functional resume with several other personnel executives and have frequently found that their feelings are generally similar to my own. In general, the consensus appears to be that if an employment candidate uses the functional resume format, he or she is probably attempting to hide one or more of the following facts:

1. *Job Hopper.* The applicant has worked for an abnormally high number of employers in a relatively short period of time and would therefore be a high employment risk.
2. *Older Worker.* The applicant is older and is attempting to hide this fact.
3. *Employment Gap.* There is an undesirable or unexplained gap or break in the candidate's employment history. Since this gap is frequently unexplained, it leads the employment executive to conjure up all kinds of undesirable explanations (for example: major illness, marital problems, alcoholism). Most of these spell high risk in the minds of the employment professional.
4. *Educational Deficit.* The applicant lacks the requisite educational credentials normally required for the position for which he or she is applying.
5. *Minimal Experience.* The candidate has little if any meaningful experience related to his or her job objective.

CHOOSING WHICH RESUME FORMAT TO USE

The rule of thumb for choosing between the chronological (both classical and linear) and the functional resume is fairly simple. If you have good credentials and a solid work history, I strongly recommend that you use a form of the chronological resume.Why cast unnecessary doubt on otherwise excellent credentials?

The reciprocal of the above statement is also true. If you have some good experience and major credentials, but also have one or more of the aforementioned problems, you may well want to seriously consider use of the functional resume format.

Resume Format Test

If you are still in doubt, perhaps the following practical test may be of assistance to you in deciding whether to use the functional resume format.

Answer each of the following questions with a yes or no.

1. Have you worked for four or more employers in the last 10 years? Yes _____ No _____
2. If employed for more than 10 years, have you averaged less than three years of service per employer? Yes _____ No _____
3. Have you been employed with more than seven companies during your professional career? Yes _____ No _____
4. Are you age 50 or older? Yes _____ No _____
5. Have you been unemployed (or substantially underemployed) for a period of more than one year in the last 10 years? Yes _____ No _____

6. Have you been unemployed (three months or longer) more than once in the last five years? Yes _____ No _____
7. Do you have the necessary educational qualifications normally required by most employers for your occupation? Yes _____ No _____
8. Do you lack the experience normally required by most employers for your profession? Yes _____ No _____

If you answered yes to one or more of the first three questions, by most standards you would be considered a job hopper. Use of the chronological resume would highlight this problem early in the resume and, in many cases, would result in your being screened out. Use of the functional resume under these circumstances is to your distinct advantage.

Although illegal, the screening out of candidates because of age continues. Some inroads have been made in this area, and it appears that most employers are now paying less attention to age than they are to the qualifications of the individual. Unfortunately, however, age 50 appears to be a bench mark of sorts for many employers. Rightly or wrongly, many employers assume that a candidate who is age 50 or older lacks the necessary energy and vitality to be a productive worker. This fact suggests that candidates who are in this age group should attempt, where possible, to conceal this fact. The functional resume format can be particularly effective in accomplishing this objective.

Questions 5 and 6 have to do with gaps in the employment history. Should you have such a gap or should you have been substantially underemployed for a period of time, you may wish to employ the functional resume to focus on your strengths and accomplishments and to draw attention away from this employment gap.

If you answered no to question 7, chances are that use of the chronological resume may cause you to be screened out on the basis of educational credentials. The functional resume, on the other hand, gives you the opportunity to highlight your strengths and accomplishments early in the resume and may generate interest sufficient to cause the prospective employer to disregard your educational deficiency.

If question 8 was answered yes, you will probably do better in adopting the functional resume. Here again the reader's attention will be focused on what you consider your strengths and accomplishments rather than on your lack of specific job experience. . . .

FUNCTIONAL RESUME COMPONENTS

Although there are several different styles or versions of the functional resume, the most frequently used format is comprised of the following components:

1. Heading
2. Qualification summary

3. Major accomplishments
4. Work history
5. Education

The order in which these components are presented on the resume can vary, but the sequence represented above is the most commonly used, accepted, and recommended format. The most frequent deviation from this recommended sequence, however, is the positioning of the education and work experience components. Note that the sample resume shown in Figure 1 has been positioned such that education follows work experience.

The general rule for deciding whether to position education before or after work experience on the functional resume is this: Always position education after work history unless it will clearly serve to enhance your marketability to do otherwise.

In cases where the candidate has exceptionally strong educational credentials and this would be clearly recognized as the case by prospective employers, it would be in the candidate's best interest to position education prior to work history on the resume. Conversely, a candidate whose work history is extensive but whose educational credentials would be considered somewhat light by most employers' standards should definitely position education following the presentation of work history.

Additionally, it is recommended that younger candidates with good educational credentials and little work experience list education first. By contrast, older workers with considerable experience and persons who do not have strong educational credentials are generally best served by placing education after work experience.

For the older worker, placing education first in the resume can often draw attention to his or her age—something that should be avoided. This makes it easy for those employers who are in defiance of federal law, still practicing age discrimination, to weed you out. If you are successful in getting such employers to review your work experience and major accomplishments first, and thereby convince them that you have something of value to contribute to their organization, perhaps less attention will be paid to your age.

You are now ready to be in the step-by-step process of preparing the functional resume. The approach that we follow examines each resume component in some detail and provides some practical exercises that will assist you in developing each of these components. . . . [If] you follow the step-by-step directions provided, you will conclude this . . . [selection] having prepared a logical, professional resume that will effectively portray and market your qualifications to prospective employers.

BRUCE B. CHAMBERS
813 Locust Lane
Lake City, Texas 09413
Phone: (217) 855-3939

SUMMARY

Marketing executive with over 20 years experience in sales and marketing management. Excellent reputation as a creative, innovative manager capable of successfully revitalizing old product lines and introducing new. Full range of marketing and sales experience to include: market research, market planning & analysis, advertising & promotion, sales, and sales management.

MAJOR ACCOMPLISHMENTS

Market Research
- Investigated and analyzed European market for U.S. lumber export with resultant successful market entry.
- Worked closely with R&D in development of consumer mini-pocket calculator. Careful design of test market and resultant market feedback assured successful product development and subsequent market entry ($5 million sales in two years).
- Developed market research computer model to forecast ten-year market projection for microwave ovens.

Advertising & Promotion
- Coordinated company efforts with major New York City consumer advertising agency to develop effective campaign to revitalize failing product line (fishing reels). Campaign expenditure ($2 million) yielded annual increase in sales of $4.1 million in one year.
- Developed creative special value coupon and supportive advertising campaign that increased annual sales volume for photographic film product line by 35% over two-year period.

Marketing & Sales Management
- Managed national sales organization of 95 employees (ten regional managers and 75 salespersons) in the sale of consumer photographic film to wholesale and retail trade.
- Directed Corporate Marketing Staff (25 employees) in the development of all marketing plans and strategies for manufacturer of consumer hardware (annual sales volume $500 million).
- Successfully organized, trained and motivated new national sales organization of 45 employees for manufacture of consumer calculators. Sales reached $20 million in four years.

FIGURE I   Sample resume

WORK HISTORY

U.S. PAPER & WOOD PRODUCTS   1997 to Present
Manager of Corporate Marketing

PHOTO FILMS INTERNATIONAL   1995 to 1997
National Sales Manager

SELF EMPLOYED   1993 to 1995
Marketing Consultant

MICRO OVENS, INC.   1992 to 1993
Manager of Market Research

SELF EMPLOYED   1991 to 1992
Marketing Consultant

NATIONAL CALCULATORS, INC.   1977 to 1991
Director of Marketing   (1989 – 1991)
National Sales Manager   (1985 – 1989)
Regional Sales Manager   (1982 – 1985)
Salesperson   (1977 – 1982)

EDUCATION

B.A., Utah State University, 1977
Major: Business Administration

FIGURE I   *(Continued)*

Heading

The resume heading consists of three parts: your full name, complete address, and home telephone number. Some sample resume headings follow:

DAVID. C. JOHNSON
18 Smith Road
Richmond, Ohio 19870
Phone: (815) 852-4310

CAROLYN A. CRISWELL
401 East 7th Street
Knoxville, Tennessee 19760
Phone: Home (315) 769-6588
Office (315) 766-4233

Your name should be typed in capital letters and in bold type so that it stands out from the rest of the heading. Your address and telephone number, on the other hand, are typed in lower case. It is advisable to exclude your office telephone number from your resume unless you are in a position to accept employment-related telephone calls at your office. Should you be on a confidential voice mail system where you are the only one who can access your messages, this is likely not a problem. Listing your office telephone number on your resume, however, sometimes raises suspicion on the part of a prospective employer that you are "on the skids" with your current employer and that your current employer is cooperating with you in your job search. It is therefore advisable to avoid listing your office telephone number on your employment resume unless by excluding it you would be making it difficult for prospective employers to reach you.

Most employment managers do not mind placing telephone calls in the evening to prospective candidates. These managers can be encouraged to do so by inclusion of a simple request to call you at home in the evenings. Such a request can easily be inserted into your cover letter. For example, you might include a statement similar to the following in the cover letter that accompanies your resume:

> I can usually be reached at my home telephone number during the evening hours after 8:00 P.M. In the event that you find this inconvenient and wish to contact me during business hours, my husband, George, will be pleased to relay your message. George can be reached at (315) 972-8051.

Using the preceding instructions, try your hand at developing your own heading.

Summary

The purpose of the Summary section of the functional resume is twofold. First, it is intended to convey the breadth and scope of your experience as a professional. Second, it is intended to provide you with the opportunity to sell your key strengths. The overall intent is to hook the resume reader so that he or she will be encouraged to read further.

Let's now examine a few summary statements.

1. *Summary.* Marketing executive with over 20 years of experience in sales and marketing management. Excellent reputation as a creative, innovative manager capable of revitalizing old product lines and introducing new ones. Full range of marketing and sales experience to include: market research, market planning and analysis, advertising and promotion, sales, and sales management.
2. *Summary.* Accomplished human resources professional with over 25 years experience in all phases of human resources management. Excellent reputation as an individual who truly understands the relationship between human resource utilization and profitability. Full range of human resources experience incudes: human resource planning, internal and external staffing, organization design and development, compensation and benefits, employee and labor relations, and public affairs.

3. *Summary.* Creative design engineer with 15 years experience in innovative design of high speed electromechanical devices. Well known for innovative, practical, and cost-effective designs that work. A real contributor to bottom-line results.

In reviewing these summary statements, you will note certain similarities. First, the initial sentence of each summary statement gives the writer's career/professional area and indicates the number of years of experience The second sentence is used to market particular strengths possessed by the writer. The third sentence, as in the case of the marketing executive and human resources professional, can be used to further communicate the breadth of the writer's experience. As an option to this, as shown in the designer's summary statement, the resume writer can use this third sentence to further market some unique skill or fact that would likely prove valuable to the prospective employer.

Another similarity between these summary statements is that they are fairly concise. In preparing your summary statement, you should avoid lengthy, rambling paragraphs that will only serve to dilute your message to the prospective employer. The key to this section is to be concise while stating sufficient positive information to entice the reader to continue. Additionally, your summary statement should convince the prospective employer that you have something of value to contribute to their organization.

Note that these sample summary statements are not written in complete sentences. These statements are simply descriptive phrases that are intended to convey strong meaning in as few words as possible. They are short "bullets" much the same as you would find in magazine or newspaper advertising. As with such advertising copy, each is intended to convey value and to compel the reader to respond favorably.

Review of the sample summary statements will also reveal the lack of the articles "a" and "the." Each sentence begins with an adjective followed by a noun. Following this type of format will force you to write concise, brisk statements that have considerably more marketing impact than the normal sentence.

Now try your hand at writing your own sample summary statements. Try writing this statement three times with the idea of bringing about quantum improvement with each successive writing.

Major Accomplishments

The Major Accomplishments section is intended to point out specific accomplishments that you have realized during your professional career and to highlight these accomplishments in discrete, functional areas. The functional areas under which these accomplishments are highlighted are normally major subfunctions or subactivities of your profession. Some examples of these functional areas would be as follows:

Marketing Executive

1. Market research
2. Advertising and promotion
3. Marketing and sales management.

Human Resources Executive

1. Human resource management
2. Labor negotiations
3. Employment
4. Compensation and benefits.

Research Engineer

1. Conceptual design
2. Prototype development
3. Prototype testing
4. Redesign
5. Reduction to practice.

In preparation for this section of your resume, make a detailed listing of all of the subfunctions relating to your career specialty. Think in terms of the major components of your present and past positions—the key functional areas for which you were accountable in each of your past positions. Having done this, you are now ready to develop a listing of your major accomplishments for each of the corresponding functional areas that you have listed.

A good exercise for stimulating your memory is to list each of your past employers in reverse chronological order, including each of the respective positions you have held. Starting with the most recent employer, make a list of the most significant results that you have realized—the most significant contributions you have made to each employer. As you list each of these major accomplishments, be sure to include, wherever possible, some kind of quantitative measurement that will serve to convey to the reader the significance of this result to your past employer. Think in terms of the savings in time and money that your accomplishments provided. Additionally, think in terms of other ways you have positively affected the profitability of your employer. In this regard, actions that you have taken to increase employee productivity, sales volume, and so on should certainly be included in the category of major accomplishments.

Although the procedure described may seem laborious, it will prove to be extremely worthwhile to you as you begin the job-hunting process. In addition to providing you with an excellent basis for the construction of your functional resume, this personal analysis of major accomplishments will provide you with the information and confidence necessary to perform well during the interview phase of your employment campaign. Take time therefore to carefully develop this inventory of major accomplishments.

After you have listed your major accomplishments by functional area, you are ready to begin to rework these for inclusion in the final draft of your functional resume. Before doing this, however, review the major accomplishments section of the sample resume shown in Figure 1. . . . You will note that this area of the resume is comprised of only three functional areas: market research, advertising and promotion, and marketing and sales management. It is important that your resume contain no more than three or four functional areas which you plan to highlight. Listing more than four areas will serve to dilute the impact that is intended when using the functional resume. This intent is to highlight and market your major areas of strength. Listing more than four functional areas with accompanying major accomplishments may serve to give the resume reader the impression that you are a jack of all trades but master of none. The reader will begin to question the credibility of your accomplishments rather than being impressed with a carefully chosen few.

A careful review of your list of past employers and positions held will highlight your major accomplishments and contributions. Scanning these accomplishments will show that several are similar in nature and can fit nicely under the key functions that you have chosen to feature on the resume. Picking what appear to be the most prominent three or four subfunctions, put each of these subfunctions as a heading on a separate piece of paper. Then systematically go back over your employment history, extract each major accomplishment and add it to the page containing the functional heading under which it fits. In following this procedure you will soon see the major accomplishments section of your functional resume taking shape.

Your next step is to decide in what order these key functions should be listed in the major accomplishments section of your resume. There are two schools of thought relative to the order in which these functions should be listed. The first school of thought focuses on the old adage "lead from strength." This adage suggests that you should list that function in which you have made the greatest contributions and have realized the most significant accomplishments. This approach reasons that your most marketable or salient qualifications should be listed first since they are your areas of greatest strength and therefore are those areas that will be of greatest value to any prospective employer. The second school would argue that you should position that function first that most closely parallels your job-hunting objective. For example, if you are a human resources professional and are seeking a position as director of labor relations, you would obviously wish to cite your accomplishments in the area of labor relations first. Prospective employers looking for a director of labor relations are probably less concerned about your expertise in the area of compensation and benefits.

Both of the above schools of thought have their point. This may leave you a bit confused as to which function to list first. As a general rule of thumb, I would suggest listing first that function that most closely relates to your current career objective. However, should your accomplishments in this area be less than stellar, I would suggest listing your area of strongest accomplishment first without respect to job-hunting objective. The remaining functional areas should be listed in a priority sequence based upon the significance of the accomplishments in each of these respective functions.

In this section of your resume, writing style can be extremely important if you wish to have maximum impact on the reader. I have extracted two functions from the major accomplishments section of the resume shown in Figure 1 for your review. The first is market research:

- Investigated and analyzed European market for U.S. lumber export with resultant successful market entry.
- Worked closely with R&D in development of consumer mini pocket calculator. Careful design of test market and resultant market feedback assured successful product development and subsequent market entry ($5 million sales in two years).
- Developed market research computer model to forecast ten-year market projection for microwave ovens.

The second function is advertising and promotion:

- Coordinated company efforts with major New York City consumer advertising agency to develop effective campaign to revitalize failing product line (fishing reels). Campaign expenditure ($2 million) yielded annual increase in sales of $4.1 million in one year.
- Developed creative special value coupon and supporting advertising campaign that increased annual sales volume for photographic film product line by 35% over two-year period.

In reviewing these accomplishments you will note that each statement begins with a verb or action word. By starting most of your statements with a verb or action word, you will be forced to be brief, concise, and to the point in your statements. Additionally, you will be forced to state specific results and accomplishments that you contributed. . . .

Another effective technique to employ in developing the Major Accomplishments section of your resume is to use quantitative descriptions, where these serve to highlight your specific accomplishments and contributions. For example, consider the following:

Wrong	Increased sales significantly in first year.
Right	Increased sales by 50% from $100 to $150 million in first year alone.
Wrong	Developed new equipment design that saved the company a lot of money.
Right	Developed new plastic extrusion equipment design resulting in $1 million company savings in 1995 alone.

Wrong	Developed and installed computerized brand costing system resulting in substantial payroll savings.
Right	Developed and installed computerized brand costing system resulting in elimination of 12 positions and annual payroll savings of $385,000.

You are now ready to develop the Major Accomplishments section of your resume. Use the format illustrated here to carefully develop each of the functions you wish to highlight with specific accomplishments. Remember to begin each of your statements with an action word and to use quantitative descriptions wherever possible. Although you will wish to highlight only three or four functional areas, rewrite each functional area at least twice. Experiment by attempting to upgrade your second writing of each function by being more concise and choosing words that have greater impact.

Functional area: _____

Accomplishments: _____

1. _____

   _____

2. _____

   _____

3. _____

   _____

4. _____

   _____

Functional area: _____

Accomplishments: _____

1. _____

   _____

2. _____

   _____

3. _____

_____

4. _____

_____

Work History

The Work History section of the resume should be organized in reverse chronological order. Shown first should be your most recent employer, second, your second most recent employer, and so forth back to your original employer. Shown
on the left should be dates of your employment, when you began and when you
left. Adjacent to these dates should be the name of the employer followed by the
position or positions held. The following work history section has been extracted
from the sample functional resume shown in Figure 1.

Work History

1997 to Present	**U.S. Paper & Wood Products** Manager of Corporate Marketing
1995 to 1997	**Photo Films International** National Sales Manager
1993 to 1995	**Self-Employed** Marketing Consultant
1992 to 1993	**Microwave Ovens, Inc.** Manager of Market Research
1991 to 1992	**Self-Employed** Marketing Consultant
1977 to 1991	**National Calculators, Inc.** Director of Marketing (1989–1991) National Sales Manager (1985–1989) Regional Sales Manager (1982–1985) Salesperson (1977–1982)

There is nothing sacred about the order of position held and employer in this
section of the resume. When the candidate has been employed by major corporations and highlighting this fact could prove to the candidate's advantage from
a marketing standpoint, Company Name should be listed first with job title to
follow on the next line. The answer to sequence is really marketability. List first
either job title or employer, whichever you feel consistently gives you the greatest advantage from a marketability standpoint.

Listing position first followed by employer will work well as long as you have held only one position with each employer. In those cases where you have held multiple positions with a single employer, however, you will need to list employer first, followed by the positions held. (Note the National Calculators, Inc. section of the sample Work History from Figure 1.) You will note that National Calculators, Inc. is listed first, with the positions held listed following the company name. You will also note that the dates of the positions held at National Calculator are shown in brackets following each of the job titles of the positions held.

The work history section of the resume is intended to be simple. No additional information other than company name and job title should be included in this section. The following space is provided for you to develop the Work History section of your own functional resume.

From: _____ To: _____   Employer: _____

_____

Job Title: _____

_____

From: _____ To: _____   Employer: _____

_____

Job Title: _____

_____

From: _____ To: _____   Employer: _____

_____

Job Title: _____

_____

From: _____ To: _____   Employer: _____

_____

Job Title: _____

_____

From: _____ To: _____   Employer: _____

_____

Job Title: _____

_____

From: _____ To: _____   Employer: _____

_____

Job Title: _____

_____

Education

The Education section of a functional resume should include the following items:

1. Degree awarded
2. School attended
3. Year of graduation
4. Major field of concentration
5. Grade point average (G.P.A.) (List only if good and education is recent.)
6. Honoraries.

Some sample education statements have been provided for your reference.

1. <u>Education</u>:   Ph.D., Massachusetts Institute of Technology, 1997
Major: Chemical Engineering

M.S., Rochester Institute of Technology, 1995
Major Chemical Engineering
G.P.A. 3.85/4.0

B.S., Rochester Institute of Technology, 1993
Major: Chemical Engineering
G.P.A. 3.75/4.0

2. <u>Education</u>:   M.S., University of Pennsylvania, 1996
Major: Statistics
G.P.A. 3.5/4.0
Dixon Mathematics Scholarship

B.S., Drexel University, 1994
Major: Mathematics
G.P.A. 3.6/4.0
Kettering Award (Mathematics Honorary Award)

3. Education:	M.S., University of Michigan, 1996
	Major: Mechanical Engineering
	G.P.A. 3.4/4.0
	Sloan Engineering Scholarships
	B.S., University of Maryland, 1995
	Major: Mechanical Engineering
	G.P.A. 3.4/4.0
	Tau Beta Pi (Engineering Honorary)
4. Education:	B.A., North Carolina State University, 1996
	Major: Business Administration

Generally, unless you are a recent college graduate, you should leave little space for the Education section of your employment resume. The general rule to be followed in developing this section of the resume is: "The further along you are in your professional career, the less important are your educational credentials and the more important are your work experience and specific accomplishments." Additionally, since you have chosen the functional resume for the purpose of highlighting your experience and accomplishments, it stands to reason that the Education section of the resume will be de-emphasized.

You now can develop the Education portion of your resume using the instructions provided.

Miscellaneous

In almost all cases, extracurricular activities and hobbies should be completely excluded from the resume unless they are directly related to your qualifications for the position you seek. On occasion, however, there may be some advantage to listing a specific extracurricular activity or hobby, that is, if it further enhances your marketability to a specific employer. For example, if you are applying for a position with a sporting goods manufacturer, your athletic interests may be to your advantage. Likewise, if you are applying to a company that places a lot of stress on providing leadership in your local community, you may want to include such information as (1) President, Parent Teachers Organization, (2) Chairperson, March of Dimes Campaign, (3) President, School Board. The preceding are carefully chosen exceptions to the rule. In general, however, hobbies and extracurricular activities are specifically excluded from the modern resume.

In the case of technical professionals, such things as publications, papers, and patents can be important. This is particularly true for items that are directly related to the professional's employment qualifications. These may help to increase the candidate's marketability to prospective employers. List only those items that are truly important and meaningful, however, and do not list minor or insignificant publications.

INDEX